FOUNDATIONS AND METHOD

FROM

MATHEMATICS TO NEUROSCIENCE

This volume would not have been possible without Charlotte Cattivera, Patrick Suppes' editorial assistant. Her expertise, attention to detail, and patience in working with authors and editors are gratefully acknowledged.

CSLI Lecture Notes Number 213

FOUNDATIONS AND METHODS FROM MATHEMATICS TO NEUROSCIENCE

ESSAYS INSPIRED BY

PATRICK SUPPES

edited by

Colleen E. Crangle, Adolfo García de la Sienra,
and Helen E. Longino

CSLI PUBLICATIONS STANFORD

Copyright © 2014
CSLI Publications
Center for the Study of Language and Information
Leland Stanford Junior University
Printed in the United States
19 18 17 16 15   1 2 3 4 5

Library of Congress Cataloging-in-Publication Data

Crangle, Colleen, 1952–.
  Foundations and methods from mathematics to neuroscience / Colleen Crangle,
Adolfo Garcia de la Sienra, and Helen Longino.
        p. cm.
  "Center for the Study of Language and Information, Leland Stanford Junior
University."
  Includes bibliographical references.
  ISBN 978-1-57586-744-1 (cloth : alk. paper) –
  ISBN 978-1-57586-745-8 (pbk. : alk. paper) –
  ISBN 978-1-57586-746-5 (electronic)
  1. Science–Philosophy.   2. Science–Philosophy–History–20th century.   3. Logic,
Symbolic and mathematical.   4. Suppes, Patrick, 1922–2014.   I. García de la
Sienra, Adolfo.   II. Longino, Helen E.   III. Center for the Study of Language and
Information (U.S.)   IV. Title.

  Q175.C8825 2015
  501–dc23

                                                                                              2015000906
                                    CIP

∞ The acid-free paper used in this book meets the minimum requirements of the American
National Standard for Information Sciences—Permanence of Paper for Printed Library
                            Materials, ANSI Z39.48-1984.

CSLI was founded in 1983 by researchers from Stanford University, SRI International, and Xerox
PARC to further the research and development of integrated theories of language, information, and
computation. CSLI headquarters and CSLI Publications are located on the campus of Stanford
                                        University.

            CSLI Publications reports new developments in the study of language,
                information, and computation. Please visit our web site at
                            http://cslipublications.stanford.edu/
                for comments on this and other titles, as well as for changes
                            and corrections by the author and publisher.

Contents

Contributors

KENNETH J. ARROW, Stanford University, Department of Economics

JOSÉ ACACIO DE BARROS, San Francisco State University, Liberal Studies

NANCY CARTWRIGHT, University of California, San Diego, Department of Philosophy, and Durham University, Department of Philosophy

CLAUDIO G. CARVALHAES, Stanford University, Center for the Study of Language and Information

COLLEEN E. CRANGLE, Stanford University, Center for the Study of Language and Information, and Converspeech LLC

ANNE FAGOT-LARGEAULT, Collège de France & Académie des sciences, Philosophe des sciences

JEAN-CLAUDE FALMAGNE, University of California, Irvine, Department of Cognitive Sciences

JENS ERIK FENSTAD, University of Oslo, Department of Mathematics

DAGFINN FØLLESDAL, Stanford University, Department of Philosophy

HARVEY M. FRIEDMAN, Ohio State University, Department of Mathematics

MICHAEL FRIEDMAN, Stanford University, Department of Philosophy

RUSSELL HARDIN, New York University, Department of Politics

STEPHAN HARTMANN, Ludwig Maximilians-Universität München, Munich Center for Mathematical Philosophy

JAAKKO HINTIKKA, Boston University, Department of Philosophy

PAUL HUMPHREYS, University of Virginia, Department of Philosophy

HANNES LEITGEB, Ludwig Maximilians-Universität München, Munich Center for Mathematical Philosophy

WILLEM J. M. LEVELT, Max Planck Institute for Psycholinguistics, The Netherlands

ELIZABETH F. LOFTUS, University of California, Irvine, School of Social Ecology

HELEN E. LONGINO, Stanford University, Department of Philosophy

R. DUNCAN LUCE, University of California, Irvine, Institute for Mathematical Behavioral Sciences

ALEXANDRE MARCELLESI, University of California, San Diego, Department of Philosophy

Gary Oas, Stanford University, Center for the Study of Language and Information

Marcos Perreau-Guimaraes, Stanford University, Center for the Study of Language and Information

Thomas Ryckman, Stanford University, Department of Philosophy

Adolfo García de la Sienra, Universidad Veracruzana, Instituto de Filosofía

Brian Skyrms, Stanford University, Department of Philosophy

Foundations and Methods from Mathematics to Neuroscience: Essays Inspired by Patrick Suppes

FOREWORD BY HELEN E. LONGINO

On March 17, 2012, Patrick Suppes reached 90 years of age. If you date from 1950, the year he completed his PhD at Columbia University, that is 62 years of amazing productivity -- surely an occasion for a big party. And that is exactly what took place.

With Suppes' advice and collaboration, the Suppes Center for History and Philosophy of Science and the Stanford Philosophy Department organized a two-day symposium, complete with banquet, featuring work in some of the many areas of science and philosophy of science to which Suppes has contributed. Participants ranged from Suppes' former and current collaborators and colleagues to former and current students. Distinguished and honored in their own fields (of physics, economics, linguistics, psychology, logic, foundations of mathematics, philosophy of science), they came from near and far to celebrate.

The Symposium opened with an overview of Suppes' contributions to science and its philosophy by Michael Friedman. This serves as the Introduction to the present volume. It was followed by papers on foundations of physics, on measurement theory, probability, and decision theory, on foundations of economics and political theory, on foundations of logic and mathematics, on linguistics and philosophy of language, and on psychology and neuroscience. The symposium concluded with a paper by Nancy Cartwright on evidence and its use and abuse in policy circles, reminding us that the work of philosophers matters far beyond the halls of the academy. This paper closes the present volume. Each 25 to 30 minute talk was followed by general discussion, opened each time by trenchant comments from Suppes.

Most of the symposium participants were able to contribute papers for this volume. Once their papers reached us, Suppes read them and transformed his off the cuff comments into perceptive commentaries on each paper that expand on its themes or mark his different take on its claims.

To those who know Patrick Suppes, many of the ideas in his commentaries will seem familiar. He still adverts to Aristotle, as having provided the most satisfying overarching metaphysical framework within which to pursue scientific investigation of the world. And his admiration of the ancient astronomers crops up several times. Suppes is still both an empiricist and a believer in the power and necessity of precise, quantitative, mathematical, representation. It would not be surprising to find that "details" is the most frequently used word in his commentaries. And, he treats physics as in some sense fundamental. But, over time, he has come to emphasize different aspects of the scientific process than he did earlier in his philosophical thinking. Perhaps the most significant is a transformation that has occurred over a long stretch of time and that has two aspects. One aspect is a shift of focus from theory to experiment. The shift of focus brings the matter of data into salience, and taking a clear-eyed look at data, at their multiplicity, diversity, resistance to representation in simple models, forces an acknowledgement of the second aspect: the fragmentary character of scientific knowledge.

While formalism played a central role in his early thinking, so did the need for quantitative characterization to be answerable to qualitative principles. Pluralism and the fragmentation of knowledge were already a part of his world-view in 1980. This conception of science is even more pronounced in 2014. The physics that is a source of models for other phenomena is not a unified theory of everything, but a conglomeration of models of well-characterized but distinct systems. These systems are themselves in many cases abstractions and simplifications compared to what is to be found in nature, offering a way to think about a phenomenon without being a complete specification of the phenomenon. For example, the abstract model of an oscillating system developed in physics can be brought into brain science to study relations among neurons. But the ways neuronal systems constitute oscillators will resemble physical oscillators in a very abstract way. One needs to study those systems directly to characterize them adequately. Suppes' comments in this volume are constantly reminding us that matters are not nearly as simple as hoped in the paper on which he is commenting. Finally, as an example of his openness to the world's complexity, we might consider his new-found appreciation of the role of emotion in learning and cognition, and the need to take emotion into account in studying both the psychology of learning and the neurobiological processes that realize human cognition (processes that play the matter to cognition's form). Typically, Suppes emphasizes that recognizing complexity is not an occasion for throwing up one's hands, but rather for modesty and focusing on making whatever progress is possible with the time and tools at hand.

Patrick Suppes does not stand still. As the commentaries show, he is continuing to think deeply about matters that have engaged his attention since he began his professional career. We are fortunate in having such a vital example of the philosophical life among us. The essays in this volume are an expression of the respect and affection in which Suppes is held globally and a partial testament to the deep and wide-ranging influence he had in 20th Century philosophy of science and continues to have in the 21st.

Near the completion of this volume, Patrick Suppes passed away, on November 17, 2014. He was delighted to see the work come to fruition and at the request of his widow, Michelle Nguyen, an advance copy of the book was laid in his casket.

An Overview: Suppes as Scientific Philosopher

Michael Friedman

I shall try in this paper to delineate Patrick Suppes' unique perspective on the relationship between science and philosophy.[1] I want to show, in particular, that Suppes' extraordinarily broad-ranging and diverse body of work amounts to a new and not yet sufficiently appreciated contribution to the ongoing development of a tradition in scientific philosophy that began with the efforts of the Vienna Circle (and related groups) in Central Europe in the 1920s and 30s, and continued in this country after the great intellectual migration that took place immediately before, during, and after the Second World War. Here it was enriched by the tradition of American pragmatism, a healthy dose of which Suppes absorbed as a graduate student in philosophy at Columbia after the War. He was there influenced especially by Ernest Nagel, who was one of the most important creators of the sub-discipline we now call philosophy of science during this period, along with Rudolf Carnap, Hans Reichenbach, and my own teacher Carl ("Peter") Hempel.

The first point I want to make, however, is that, although Suppes has of course made many fundamental contributions to this sub-discipline, he is not a philosopher of science in the standard sense. The concept of scientific philosophy fits him better, in so far it implied (in the late nineteenth and early twentieth century context in which it was developed) a revolutionary effort to replace the entire academic discipline of philosophy as it had come to be practiced with a new orientation devoted to much closer cooperation with the sciences themselves.[2] Yet, even though it makes good sense to take Suppes to be an inheritor of this tradition, the concept of scientific philosophy, as it was used by the Vienna Circle, for example, may still seem too narrow in his case. For Suppes is even more devoted to making original cutting-edge contributions to science itself—and, indeed, in a number of scientific disciplines.

Thus, for example, he is one of only three contemporary philosophers (together with Allan Gibbard and Brian Skyrms) who are members of both the American Academy of

---

[1] I originally prepared this overview, at Suppes' request, for the celebration of his 90th Birthday at Stanford in March 2012. I was more than happy to comply, but I was also grateful, in particular, for his wise and helpful suggestion that I use his Intellectual Autobiography (2006) as the basis for my discussion. I rely heavily on this Autobiography in what follows.

[2] For an extended discussion of the character and development of this concept see Friedman (2001).

Foundations and Methods from Mathematics to Neuroscience.
Colleen E. Crangle, Adolfo García de la Sienra and Helen Longino.
Copyright © 2014, CSLI Publications.

Arts and Sciences and the National Academy of Science. Moreover, one of the things of which Suppes is justly most proud is the award of the National Medal of Science in 1990. He cites the statement of this award at the end of his Intellectual Autobiography (2006, p. 75): "For his broad efforts to deepen the theoretical and empirical understanding of four major areas: the measurement of subjective probability and utility in uncertain situations; the development and testing of general learning theory; the semantics and syntax of natural language; and the use of interactive computer programs for instruction." Although an excellent summary of some of Suppes' most important contributions, this certainly does not sound like the profile of a typical philosopher of science—or, even, if there is such a thing, of a typical scientific philosopher.

Suppes is well aware of the unique nature of his enterprise. He explains it this way several pages earlier (2006, p. 70):

> I had worked both in philosophy and science, though from the standpoint of research, I primarily thought of myself as a philosopher of science. I think this is probably not really the most accurate characterization. I continue to do and continue to have great interest in the philosophy of science, but it is certainly also true that, in many respects, more of my energy in the last quarter of a century has been devoted to scientific activities. . . . The two main areas are psychology, above all, and physics. . . . I could take another line and say that a distinction between philosophy of science and science is in itself incorrect. In many ways I am sympathetic with such a summary of Quine's view, namely, that philosophy should mainly be philosophy of science and philosophy of science should mainly itself be science. This is a way of saying that philosophy is not privy to any special methods different from the methods used in the sciences.

I shall return to Suppes' relation to Quine below, and also to his relation to Quine's teacher and later great adversary, namely, Carnap.

My second point concerns Suppes' distinctive version of philosophical empiricism, which, once again, is also quite unique (2006, p. 37): "I think the influence of [my] scientific work on my philosophy has been of immeasurable value. I sometimes like to describe this influence in a self-praising way by claiming that I am the only genuinely empirical philosopher I know. It is surprising how little concern for scientific details is to be found in the great empirical tradition in philosophy. It has become a point with me to cite scientific data and not just scientific theories whenever it seems pertinent." Suppes' philosophical empiricism thus involves a commitment to empirical science as a paradigm of knowledge and a consequent commitment to attending very closely to the details of this science, both theoretical and empirical. And here Suppes' philosophical practice could not be more different from Quine's. For, although Quine is certainly the most celebrated recent proponent of philosophical empiricism, he appears, paradoxically, to be almost wholly uninterested in the details of any real empirical science—with the possible exception of a rather superficial fragment of behavioristic psychology.

However, and this is my third point, although Suppes' philosophical practice is unusually attentive to empirical data, and his scientific practice, moreover, is centrally concerned with developing detailed ways of empirically testing theoretical hypotheses, an unusually large part of Suppes' activities—both philosophical and scientific—revolves around formal mathematical work in axiomatic foundations. Much of his earliest work in the 1950s, for example, concerned the axiomatic foundations of mechanics, including axiomatic foundations for special relativity in the tradition that we now identify as beginning with A. A. Robb.[3] Suppes then turned in the 1960s to the foundations of probability in connection with the foundations of quantum mechanics, and embarked on a program for analyzing the phenomenon of non-commutativity or incompatibility between conjoint physical magnitudes or "observables" (such as position and momentum, for example, or different directions of spin) in terms of the non-existence of joint distributions for the associated random variables.[4] During this same period, moreover, he embarked on an ambitious and wide-ranging program in the foundations of measurement theory, which culminated in the now classic three-volume treatise, *Foundations of Measurement*, written with David Krantz, Duncan Luce, and Amos Tversky.[5] This work both created the formal axiomatic theory of measurement and developed it in extraordinary detail.

The central concept of this theory is that of a representation theorem, which formally explains the connection between the qualitative relations characterizing a given empirical domain (such as the relation of heavier-than between different masses or weights, warmer-than between differently heated objects, and so on) and scales of numerical measurement that may then be applied to this domain. The relevant representation theorem says (roughly) that, for any model satisfying the appropriate axioms for these relations, there is a function from the objects in the domain into the real numbers that is unique up to a certain class of transformations (characteristic of a ratio scale, or an interval scale, and so on)—or, to put it another way, any model for these axioms is isomorphic to a purely numerical model (unique up to the same class of transformations). The historical source for this type of axiomatic foundation was Hilbert's celebrated work on the foundations of geometry, which aimed, among other things, thereby to explain the connection between synthetic and analytic geometry, see Hilbert (1899). Thus, any model for Hilbert's axioms of incidence, order, congruence, and so on (in general, a qualitative or synthetic model) is isomorphic to a model in the domain of pairs of real numbers (a numerical or analytic model), where the appropriate class of transformations under which the relations in the model are invariant is given by the Euclidean group (rotations and translations) plus dilations (where the latter signify an arbitrary choice of unit of measurement).

This extension and generalization of the Hilbertian style of formal axiomatics to a formal theory of measurement for a great variety of possible physical magnitudes—including not only classical extensive magnitudes like length but also what we now call intensive magnitudes like (phenomenological) temperature and, most importantly, psychological magnitudes such as subjective probabilities and utilities—was a magnificent mathematical achievement. Moreover, the fundamental ideas of representation, isomorphism (and its generalizations), and invariance lie at the core of all of Suppes' scientific and philosophical work. Indeed, his encyclopedic later book on representation and invariance uses precisely these ideas as a framework for systematically organizing virtually all of his work, from his earliest work on mechanics and the foundations of physics, through the foundations of probability, language, and linguistics, up to his most recent (and very empirically oriented) work on the representations of language understanding via electro-magnetic waves in the brain, see Suppes (2002).

---

[3]The best known of these efforts from the 1950s is the axiomatic treatment of classical particle mechanics, McKinsey et al. (1953). The earliest treatment of the relativistic case is Rubin and Suppes (1954). Suppes cites Robb (1936) in his Intellectual Autobiography (2006, p. 7).

[4]The first paper in this now long series is Suppes (1961).

[5]See Krantz et al. (1971); Suppes et al. (1989); Luce et al. (1990).

This predominance of formal axiomatics in Suppes' thinking is extremely striking, and it appears to be especially striking, in particular, against the background of Suppes' avowed commitments to philosophical pragmatism and empiricism. This is not the kind of framework that one typically finds in the work of philosophical pragmatists and empiricists—or, for that matter, in the writings of typical working scientists, neither experimental nor theoretical. Suppes recently told me a delightful anecdote in this connection concerning another of his scientific contributions of which he is justly most proud—the publication of his work, co-authored with Acacio de Barros, on the extension of the no-joint-distribution approach to quantum non-commutativity to Bell-type inequalities involving three entangled particles in *Physical Review Letters*, see de Barros and Suppes (2000). The center-piece of this work involved two rigorous theorems, labeled as such, and the editors had only one request for changes—that the theorems be removed and that the authors merely state their results without giving proofs or even using the word 'theorem'. The point is that not even theoretical physicists, as such, are interested in these kinds of attempts at mathematical rigor.

Suppes' work, in this respect, resembles that of no one in the previous tradition of scientific philosophy so much as Carnap's. For Carnap, too, devoted himself to the formal analysis of scientific concepts and principles, and Carnap, too, was deeply influenced by the Hilbertian axiomatic method (as refracted, in Carnap's case, through the lens of Frege's *Begriffsschrift* and Whitehead's and Russell's *Principia Mathematica*). Moreover, Carnap, like Suppes, began his career with a strong interest in the axiomatic foundations of relativity theory and then spent a large portion of his later career on the foundations of probability and statistics.[6] Yet the formal framework within which Suppes constructs his analyses of scientific concepts and principles is very different from Carnap's. Whereas Carnap focusses on formal languages, Suppes focuses on set-theoretical models, and, accordingly, Carnap's approach is paradigmatic of what one now calls the syntactic view of scientific theories while Suppes' is paradigmatic of what one now calls the semantic view.[7]

The crucial question, however, is why is this important? The point is not, as it is sometimes suggested, that Carnap is committed to the restricted resources of first-order logic while Suppes (at least for this purpose) presupposes all the set theory that is needed. For Carnap is not committed to first-order logic at all, and his preferred formal language is a strong version of higher-order type theory essentially equivalent to Zermelo-Fraenkel set theory.[8] The point, I believe, is that Suppes' distinctive generalization of Hilbert's axiomatic approach is one significant step closer to the style of mathematics used by actual working mathematicians, whereas Carnap's approach remains at an excessive level of abstraction and, accordingly, yields very few results that are useful to such mathematicians.

[6]Carnap (1925) is an early discussion of the axiomatic foundations of relativity theory, based also on the approach that we now identify with A. A. Robb. I shall briefly touch on Carnap's later work on the foundations of probability below.

[7]A useful collection of philosophy of science literature on the contrast between these two approaches is Suppe (1974); van Fraassen (1980) is an influential defense of the semantic approach. As we shall see, however, Suppes' own approach has distinctive intellectual motivations that are concerned more with mathematical foundations than with philosophy of science in the standard sense.

[8]Carnap in fact worked closely with Fraenkel during the late 1920s and early 1930s. For illuminating discussion of Carnap's essentially higher-order conception of logic, and his interactions with Fraenkel, in particular, see Reck (2007).

Suppes illustrates this (without naming Carnap explicitly) by contrasting the perspectives on the empirical interpretation of a theory resulting from the two different approaches. While Carnap appeals to the very abstract and general notion of a correspondence rule (namely, any sentence containing both observational and theoretical terms), Suppes appeals to the representation theorems that one may attempt to provide for qualitatively formulated theories, which explain (at least in principle) how quantitative measurement of the corresponding empirical magnitudes is possible. Indeed, since the qualitative relations in question correspond to idealized experimental procedures (such as placing weights in a balance and the like), we thereby obtain a useful and unproblematic perspective on how mathematics applies to the empirical world. This is a perspective, moreover, that is entirely missing, not only from Carnap's excessively abstract and schematic account, but even more from currently popular model-theoretic accounts according to which purely mathematical models are supposed to be directly correlated with purely empirical models by an abstract isomorphism—thereby making the application of mathematics a complete mystery. On Suppes' approach, by contrast, there is no sharp distinction at all between "pure" mathematical theories (like Euclidean geometry) and empirical theories—both are equally in need of appropriate representation theorems.[9]

A further and perhaps even more important point is that Suppes' axiomatic approach is directed towards the concepts and principles of developing theories at the cutting edge of scientific research—theories whose conceptual, mathematical, and empirical foundations are still quite unsettled. This, for example, is why psychological concepts and principles are especially important in his work. For there has been a strong tendency to question even the possibility of fundamental measurement of intensive (as opposed to extensive) magnitudes and thereby to question the possibility of a proper application of mathematics to psychology. This point of view, as Suppes emphasizes, simply ignores the strong tradition in nineteenth-century psychology, beginning with Weber and Fechner and continuing through Helmholtz and beyond, focusing on the experimental measurement and mathematical modeling of psychological phenomena.

The approach to fundamental measurement taken by Suppes and his co-workers, by contrast, is explicitly intended to right this wrong and thereby to contribute to the articulation and development of a scientific psychology (2006, p. 11):

> The theory of measurement provides an excellent example of an area in which real progress has been made in the foundations of psychology. In earlier decades psychologists accepted the mistaken beliefs of physicists like Norman Campbell that fundamental measurement in psychology was impossible. Although Campbell had some wise things to say about experimental methods in physics, he seemed to have only a rather dim grasp of elementary formal methods, and his work in measurement suffered accordingly. Moreover, he did not even have the rudimentary scholarship to be aware of the important earlier work of Helmholtz, Hölder, and others.[10]

---

[9]See (2002, p. 33): "It is one of the theses of this book that there is no theoretical way of drawing a sharp distinction between a piece of pure mathematics and a piece of theoretical science.... From a philosophical standpoint there is no sharp distinction between pure and applied mathematics, in spite of much talk to the contrary. The continuity between pure and applied mathematics, or between mathematics and science, will be illustrated here by many examples drawn from both domains."

[10]Suppes' targets here are Campbell (1920, 1928). As he makes clear on the preceding page, the axiomatization of extensive measurement was the subject of Suppes' first published paper (1951), building on earlier work by Otto Hölder (1924).

In sum, Suppes' use of formal axiomatic measurement theory is intended not only to dispel philosophical mystery mongering concerning how mathematics can possibly apply to the empirical world, it is also intended to help set a still embryonic theoretical and empirical discipline like psychology on the secure path of a mature science.[11]

Yet Carnap has an overarching interest in furthering the progress of empirical science as well. Indeed, Carnap's commitment to philosophical empiricism, like Suppes', rests simply on the conviction that ongoing empirical science is the best example of genuine knowledge that we have. Moreover, Carnap intends the formal discipline that he recommends as the replacement for traditional philosophy—which he calls *Wissenschaftslogik*—to provide the best means for fostering fruitful cooperation between philosophers in this sense and ongoing empirical science. In particular, such logicians of science are to cooperate with applied mathematicians in constructing and formally investigating a variety of possible axiomatic foundations so that the most fruitful ones for actual empirical research can then be chosen on basically pragmatic grounds. Thus far, therefore, Carnap's enterprise does not sound so very different from Suppes'.

The crucial difference emerges in what comes next. For it is centrally important to Carnap that what he calls *Wissenschaftslogik* takes place at a fundamentally different level from empirical science itself—that it constitutes a meta-discipline relative to empirical science rather than a part of this science. In particular, whereas the sentences of empirical science are paradigmatically contentful and synthetic, for Carnap, those of *Wissenschaftslogik* are contentless and analytic. They make no substantive assertions about reality but concern either what follows from what in a given linguistic framework or the purely practical and pragmatic decision to adopt one such framework rather than another. Carnap's insistence on a sharp distinction between synthetic and analytic questions—or, more-or-less equivalently, between internal questions decidable within a given linguistic framework and external questions involving the choice of one such framework over another—does not rest (following Quine's picture of Carnap) on the desire to explain or secure the special epistemic status of logic and mathematics. It rests, rather, on the desire systematically to avoid metaphysical pseudo-problems once and for all by appealing to precisely the distinction between internal and external questions. Thus, for example, the choice between classical and intuitionist logic for the language of science involves no substantive issue at all, for Carnap, but merely a practical choice that can only be settled, in turn, on the basis of long-run scientific fruitfulness balanced against the desire to avoid possible inconsistencies.[12]

At the end of "Two Dogmas of Empiricism," while applauding Carnap for this pragmatic attitude, Quine famously goes on to criticize him for continuing to maintain a boundary between the analytic and the synthetic. "In repudiating such a boundary," Quine says (1951/1953, p. 46), "I espouse a more thorough pragmatism." And there is no doubt that Suppes agrees with Quine on this point. This is clear, for example, in the passage quoted above (2002, p. 33), maintaining that "there is no theoretical way of drawing a sharp distinction between a piece of pure mathematics and a piece of theoretical science." But a more illuminating example of his fundamental divergence

---

[11] Judging from the statement of the award of the National Medal of Science, quoted above from the end of Suppes' Intellectual Autobiography (2006, p. 75), it is clear that he has indeed made substantial progress in this direction.

[12] For discussion of Carnapian *Wissenschaftslogik* and its anti-metaphysical ambitions see Friedman (2007). As explained, the project is initiated in Carnap (1934) and receives its canonical mature expression in Carnap (1950a/1956).

SUPPES AS SCIENTIFIC PHILOSOPHER / 7

from Carnap here emerges in Suppes' detailed and substantive discussion (2002, §5.5) of Carnap's "logical" approach to the foundations of probability—which discussion culminates, not surprisingly, in a characterization of Carnap's originally unique confirmation function (based on the measure m*) in a representation theorem (2002, p. 194). Suppes concludes with a number of critical comments (2002, pp. 195–198), the most important of which, for our purposes, is the last:

> Carnap ... makes a point of emphasizing that in his view there are two distinct concepts of probability. One concept is that of empirical relative-frequency, and the other is the concept of probability expressed in confirmation theory. He emphasizes most strongly that the statistical concept is empirical, while the confirmation-theory concept is analytic or logical in character. Although this distinction seems to have been accepted by a great many philosophers, it is not at all clear that a persuasive case has been made by Carnap for this dualistic approach. (2002, p. 197)

Indeed, this is perhaps Suppes' most incisive criticism of Carnap's whole logico-linguistic approach to the subject.

Carnap had initially introduced a fundamental distinction between two concepts of probability in the mid 1940s as a way of attempting to dissolve once and for all what he saw as a typically fruitless metaphysical dispute—about the "true nature" of probability—between Harold Jeffreys (representing degree of confirmation) and Richard von Mises (representing long-run relative frequency), see Carnap (1945). Although Carnap did expand his frame of reference in the 1950s to include (among others) both the statistical-likelihood approach of R. A. Fisher and the Bayesian approach of Bruno De Finnetti, the point of Suppes' skepticism, as I understand it, is that some of the most interesting more recent work in the field involves *connecting* the two concepts in ways that entirely escape Carnap's framework.[13] For, on the one hand, the deep mathematical investigations inspired by Poincaré's "method of arbitrary functions" for analyzing the chaotic behavior of deterministic dynamical systems shows precisely how certain classical long-run relative frequencies—such as those underlying the "probability of heads" in tossing a coin—emerge from the empirically given symmetries of the physical situation.[14] And, on the other hand, some of the most interesting modern Bayesian approaches involve interpreting certain privileged prior probabilities (generated by measures like Carnap's m*) as objective probabilities reflecting presupposed *de facto* symmetries.[15]

These considerations undermine the philosophical motivations for Carnap's logico-linguistic approach. For Carnap's guiding ambition was to develop a platform upon which the scientific philosopher could stand while simultaneously carrying out two distinct aims: to cooperate with applied mathematicians in developing appropriate mathematical frameworks for articulating and testing empirical scientific theories, and, at the same time, to ward off fruitless metaphysical disputes by systematically distinguishing between internal and external questions in the theory of linguistic frameworks. This is why, in the present case, Carnap insists on a logico-linguistic approach to probability.

---

[13]See Carnap (1950b/1962) for his expanded frame of reference. For a detailed and sympathetic discussion of how Carnap's work does interact that of working statisticians see Zabell (2007).

[14]Suppes discusses this approach in the following section of his long chapter on probability concerned with "propensity representations" of objective probabilities (2002, §5.6). In particular, he illustrates Poincaré's method of arbitrary functions by the treatment of coin tossing in Keller (1986, pp. 214–218) of this section.

[15]Suppes mentions this point at the end of his discussion of Carnap's approach (2002, p. 198), referring to Fitelson (2001) for further discussion.

It now appears, however, that it is for precisely this reason that Carnap's approach interacts only weakly with ongoing research in applied mathematics—and, in connection with his sharp distinction between two fundamentally different concepts of probability, thereby fails to engage with some of the most interesting recent developments. One might then reasonably harbor the suspicion, more generally, that this is also true of other attempted applications of Carnapian *Wissenschaftslogik*—including his attempted dissolution of the so-called "crisis" in the foundations of mathematics of the late 1920s and early 1930s to which Carnap (1934) was originally responding.

My final question, then, is: What does Suppes put in place of Carnapian *Wissenschaftslogik*? The answer is a version of pragmatism, but one that is again completely unprecedented and unique. In particular, it has very little in common, in the end, with what Quine calls his "more thorough pragmatism." For what Quine comes up with is an extremely broad-brush version of philosophical naturalism and empiricism, featuring a physicalist ontology and a vaguely holistic picture of empirical scientific method. And this version of philosophical naturalism and empiricism, as I suggested, appears (paradoxically) to be wholly uninterested in the details of virtually all extant empirical science.

What Suppes puts in place of Quine's broad-brush philosophical program is an astonishingly detailed engagement with real empirical science in all of its variety and complexity—from the level of abstract mathematical foundations to that of the most concrete empirical applications. Suppes' focus, to be sure, has thus been, like Carnap's, on the mathematical sciences, beginning with the physics in which he majored as an undergraduate. But his view of physics was always of a science distinguished more by the complexity and subtlety of its interaction with empirical data than any special foundational character—whether ontological or epistemological. As explained at the very beginning of Suppes' Intellectual Autobiography, his position as a military meteorologist during the War naturally encouraged such an attitude (2006, p. 2): "Knowledge of meteorology has stood me in good stead throughout the years in refuting arguments that attempt to draw a sharp distinction between the precision and perfection of the physical sciences and the vagueness and imprecision of the social sciences. Meteorology is in theory a part of physics, but in practice more like economics, especially in the handling of a vast flow of non-experimental data." As I understand him, therefore, Suppes' sees even our best mathematical science as proceeding wholly without philosophical foundations—whether ontological or epistemological—and also without any claim to ultimate truth. Mathematical or axiomatic foundations, by contrast, are quite a different story. For this enterprise, from Suppes' point of view, has less to do securing our situation against future uncertainty than with developing entirely new concepts at junctures where we are quite unsure of the right direction in which to go next—as in the attempt to develop a satisfactory mathematical psychology, for example, or to appreciate the tension between quantum mechanics and relativity created by the phenomenon of Bell-type entanglement.

Suppes' version of pragmatism is thus essentially pluralistic, with no particular approach singled out as definitive. He explains in his Intellectual Autobiography that the enterprise of modern science, accordingly, is entirely tentative and continually open-ended towards the future (2006, p. 39): "[I]t is my conviction that an important function of contemporary philosophy is to understand and to formulate as a coherent world view the highly schematic character of modern science and the highly tentative character of the knowledge that is its aim. The tension created by a pluralistic attitude toward knowledge and skepticism about achieving certainty is not, in my judgment, removable. Explicit recognition of this tension is one aspect of recent historically oriented work in the philosophy of science that I like." The reference to "historically oriented work" leads me to my final observation: Suppes' version of scientific pragmatism contains an important place for a detailed study of the history of science and, indeed, of the relationship between the history of science and the history of philosophy. Suppes' conception thereby essentially involves a very long-term historical perspective, including an appreciation of the surrounding intellectual context.

This comes out most clearly, perhaps, towards the end of his Intellectual Autobiography, summarizing a conversation with Noel Swerdlow (one of Suppes' favorite historians). Suppes reports (2006, p. 74):

> I tried to push a pragmatic theory, looking at the broad history from ancient astronomy to Kepler, to show how many concepts that were important to Babylonians for making omens and the like, and later, many aspects of Greek thought as well, were simply pushed out of the way and ignored. But the varied and detailed observations made by the Babylonian astronomers and used by Ptolemy more than five hundred years later, are even of some use today. Ptolemy's own central work was preserved in the tradition of a millennium and a half span leading up to Kepler and including, of course, the less important work of Copernicus. In this long period two important things were preserved: the observations reaching back to Babylonian times, and many of the Ptolemaic methods of computation, which Copernicus himself continued and were only changed by the new astronomy, as Kepler called what he introduced. The whole subject was then given a much greater perfection by Newton, with the introduction of gravitational dynamics. But much of what Kepler and Newton did rested on the shoulders of these observational and calculational giants of the distant past. It is this that is pragmatic—keeping the useful and letting go of the rest.

Note that there is no suggestion of convergence towards any ultimate truth in this description—just the ever richer accumulation of empirical data and the development of ever more powerful methods of computation for predicting and accommodating such data based on whatever mathematical models prove most useful at the time. Nor is there any systematic way of drawing the limits of scientific knowledge and placing fruitless metaphysical controversies forever outside this boundary. Such controversies, if Suppes is right, will simply be let go, on pragmatic grounds, as the observational and mathematical progress distinctive of modern science continues. Yet scientific philosophy still remains: a scientific philosophy for our own time.

References

Campbell, N. R. 1920. *Physics: The Elements*. Cambridge: The University Press.

—. 1928. *An Account of the Principles of Measurement and Calculation*. New York: Longmans, Green & Co.

Carnap, R. 1925. Über die Abhängigkeit der Eigenschaften des Raumes von denen der Zeit. *Kant-Studien* 30:331–345.

—. 1934. *Logische Syntax der Sprache*. Wien: Springer. Translated by A. Smeaton as *The Logical Syntax of Language*, London: Kegan Paul, Trench, Trubner & Co., 1937.

—. 1945. The two concepts of probability: The problem of probability. *Philosophy and Phenomenological Research* 5(4):513–532.

—. 1950a/1956. Empiricism, semantics, and ontology. *Revue Internationale de Philosophie* 4:20–40. Reprinted in *Meaning and Necessity*, Chicago: University of Chicago Press, 2nd edition, 1956, pp. 205-221.

—. 1950b/1962. *Logical Foundations of Probability*. Chicago: University of Chicago Press. 2nd edition, 1962.

de Barros, J. A. and P. Suppes. 2000. Inequalities for dealing with detector inefficiencies in Greenberger-Horne-Zeilinger-type experiments. *Physical Review Letters* 84(5):793–797.

Fitelson, B. 2001. *Studies in Bayesian Confirmation Theory*. Ph.D. thesis, University of Wisconsin-Madison.

Friedman, M. 2001. *Dynamics of Reason: The 1999 Kant Lectures at Stanford University*. Stanford: CSLI Publications.

—. 2007. Introduction: Carnap's revolution in philosophy. In R. Creath and M. Friedman, eds., *The Cambridge Companion to Carnap*, pages 1–18. Cambridge: Cambridge University Press.

Hilbert, D. 1899. *Grundlagen der Geometrie*. Leipzig: B. G. Teubner. Translated from the 10th German edition by L. Unger as *Foundations of Geometry*, LaSalle: Open Court Publishing Company, 1971.

Hölder, O. 1924. *Die mathematische Methode: logisch-erkenntnistheoretiche Untersuchungen im Gebiete der Mathematik, Mechanik und Physik*. Berlin: Springer.

Keller, J. B. 1986. The probability of heads. *The American Mathematical Monthly* 93(3):191–197.

Krantz, D. H., R. D. Luce, P. Suppes, and A. Tversky. 1971. *Foundations of measurement, Volume I: Additive and polynomial representations*. New York: Academic Press.

Luce, R. D., D. Krantz, P. Suppes, and A. Tversky. 1990. *Foundations of Measurement, Vol. III: Representation, Axiomatization, and Invariance*. New York: Academic Press.

McKinsey, J., A. Sugar, and P. Suppes. 1953. Axiomatic foundations of classical particle mechanics. *Journal of Rational Mechanics and Analysis* 2(1):253–272.

Quine, W. V. O. 1951/1953. Two dogmas of empiricism. *Philosophical Review* 60:20–43. Reprinted in *From a Logical Point of View*, New York: Harper & Row, 1953, pp. 20-46.

Reck, E. 2007. Carnap and modern logic. In R. Creath and M. Friedman, eds., *The Cambridge Companion to Carnap*, pages 176–199. Cambridge: Cambridge University Press.

Robb, A. A. 1936. *Geometry of Space and Time*. Cambridge: Cambridge University Press.

Rubin, H. and P. Suppes. 1954. Transformations of systems of relativistic particle mechanics. *Pacific Journal of Mathematics* 4(4):563–601. doi:10.2140/pjm.1954.4.563.

Suppe, F., ed. 1974. *The Structure of Scientific Theories*. Urbana: University of Illinois Press.

Suppes, P. 1951. A set of independent axioms for extensive quantities. *Portugaliae Mathematica* 10:163–172.

—. 1961. Probability concepts in quantum mechanics. *Philosophy of Science* 28:378–389.

—. 2002. *Representation and Invariance of Scientific Structures*. Stanford, CA: CSLI Publications.

—. 2006. Intellectual autobiography: Part I (1978); Part II (2006). URL `http://suppes-corpus.stanford.edu/autobiography.html`.

Suppes, P., D. Krantz, R. D. Luce, and A. Tversky. 1989. *Foundations of Measurement, Vol. II: Geometrical, Threshold, and Probabilistic Representations*. New York: Academic Press.

van Fraassen, B. C. 1980. *The Scientific Image*. Oxford: Clarendon Press.

Zabell, S. L. 2007. Carnap on probability and induction. In R. Creath and M. Friedman, eds., *The Cambridge Companion to Carnap*, pages 273–294. Cambridge: Cambridge University Press.

Part I

Foundations of Mathematics

1

What Numbers Are

JENS ERIK FENSTAD

Patrick Suppes has in a recent paper, "Why the effectiveness of mathematics in the natural sciences is not surprising" Suppes (2011), discussed the emergence of geometry. He sees an early beginning in the perceptual processes necessary for humans to survive. He further notes in his paper how human culture developed the skill to represent "the structure of the external world in remarkable paintings and drawings", and concludes that the key to the early development of geometry lies in the gradually emerging "structural isomorphism between, in one case a perception and an object or process in the world, and in another, between a mental image and an 'abstract' structure". Modern geometry is thus a complex product of nature, mind and culture. Except for one brief mention—"the non-geometric arithmetic methods of calculation of the Babylonians were clearly earlier than those of the Greeks"—numbers are missing from Suppes' account. But there is no mathematics without numbers and counting; we need to know what numbers are.

1 On What There Is

The ontological status of numbers is controversial. Few would deny that our "world" allows for both physical objects and states of consciousness. Some would further argue for a reduction of the mental to the physical, others would maintain a strict dualism. But a "world" consisting of only material and/or mental objects is, in my view, not sufficient for an understanding of what numbers are. A third kind of objects is necessary to give an explanation of mathematical existence.

The question of the existence of "ideal" objects has a long history. We shall make use of Popper's theory of objective knowledge as a convenient entering point for our few remarks. His philosophy with roots both in Frege and Husserl recognizes three ontologically distinct domains:

> "or, as I shall say, there are three worlds: the first is the physical world or the world of physical states; the second is the mental world or the world of mental states; the third is the world of intelligibles, or of *ideas in the objective sense*; it is the world of possible objects of thought: the world of theories in themselves, and their logical relations; of arguments in themselves; and of problem situations in themselves."

> Popper (1972, p. 154)

Foundations and Methods from Mathematics to Neuroscience.
Colleen E. Crangle, Adolfo García de la Sienra and Helen Longino.

Even if the so-called World 3 is a "man-made product", Popper argues for the objective nature of what it contains, and he highlights the causal interaction between this world and the other two. His discussion ranges over a number of themes, among them the possibility of how a World 3 theory could provide an epistemology for the humanistic sciences. But rather than continue with Popper's World 3, I shall briefly turn my attention from philosophy to anthropology.

In a paper published some years ago, Fenstad (1985), I pointed to a striking similarity between the Popperian ontology and the views on mathematical existence expressed by the anthropologist L. A. White in an essay, *The Locus of Mathematical Reality: An Anthropological Footnote*, White (1947): Do mathematical truths reside in the external world, there to be discovered by man, or are they man-made inventions? Does mathematical reality have an existence and validity independent of the human species or is it merely a function of the human nervous system?

White provides an answer to these questions by adding to the two existential categories of *matter* and (individual) *mind* the category of the *mind of the species*. The latter is a term sometimes used by anthropologists in their general discussion of *human culture*. For our purposes we may note the close relationship of this concept to the Popperian notion of World 3. It is in this notion of culture that our anthropologist sees a solution to the problem of mathematical existence – in his words, there is no mystery about mathematical reality. Mathematics is a kind of primate behavior as languages, musical systems and the penal code are, and mathematical concepts are man-made just as ethical values, traffic rules, and bird cages are:

> "Mathematical truths exist in the cultural tradition into which the individual is born, and so enters his mind from the outside. But apart from cultural tradition, mathematical concepts have neither existence nor meaning, and of course, cultural traditions have no existence apart from the species. Mathematical realities thus have an existence independent of the individual mind, but are wholly dependent upon the mind of the species".

> White (1947)

Our anthropologist is not an isolated phenomena. The cultural view on mathematics has a significant support within the mathematical community. A classical reference is "The Mathematical Experience" by P. J. Davies and Reuben Hersh (1981), see also Hersh (1997). A recent voice in support is David Mumford, see "the taxonomy of mathematics" in Mumford (2000). For a carefully argued account of mathematics as a "social construct", see Cole (2008) and Cole (2012). We should also mention the earlier contribution of Kitcher (1984). Philosophers of mathematics have not always been convinced by this approach, see the critical analysis in M. Leng (2007).

The arguments of the anthropologist may well convince us as long as the discussion proceeds on this general level. But we need to understand with much greater precision how matter, mind and culture interact, how mathematical objects come into existence, and how they can be known by man. Some would at this point turn back to philosophy in the tradition of Frege, see Linnebo (2008, 2009), others would turn to the phenomenology of Husserl, a noteworthy example is Gödel (1995). For a general discussion of the relationship between Husserl and Gödel see Føllesdal (1995) and Hauser (2006). A useful introduction to Husserl's philosophy of mathematics is the survey article written by Richard Tieszen for the *Cambridge Companion to Husserl*; see Tieszen (1995). For a post-Gödel development of these ideas see Tieszen (2011). Not being a Husserl scholar I shall refrain from going into the details of his theory of *intentionality* and how it can be used both to elucidate the nature of mathematical objects and to explore how we obtain knowledge of them.

One noteworthy aspect of Tieszen's survey is his sharp criticism of much of current philosophy of mathematics from a Husserlian perspective. In many respects this criticism is similar to the one offered by our anthropologist. But there is one crucial difference, Husserl, according to Tieszen, holds that "mathematical objects cannot be assimilated to social or cultural objects, because social and cultural objects are bound to times and places", Tieszen (1995, p. 448).

But is this necessarily so? How could the cultural view be strengthen to meet the criticisms of Husserl? This is a question I would like to explore further. We agree that numbers are abstract objects, but how are they related to brain and mind? I shall approach this question by first discussing the evolution of language and how syntactic categories such as noun phrases NP and verb phrases VP became objects of the collective mind, see chapter three of *Grammar, Geometry and Brain*, Fenstad (2010). Highly relevant are the results by Nowak et al. (2000), based on a general model of evolutionary dynamics. These results show how an elementary form of "protosyntax", the decomposition a message into a NP + VP form, is the outcome of an evolutionary process. Note that this can be seen as a starting point for the iterative or recursive property characteristic of human languages. The evolutionary process explains how a NP + VP structure which first occurs in the mind of individuals, evolves into an object in the collective mind of the speech community. And since the structure of the world which lies at the base of the Nowak model, is simple and almost universally shared, we also come to understand how syntactic categories will, as many grammarians strongly insist, be the same across the many existing and seemingly different speech communities, and thus become "true" objects in the collective mind of the human species.

There has been much current interest among neuroscientists on how mathematics is related to brain structure, see *The Number Sense. How the mind create mathematics*, S. Dehaene (1997). His starting point is clear: "Our abstract mathematical constructions originate in the coherent activity of our cerebral circuits and of the millions of other brains preceding us that helped shape and select our current mathematical tools", Dehaene (1997). The author thus clearly recognizes that both the mind of the individual and the collective mind of the species are necessary for an understanding of the nature of mathematics. Part of his work reviews brain mechanisms that are believed to underlie mathematical activities. This is no easy task. Experimental techniques can tell us the location of cognitive activities in the brain, but rarely reveal the exact mechanisms, i.e. the equations, that govern these activities. Sometimes neural network simulations can be used to suggest the actual mechanisms, one interesting example is the so-called "accumulator model" of Dehaene and Changeux (1993) which simulates how animals extract numbers from their environment. The constructed network can be used to understand rudimentary number processing by man and animals. Another early contribution is Amit (1989). In the section *Tentative steps into Abstract Computations* he constructs a network for counting chimes and writes down the exact equation determining its dynamics. In discussing the cognitive relevance of the model he notes: "Once again, one should recall the methodological disclaimer. We do not, by any stretch of the imagination, purport to describe children by the little model counting chimes. Yet it is of interest to observe that a number of features, which one usually associates with human cognitive psychology, do find a simple and well defined echo in this version of a neural network", Amit (1989). Much have happened since Amit and Dehaene; for a current update concerning our understanding of the cognitive basis of mathematical knowledge, see Cappelletti and Giardino (2007).

But is this sufficient for an invariant foundation? Will an evolutionary perspective also be needed? In the same way as the Nowak model suggested an objective basis for some syntactic categories, there could be a similar story to tell concerning numbers and other mathematical objects. The Nowak model presupposed a simple "world structure" consisting of individuals and events as an input to the evolutionary process. In analogy to this I suggest that some elementary skills and intuitions about counting and measurement acted as a necessary input to a similar development of mathematics. This does not exclude that some rudimentary part of numerical competence is innate. This is the conclusion in Cappelletti and Giardino (2007): "We have addressed the issue of whether numerical knowledge represents a domain-specific, innate semantic category and we supported this view within the framework of cognitive science by presenting evidence from experimental, developmental and animal psychology, from neuropsychology and neuroimaging studies." But this is not necessarily the whole story. We also need to understand how mathematics evolved: First, a survey of the non-European roots of mathematics, G. G. Joseph (1991), reveals a long and complicated history of how numbers and systems of notation developed out of "bones, strings and standing stones". Next, we note the lessons learned from the many studies of numerical cognition in "primitive" societies today, Gordon (2004); Beller and Bender (2008). Gordon concludes his study of the Piraha tribe in the Amazon Basin that what is innate is at most the ability to see specific numbers up to three, the rest is culturally determined. We also note that a considerable body of "trade and marketplace" mathematical skills, independent of abstract mathematical knowledge, survived for a long time in our culture, see Ferguson (1992). But mathematics as known today did emerge. It is a difficult and still unresolved question what is *ab initio* hard wired and what is the result of cultural evolution. The relationship between culture and the human genome is complicated, see *How culture shaped the human genome: bringing genetics and the human sciences together*, Laland et al. (2010), but it is definitively a two-way and highly interactive process. There is still much to learn and many sciences will be called upon for help, but I would claim that enough is now known to assert that *number* and some basic *syntactic categories* did emerge as invariants across human cultures, hence became true objects of the collective mind of the species.

Not everyone will agree to this cultural view. It has been a widely shared opinion that the absoluteness of simple arithmetical facts, such as $5 + 7 = 12$, is a strong argument against the cultural dependence of mathematics: If mathematics is culture-dependent, there could be cultures where the meaning of terms are the same, but at the same time $5 + 7$ and 12 are not equal. If this is not possible, the "fact" that $5 + 7 = 12$ would necessarily be both culture- and mind-independent. Some would see this as an argument in favour of some form of platonism, Tieszen (2010). I do not agree with this line of reasoning. I shall argue below that structure is the basic entity, syntax is derived and adapted. Thus if structure is given and a suitable structure-dependent syntax has emerged, the arithmetical "facts" will follow with necessity. Therefore, if I say that 5 + 7 equals 12 and you say no, we have to go back to basic structures. I have argued above for the cultural invariance of numbers and of counting / addition based on what we currently know about neuroscience and anthropology. I do not deny that there could be cultures, past or future, where basic structures and therefore "facts" are otherwise. But for cultures in our invariance class numbers are true objects in the "mind of the species", and elementary arithmetical facts are absolute. This is what is needed for the applicability of mathematics.

2   On How We Know

Much should still be said about the complex relationship between brain, mind and culture. And many sciences must contribute to this story. I have pointed to some facts from evolutionary theory and neuroscience, others will stress the importance of a psychological theory of learning in relating the individual learning to the gradual acquisition of knowledge in a culture, see Suppes (2011). But in addition to "what there is" we also need to have an answer to the question "how we know".

Against Wigner's "unreasonable effectiveness", Wigner (1960), we shall argue for an "obvious applicability" - it is possible to understand how we know. Let us seek a first clue from studies on the origin of language. In a previous section we mentioned an evolutionary model for how syntactic communication evolved. The starting point was a structural awareness or understanding of an environment consisting of both individuals/objects and properties/events. The result of the proposed evolutionary process was a transition from the initial communication in whole messages to a language showing some proto-syntactic form. And no one would question that this language was quite appropriate for what was seen in the environment.

We could also point to numbers and geometry to see some early examples of "obvious applicability", Suppes (2011). Measurements and constructions relied on some structural understanding of what we "see", and a "proto-mathematics" developed as a tool to make use of these structures, in particular, in the building and construction trades. It is interesting to note that an implicit knowledge of geometric structures survived up through and well beyond the middle ages within these trades; the Gothic cathedrals being an outstanding example; Ferguson (1992).

In these examples structure is prior to syntax. The most general analysis of the notion of structure is today found in mathematical logic. This analysis has an important root in the work of the Norwegian logician Thoralf Skolem (1920), see Fenstad and Wang (2009). But the logic approach may at times be somewhat too restrictive. An alternative would be to see the notion of structure in light of prototype theory, which in a sense is to understand structure as a kind of "fixed point" for what is seen; see Rosch (1983) and Gärdenfors (2000). The reader should also at this point consult Mumford (2002) on pattern theory and the mathematics of perception. In insisting that that structure is what we "see" and that syntax is what we introduce to make sense of what we "see", we come close to the idea of intentionality as used in phenomenology.

The art and science of *mathematical modeling* will be our bridge from what we see to how we know. It consists typically of three stages: First an analysis of a given scientific or technological problem leading up to a well-defined *structure* (notice that we in this one sentence have hidden all we have to say about *observations* and *experiments*); next comes *syntax* and "equations", which are used to analyse the properties of the structure; finally the development of *algorithms*, which in a scientific context can be used to make specific predictions and to determine its long time development, and which in a technological context used to produce "blueprints" for construction and production. But equations normally have their limitations, they cannot always be exactly solved. This adds a further ingredient to the practice of mathematical modeling, *simulations* and graphic displays.

Much need to be added to these few words about mathematical modeling as a general model. We need, in particular, to understand the notion of algorithms in a suitable general form. There are also many questions concerning explicit/conscious versus implicit/sub-conscious levels of both structure, syntax and algorithms. I have taken some notice of this in my reference to the cathedral builders of the middle ages, but more needs to be said. At this point let me just stress that all three parts, structure, syntax, algorithms, are abstract objects. A structure is an object of our culture which describes or pictures some part of the material and/or the mental, and even of the cultural world. Syntax and algorithms are tools created to understand and make use of structures. They are "obviously applicable". And if this is accepted, we are, in my view, a step closer to an understanding of "how we know".

This is the simple picture. But Wigner (1960) points to examples where some pre- existing theories, developed deep inside mathematics, have important applications. Today he would probably refer to the interaction between algebraic geometry, topology and quantum field theory, surely in his view this "is something bordering on the mysterious". We are not so convinced. We believe that a first step to understand this mystery is to note that "pure" mathematics evolved out of applied mathematics *when syntax became structure*. The first example is, perhaps, the elementary theory of numbers. We speculated above on the neuronal and cultural origin of some rudimentary forms of arithmetic based on numbers, counting/addition and measurements/multiplication. But when this syntax for counting and measurement is itself seen as a new structure, we understand how the "rekurrirende Denkweise" of Skolem (1923) and higher forms of mathematics could develop.

One final note, our discussion is seen to strongly support a semantic or model- theoretic view of general scientific methodology, different from the standard HD (hypoteticodeductive) approach. The HD methodology is dominated by the sequence: hypothesis, deduction, test and falsification. This is a methodology dominated by syntax and logic. We have insisted on starting with what we "see", which in a sense is a structural alternative to the standard HD method, see Suppes (2002). In this approach structures and patterns are the foundation. We find a forceful expression of a similar view in Barrow (2010): "Mathematics is simply the catalogue of all possible patterns. Some of those patterns are especially attractive and are studied for decoration, others are patterns in time or in chains of logic. Some are described solely in abstract terms, while others can be drawn on paper or carved in stone. Viewed in this way, it is inevitable that the world is described by mathematics."

References

Amit, D. J. 1989. *Modeling Brain Function: The World of Attractor Neural Networks*. Cambridge, UK: Cambridge University Press.

Barrow, J. D. 2010. Simple really: From simplicity to complexity—and back again. In B. Bryson and J. Turney, eds., *Seeing Further: The Story of Science and the Royal Society*. London: HarperPress.

Beller, S. and A. Bender. 2008. The limits of counting: Numerical cognition between evolution and culture. *Science* 319(5860):213–215. doi:10.1126/science.1148345.

Cappelletti, M. and V. Giardino. 2007. The cognitive basis of mathematical knowledge. In M. Potter, A. Paseau, and M. Lang, eds., *Mathematical Knowledge*, pages 74–83. Oxford: Oxford University Press.

Cole, J. C. 2008. Mathematical domains: Social constructs? In B. Gold and R. A. Simon, eds., *Proof and other Dilemmas: Mathematics and Philosophy*, pages 109–128. Mathematical Association of America.

—. 2012. Toward an institutional account of the objectivity, necessity, and atemporality of mathematics. *Philosophia Mathematica* doi:10.1093/philmat/nks019.

Davies, P. J. and R. Hersh. 1981. *The Mathematical Experience*. Boston: Birkhäuser Boston.

Dehaene, S. 1997. *The Number Sense: How the Mind Creates Mathematics*. London: Allen Lane The Penguin Books.

Dehaene, S. and J. P. Changeux. 1993. Development of elementary numerical abilities: A neuronal model. *Journal of Cognitive Neuroscience* 5(4):390–407.

Fenstad, J. E. 1985. Is nonstandard analysis relevant for the philosophy of mathematics? *Synthese* 62(2):289–301.

—. 2010. *Grammar, Geometry, and Brain*. Stanford: CSLI Publications.

Fenstad, J. E. and H. Wang. 2009. Thoralf Albert Skolem. In D. M. Gabbay and J. Woods, eds., *Handbook of the History of Logic*, vol. 5, pages 127–194. Amsterdam: North Holland. doi:10.1016/S1874-5857(09)70008-1.

Ferguson, E. S. 1992. *Engineering and the Mind's Eye*. Cambridge, MA: MIT Press.

Føllesdal, D. 1995. Introductory note to Gödel 1961. In S. Feferman, J. J. W. Dawson, W. Goldfarb, C. Parsons, and R. N. Solovay, eds., *Kurt Gödel. Collected Works, Vol III*, pages 364–373. New York and Oxford: Oxford University Press.

Gärdenfors, P. 2000. *Conceptual spaces: The geometry of thought*. Cambridge, MA: MIT Press.

Gödel, K. 1995. The modern development of the foundations of mathematics in the light of philosophy. In S. Feferman, J. J. W. Dawson, W. Goldfarb, C. Parsons, and R. M. Solovay, eds., *Kurt Gödel. Collected Works: Unpublished Essays and Lectures*, vol. III, pages 374–387. New York and Oxford: Oxford University Press.

Gordon, P. 2004. Numerical cognition without words: Evidence from amazonia. *Science* 306(5695):496–499. doi:10.1126/science.1094492.

Hauser, K. 2006. Gödel's program revisited, part i: The turn to phenomenology. *The Bulletin of Symbolic Logic* 12(4):529–590.

Hersh, R. 1997. *What is Mathematics, Really?* Oxford: Oxford University Press.

Joseph, G. 1991. *The Crest of the Peacock: Non-European Roots of Mathematics*. London: Penguin Books.

Kitcher, P. 1984. *The Nature of Mathematical Knowledge*. New York: Oxford University Press.

Laland, K. N., J. Odling-Smee, and S. Myles. 2010. How culture shaped the humen genome: Bringing genetics and the human sciences together. *Nature Reviews/Genetics* 11(2):137–148. doi:10.1038/nrg2734.

Leng, M. 2007. Introduction. In M. Leng, A. Paseau, and M. D. Porter, eds., *Mathematical Knowledge*, pages 1–15. Oxford: Oxford University Press.

Linnebo, O. 2008. The nature of mathematical objects. In B. Gold and R. A. Simon, eds., *Proof and Other Dilemmas: Mathematics and Philosophy*, pages 205–219. Washington D.C.: Mathematical Association of America.

—. 2009. The individuation of the natural numbers. In O. Bueno and O. Linnebo, eds., *New Waves in Philosophy of Mathematics*, pages 220–238. Hampshire: Palgrave Macmillan.

Mumford, D. 2000. The dawning age of stochasticity. In V. I. Arnol'd, ed., *Mathematics: Frontiers and Perspectives*, pages 199–218. Providence, R.I.: American Math Society.

—. 2002. Pattern theory: The mathematics of perception. In *Proceedings of ICM*, vol. I, pages 401–422. Beijing, China.

Nowak, M. A., J. B. Plotkin, and V. A. A. Jansen. 2000. The evolution of syntactic communication. *Nature* 404(6777):495–498. doi:10.1038/35006635.

Popper, K. R. 1972. *Objective Knowledge: An Evolutionary Approach.* Oxford: Clarendon Press.

Rosch, E. 1983. Prototype classification and logical classification: The two systems. In E. K. Scholnick, ed., *New Trends in Conceptual Representation: Challenges to Piaget's theory?* Hillsdale, NJ: Lawrence Erlbaum Associates.

Skolem, T. A. 1920. Logisch-kombinatorische untersuchungen über die erfüllbarkeit und beweisbarkeit mathematischen sätze nebst einem theoreme über dichte mengen. *Skrifter uget av Videnskabsakademiet i Kristiania* Skrifter I(3):1–36.

—. 1923. Begründung der elementären arithmetik durch die rekurrierende denkweise ohne anwendung scheinbarer veränderlichen mit unendlichem ausdehnungsbereich. *Skrifter uget av Videnskabsakademiet i Kristiania* Skrifter I(6):38 pp.

Suppes, P. 2002. *Representation and Invariance of Scientific Structures.* Stanford, CA: CSLI Publications.

—. 2011. Why the effectiveness of mathematics in the natural sciences is not surprising. *Interdisciplinary Science Reviews* 36(2):244–254. URL 10.1179/030801811X13082311482645.

Tieszen, R. 1995. Mathematics. In B. Smith and D. W. Smith, eds., *The Cambridge Companion to Husserl*, pages 438–462. Cambridge: Cambridge University Press.

—. 2010. Mathematical problem-solving and ontology: An exercise. *Axiomathes* 20(2-3):295–312.

—. 2011. *After Gödel: Platonism and Rationalism in Mathematics and Logic.* Oxford: Oxford University Press.

White, L. A. 1947. The locus of mathematical reality. *Philosophy of Science* 14(4):289–303. Reprinted in Newman (1956), *The World of Mathematics*, New York: Simon and Schuster.

Wigner, E. P. 1960. The unreasonable effectiveness of mathematics in the natural sciences. *Communications in Pure and Applied Mathematics* 13(I):1–14. Reprinted in Wigner (1979), *Symmetries and Reflections*, Woodbridge, Ct: Ox Bow Press, pp. 222-237.

2

The "Axiom" of Choice is not an Axiom of Choice

Jaakko Hintikka

Patrick Suppes has for a long time been one of my best friends, literally the best man at my wedding to Merrill in 1978. I was delighted to have a chance at honoring him by participating in the Suppes symposium at Stanford in the spring 2012. I was so enthusiastic that the paper I gave there, entitled "There is no Set Theory, but there are Set Theoretical Problems" soon became far too long for inclusion in the proceedings volume. What is found here is a fragment of that self-contained longer paper in Hintikka (forthcoming).

1 The ZF Set Theory is not What Zermelo Intended

One of Zermelo's main aims in axiomatizing set theory was to justify the principle ("axiom") of choice that he had earlier relied on in his well-ordering theorem. This assumption is the most interesting axiom in the usual axiomatization of set theory. Its precise status in axiomatic set theories is nevertheless ambiguous.

It is usually said that the axiom of choice is one of the axioms of ZF set theory. This is inaccurate, however, for the unlimited form of the axiom of choice is not only not among the axioms of ZF but is incompatible with it. For what is the general form of this "axiom"? It is supposed to say that for any set of sets there exist a function (choice function) that picks out precisely one member of each member set. Thus if member sets are distinguished from each other by the variable x, each of them can be thought of as consisting of elements y that satisfy the condition $F[x, y]$. Then a choice function f is one that satisfies the formula

$$(\forall x)F[x, f(x)] \tag{1}$$

The axiom of choice then says in this case that

$$(\forall x)(\exists y)F[x, y] \supset (\exists f)(\forall x)F[x, f(x)] \tag{2}$$

The deductive force of (2) depends on the (usually complex) formula $F[x, y]$ Accordingly, we have a large family of varieties of axiom of choice type assumptions. What is common to them is that they all assert the existence of Skolem functions. Thus the most general form of the axiom of choice would assert that there exists a full set of Skolem functions for each true sentence. An outright contradiction is avoided in ZF by

Foundations and Methods from Mathematics to Neuroscience.
Colleen E. Crangle, Adolfo García de la Sienra and Helen Longino.

restricting the force of the axiom. In the jargon of von Neumann-Bernays set theory, it holds only for sets but not in general for classes.

This is already puzzling; for whatever intuitions we are supposed to have about the axiom of choice seem to support the unlimited form quite as much as the more limited forms. They do not tell us anything about why such restrictions are needed.

This shows that the ZF set theory does not after all implement the intended force all of Zermelo's axioms, Moore (1982). It incorporates only a limited version of what the axiom of choice should do. This limits the usefulness of ZF as a tool for model theory, including its own model theory.

2 Axiom of Choice and Existential Instantiation

The confusions about the axiom of choice can be said to be reflected in its name. It is not, it will be argued, one axiom among others in some particular mathematical theory, and it is not about choices but about existence. It is a first-order logical principle that can be considered a generalization of the familiar rule of existential instantiation that authorizes us to move in a proof of logical truth from a formula of the form $(\exists x)F[x]$ to $F[b]$, where b is a new individual constant.

The meaning of this rule is obvious. It is a kind of "John Doe" rule. When we know or are assuming that there exists individuals x that satisfy $F[x]$, we simply decide to consider one of them and, literally for the sake of argument, assign a name for it in order to be able to argue about it. In a court of law we might call such a person John Doe or Jane Roe, and in algebra we could use letters for "unknowns" $x, y, z, \dots$. (One of the pioneers of modern algebra, John Wallis, claimed that the latter practice was modeled on the former.) As long as this rationale for existential instantiation is kept in mind, the rule is totally obvious and intuitive.

The only main formal restriction is that the rule can only be applied to sentence-initial existential quantifiers. It is a fallacy to apply to an existential quantifier within the scope of other quantifiers or strictly speaking to a quantifier that depends on other quantifiers. One cannot infer $(\forall x)F[x, b]$ from $(\forall x)(\exists y)F[x, y]$ If a student does so, you correct him or her. But what if the student tries to defend himself and appeals to the Jane Roe idea? What an existential quantifier expresses is existence, whether or not it occurs within the scope of other quantifiers. Why cannot one simply dub one of the acknowledgedly existing ones? How the instructor should respond is to point out that the choice of the existing individual depends on the values of other quantifiers. Hence a simple dubbing step is impossible.

But if the student's name were Kurt or Alfred, we can imagine him to push the instructor further and to say, "But, Sir, even if so, I can surely still use an instantiating term. It's only that now it cannot be a name. But it can be a function term that spells out those dependences."

Such a student would be right. We can extend the rule of existential instantiation to existential-force quantifiers occurring inside a formula. Assuming that the formula is in a negation normal form, the rule will authorize us to move from a formula of the form

$$S(-(\exists x)F[x]-) \tag{3}$$

to the formula

$$(S-F[f(y_1, y_2, \dots)]-) \tag{4}$$

where $(Q_1y_1), (Q_2y_2)\dots\dots$ are all the quantifiers on which $(\exists x)$ depends on in S. In the received first-order logic quantifier dependencies are expressed by the nesting of scopes.

The generalized rule applies much more widely than in the received first-order logic as soon as the dependence relations are somehow explicitly expressed.

Strictly speaking, we must also consider the possible dependencies of $(\exists x)$ on constant terms in S. They nevertheless do not affect the line argument in this paper.

In (4) f is a new function constant. But no quantification over functions is involved. The generalized rule is completely first-order; no higher-order quantification is involved. The result of the generalization is simply a new formulation of the first-order logic. The new system of first-order logic is obviously equivalent with the original one. Or, rather, it is obviously semantically (model-theoretically) equivalent. Hence, its deductive equivalence follows from any completeness proof for the usual first-order logic.

Although equivalent with the received first-order logic, the thus extended traditional logic has a great deal of interest in its own. In it we can eliminate existential quantifiers in terms of their Skolem functions and obtain a streamlined logic in which quantificational reasoning is replaced by a logic of equations, in other words, by a kind of generalized algebra. This often simplifies greatly actual proofs. (If you do not see this, try to formulate group theory in terms of quantifiers as in Suppes' uniquely applicable textbook Suppes (1957)).

Also proof theory becomes easier. This is because in proof theory the intricate questions typically concern the order of different proof rules. This can be complicated because the usual rules of reference can be applied only to the governing ("outmost") quantifier or connective of a formula. This means that in order to apply a rule to a formula inside a larger one we first have to dredge it out to the beginning of an entire formula by means of other rule applications, which complicates attempts to vary the order of rules. The generalized existential instantiation is not dependent for its applicability in this way or other rule applications.

3   The "Axiom" of Choice is a Logical Principle

The same formulas will be provable and the same inferences valid after the generalization as before in first-order logic. The differences are manifested when the reformulated first-order logic is used as the logic of set theory or as a part of a higher-order logic, Jech (1973). Then suddenly all the applications of the axiom of choice become provable. For instance, (1) becomes provable, perhaps as follows:

$$
\begin{array}{lll}
\text{(i.)} & (\forall x)(\exists y)F[x,y] & \text{assumption} \\
\text{(ii.)} & (\forall x)F[x,g(x)] & \text{generalized existential instantiation} \qquad (5) \\
\text{(iii.)} & (\exists f)(\forall x)F[x,f(x)] & \text{second-order existential generalization}
\end{array}
$$

This ought to remove all doubts about the validity of the "axiom" of choice as a principle of mathematical reasoning. The axiom of choice is a completely obvious first-order logical truth. It can be obtained from the usual first order logic by an innocuous liberalization of one of its rules. This liberalization does not make first-order logic any less obvious semantically. The truth of a quantificational sentence means the existence of suitable "witness individuals". But these individuals do not come one by one, but in combination with each other and hence dependent upon each other. Skolem functions codify these dependencies, and for that reason their existence is the natural truth condition of quantified sentences. The "axiom" of choice merely asserts the existence of such Skolem functions for all true sentences. To deny the axiom of choice is therefore to change what we mean by truth in our quantificational discourse.

Doubts about the axiom of choice seem in some cases caused by a confusion between the truth of a quantificational proposition S and its being known, see Howard and Rubin (1998). For the truth, it suffices that a full set of Skolem functions for S exists. Here existence has to be taken in a totally abstract sense. It is not needed for understanding such a sentence that the speaker (or some other kind of statement-maker) or the hearer knows what those functions are or may be, although there undoubtedly is a conversational expectation that the speaker has some idea of what they are. But of course for the knowledge of the proposition some acquaintance with the relevant Skolem functions (roughly speaking, methods of verification) is of course needed.

In any case, the axiom of choice is not needed in set theory as a special axiom or other separate assumption. It can be taken as a part of the logic that is being used. Questions about its dependence or independence of other axioms do not arise. Indeed, no other set-theoretical axioms need to be used either, which is yet another example of the vacuousness of a first-order axiomatic approach to set theory.

In a historical perspective, this shows that a ZF type system of set theory fails to fulfill Zermelo's second main aim in axiomatizing set theory. Instead of justifying the "axiom" of choice, such an axiom system turns out to be incompatible with the unrestricted principle of choice. What is worse, no axiomatization is needed to buttress the "axiom" of choice, for correctly understood it turns out to be purely logical truth completely independently of all set-theoretical questions. That such a purely logical principle cannot be incorporated in an axiom system like ZF without restriction shows that there is something deeply wrong with the system.

References

Hintikka, J. forthcoming. Axiomatic set theory in memoriam. Philosophy, Boston University.

Howard, P. and J. E. Rubin. 1998. *Consequences of the Axiom of Choice*. Providence, R.I.: American Mathematical Society.

Jech, T. J. 1973. *The Axiom of Choice*. Amsterdam; New York: North-Holland Pub. Co.; American Elsevier Pub. Co.

Moore, G. H. 1982. *Zermelo's axiom of choice: Its origins, development, and influence*. New York: Springer-Verlag.

Suppes, P. 1957. *Introduction to Logic and Axiomatic Set Theory*. New York: Van Nostrand Reinhold Co. Reprinted Dover, New York, 1999.

3

Invariant Maximality and Incompleteness

Harvey M. Friedman

I had the honor of presenting a lecture at the Symposium on the Occasion of Patrick Suppes 90th Birthday. This contribution to the Proceedings is the result of many improvements and further developments since my talk.

Abstract

We present new examples of discrete mathematical statements that can be proved from large cardinal hypotheses but not within the usual ZFC axioms for mathematics (assuming ZFC is consistent). These new statements are provably equivalent to Π_1^0 sentences (purely universal statements, logically analogous to Fermat's Last Theorem)—in particular provably equivalent to the consistency of strong set theories, including one that is in explicitly Π_1^0 form. The examples live in the rational numbers, with only order, where the nonnegative integers are distinguished elements. The statements take the general form: every order invariant $W \subseteq Q^{2k}$ has a maximal subset S^2, with an invariance condition. Certain statements of this form are shown to be provably equivalent to the widely believed Con(SRP), and hence unprovable in ZFC (assuming ZFC is consistent). Modifications are made, involving a simple cross section condition, which propels the statement beyond the huge cardinal hierarchy, to attain equivalence with Con(HUGE). We also present some nondeterministic constructions of infinite and finite length with some of the same metamathematical properties. These lead to practical computer investigations designed to provide arguable confirmation of Con(ZFC) and more.

1 Preface

Patrick Suppes is the most successful person that I have interacted with. Looking at Suppes' spectacular output at Suppes (2002), it is clear that at least by some measures, He could be the most successful general intellectual alive today. At Suppes (2002), Suppes' spectacular output is divided into five categories: Computers and Education; Language and Logic; Methodology, Probability, and Measurement; Physics; and Psychology and Neuroscience. His work in any single one of these five areas alone would represent at least a distinguished career at major intellectual centers, and probably much more.

I can only give a bare indication of the range and depth of Suppes' work here, based on Suppes (2002), and encourage the reader to look at Suppes (1992, 1999) and

Foundations and Methods from Mathematics to Neuroscience.
Colleen E. Crangle, Adolfo García de la Sienra and Helen Longino.
Copyright © 2014, CSLI Publications.

Ferrario and Schiaffonati (2012). There are many overlapping themes and cross connections between his work in these five categories, and so we found the placement of some of his honors and books into a single chosen category to be somewhat artificial.

Computers and Education

At the beginning of his work in Computers and Education, Suppes was involved in actual experimentation with 1st graders in 1959. This represents a constant theme throughout his long career—to become immersed in real world data before, during, and after developing theories. He is universally recognized as a—and perhaps the—major pioneer in Computers and Education, both on the entrepreneurial side and the academic side. The emerging MOOC movement was fully envisioned by Suppes many decades ago, and he is arguably the person who is more responsible for the MOOC movement than anyone else on the planet.

For work in the general area of Computers and Education, Suppes's honors include:

- Palmer O. Johnson Memorial Award, American Educational Research Association, 1967.
- John Smyth Memorial Lecturer, Victorian Institute of Educational Research, Melbourne, Australia, 1968.
- American Educational Research Association, Phi Delta Kappa Meritorious Researcher Award, 1971.
- E. L. Thorndike Award for Distinguished Psychological Contribution to Education, American Psychological Association, 1979.
- Henry Chauncey Award for Distinguished Service to Assessment and Educational Science, Educational Testing Service, 2003.
- Lifetime Achievement Award in Education Technology, Software and Information Industry Association, San Francisco, CA, 2012.

Suppes has authored several books in Computers and Education, including:

- Geometry for Primary Grades (with N. Hawley).
- Sets, Numbers, and Systems (with B. Meserve, P. Sears).
- Computer-assisted instruction at Stanford, 1966–68.
- Data, Models, and Evaluation of the Arithmetic Programs (with M. Morningstar).

Language and Logic

Suppes' work in Language and Logic not only lays the theoretical basis for the online education revolution, but also represents seminal advances in semantics of natural language, machine learning theory, language aspects of neuroscience, and the active field of formal verification of mathematical proofs. Many of these works were way ahead of their time. There is also much work along of the lines of finitistic and quantifier free foundations of infinitary mathematics that has not been properly followed up by the mathematical logic community, and which may be equally prescient.

For work in the general area of Language and Logic, Suppes' honors include:

- President, Division of Logic, Methodology and Philosophy of Science, International Union of History and Philosophy of Science, 1975–79.
- Eleventh Annual Alfred Tarski Lectures, University of California, Berkeley, 1999.
- Barwise Prize, Committee on Philosophy and Computers, American Philosophy Association, 2002.

Suppes has authored several books in Language and Logic, including:

- Decision Making: An Experimental Approach (with D. Davidson and S. Siegel).
- Introduction to Logic.
- Axiomatic Set theory.
- First Course in Mathematical Logic (with S. Hill).
- Probabilistic Metaphysics, Language for Humans and Robots, Language and Learning for Robots (with C. Crangle).

Methodology, Probability and Measurement

Suppes' work in Methodology, Probability and Measurement includes seminal works on utility theory, preference theory, foundations of probability theory, causality, subjective probability, confirmation, belief, and measurement theory. He is a preeminent pioneer in the area of mathematical psychology.

For work in the general area of Methodology, Probability and Measurement, Suppes' honors include:

- Distinguished Scientific Contribution Award, American Psychological Association, 1972.
- Member, National Academy of Sciences, 1978.
- National Medal of Science, 1990.

Suppes has authored several books in Methodology, Probability and Measurement, including:

- Studies in the Methodology and Foundations of Science: Selected Papers from 1951–1969.
- Markov Learning Models for Multiperson Interactions (with R. Atkinson).
- A probabilistic Theory of Causality.
- Models and Methods in the Philosophy of Science: Selected Essays.
- Foundations of Probability with Applications: Selected Papers, 1974–1995.
- Foundations of Measurement, Vols 1–3 (with R. Krantz, E. Luce, and A. Tversky).

Physics

Suppes' work in Physics concerns the foundations of classical mechanics, relativity, and quantum mechanics, and has been deeply influential in the philosophy of science community.

Suppes' honors for work in the general area of foundations of physics, include:

- President, Pacific Division, American Philosophical Association, 1972–73.
- President, International Union of History and Philosophy of Science, 1976, 1978.
- Lectures "Determinism and Prediction", "Determinism, Chaos and Randomness", "Free Will" and "Determinism and Free Will", Collège de France, Paris, 1988.
- Ernest Nagel Memorial Lecture, "Determinism, Computation and Free Will", Columbia University, 1988.
- Evert Willem Beth Lecture, "Physical Determinism and Biological Computation", Nijmegen, The Netherlands, 1989.
- Member, American Philosophical Society, 1991.

- First Annual Karl Popper Visiting Lecture, "The Nature of Freedom", London School of Economics and Political Science, University of London, 1995.
- Reichenbach Lecture, "Freedom, Determinism and Biological Computation", University of California, Los Angeles, 1997.
- Lauener Prize in Philosophy, Lauener Foundation, Basel, Switzerland, 2004.
- Lakatos Award Prize for book in philosophy of science, London School of Economics and Political Science, 2003.

Suppes has published the monumental Representation and Invariance of Scientific Structures, in 2002, for which he won the above cited Lakatos Award Prize.

Psychology and Neuroscience

Suppes' long interest in psychology has gradually moved over the years to intense ongoing path breaking research in neuroscience, with a particular focus on the processing of language. He began serious research in neuroscience only ten years after his retirement from Stanford.

Suppes's honors for work in psychology and neuroscience include the William Lowe Bryan Memorial Lecture, "Brain-wave Recognition of Words and Sentences", Indiana University, 1998. Suppes (1999) lists a very large number of publications in this area, including numerous publications in the prestigious Proceedings of the National Academy of Sciences.

2 Introduction

Mathematicians view mathematics as a special subject with singularly attractive features. Most intuitively feel that the great power and stability of some "rule book for mathematics" is an important component of their relationship with mathematics. The general feeling is that there is nothing substantial to be gained by revisiting the commonly accepted rule book, and they can continue to do all truly significant mathematics without any foundational concerns.

We approach 100 years of ZFC as the accepted rule book for mathematics. Nearly all mathematicians are aware of its existence—if not its details—and accept that they are implicitly working within its scope. Incompleteness is an attack on the power and stability of the rule book. As power and stability are integral parts of the mathematician's relationship with mathematics, Incompleteness is arguably the unique theme in mathematics today that has the real potential of profoundly altering the mathematician's relationship with mathematics at the most fundamental level. This puts Incompleteness in a category all by itself among all mathematical research.

Incompleteness, in the modern sense, was initiated by Kurt Gödel through his first and second incompleteness theorems, and his relative consistency of the axiom of choice and the continuum hypothesis—with the follow up development of Paul Cohen on AxC and CH (Gödel (1986, 1989, 1986-2003); Cohen (1963, 1964); Friedman (2011)). The mathematician's basic instinct is to defend against Incompleteness by requiring that any statement that is known to be beyond the scope of the rule book be "real mathematics". The exact nature of the "real mathematics" requirement for Incompleteness has evolved over time according to the evolving nature of Incompleteness. It was already clear soon after Cohen that mathematicians were at least implicitly putting up a general defense against Incompleteness that was going to hold until there was a radical change in the examples of Incompleteness. An informal notion of "pathological mathematical object" had emerged during the 20th century that has shaped, and continues to shape, the general course of mathematics. Informal discussion has even made it to Wikipedia with the pages "Pathological (mathematics)" and "Well Behaved".

In the late 1960s we formulated what we now call Concrete Mathematical Incompleteness (CMI), aimed at developing examples of Incompleteness not involving, directly or indirectly, pathological objects. Here we are 45 years later. The CMI examples have the important feature that they become provable if we expand the rule book in certain well studied ways (large cardinal hypotheses, or at least their consistency). Furthermore, there appears to be no alternative way to expand the rule book to remove the CMI (for many examples we have outright equivalence with their consistency or variants of consistency). With traditional set theoretic incompleteness, $V = L$ is an alternative way to remove the mathematical incompleteness.

We regularly discuss the continually evolving examples of CMI with a range of mathematicians, including top luminaries such as A. Connes, C. Fefferman, H. Furstenburg, T. Gowers, M. Gromov, D. Kazhdan, Y. Manin, B. Mazur, D. Mumford, and looking forward to continuing discussions. All of these top luminaries are fully aware of what is at stake, and had interesting reactions. I thank them all for their valuable feedback.

It is obvious that these mathematicians do not identify, in any way, "fundamental or important mathematics" with "mathematics people have done or are doing now". Our examples are judged by such top luminaries strictly on fundamental mathematical standards of simplicity, naturalness, intrinsic interest, concreteness, and depth. This is fortunate since the integration of CMI with existing concrete mathematical developments will likely occur only at later stages of CMI. On the other hand, very strong mathematicians, even some at the next level below such top luminaries, do tend to identify "fundamental or important mathematics" with "mathematics people have done or are doing now", and ironically with the mathematics that such top luminaries have done or are doing now.

The outer limits of CMI live in the Borel measurable sets and functions between Polish spaces. Some historical highlights of CMI are: long finite sequences, continuous comparability of countable sets of reals, Kruskal's and Higman's theorem, Graph Minor Theorem, Borel Diagonalization, Borel determinacy, Borel selection, and Boolean Relation Theory. A detailed overview of CMI is given in the Introduction of Friedman (forthcoming(a)).

In Friedman (forthcoming(a)), we present an earlier development called Boolean Relation Theory (BRT), which represents an earlier state of the art in CMI. The present development has many advantages over BRT, and BRT has some advantages over CMI. We do not view CMI as replacing BRT, as they both have their strengths.

The simplest statements that we prove here from large cardinals, but do not know how to prove in ZFC, are IMR(Ntsf) and IMC(Ntsf) of Section 4.1 (proved from large cardinals in Section 10.2).

The simplest statements that we prove here from large cardinals, and know cannot be proved in ZFC (assuming ZFC is consistent), are IMR($\forall Q \leq n$, Ntsf) and IMC($\forall Q \leq n$, Ntsf) of Section 4.2 (proved from large cardinals in Section 10.2).

In Section 8, we present a program of clear mathematical interest of a combinatorial nature. This is the Order Theoretic Invariance Program. The goal of the program is to determine all order theoretic invariance properties that can be required of a maximal root (clique) for an arbitrarily given order invariant relation (graph) on $Q[0,1]^k$. At this time, we have only partial results. In particular, IMR($\forall Q \leq n$, Ntsf) and IMC($\forall Q \leq n$,

Ntsf) represent partial results, as they can be readily transferred to the $Q[0,1] = Q \cap [0,1]$ setting.

These partial results establish that any carrying out of the Order Theoretic Invariance Program must involve considerably more than ZFC. This is because particular instances have been established using large cardinal hypotheses, that cannot be established using just ZFC. We conjecture that the Order Theoretic Invariance Program will be carried out using the same large cardinal hypotheses that we use here (see Section 10). The solution will take the form of an (intelligible) algorithm that, provably using large cardinal hypotheses, determines whether a given order theoretic relation $R \subseteq Q[0,1]^{2k}$ has the property in question: that all order invariant relations (graphs) on $Q[0,1]^k$ have an R invariant maximal root (clique).

IMR($\forall Q \leq n$, Ntsf) and IMC($\forall Q \leq n$, Ntsf) are implicitly arithmetic statements—even implicitly Π_1^0—because they can rather straightforwardly be shown to be equivalent to the satisfiability of a readily constructed effectively given sequence of sentences in the first order predicate calculus with equality, so that Gödel's completeness theorem applies.

Nevertheless, it is desirable to have explicitly arithmetic – even Π_1^0 – forms. In Section 6, we present an utterly straightforward infinite sequential form of INMC(nsh), an equivalent variant of IMR($\forall Q \leq n$, Ntsf) and IMC($\forall Q \leq n$, Ntsf) that has some advantages and disadvantages. The main disadvantage of INMC(nsh) is gone in the finite form which is presented in Section 6. First the infinite sequential form is presented which asserts that a certain infinite construction can be successfully carried out. The most obvious finite form is obtained by truncation, and therefore is an immediate consequence, and is explicitly Π_2^0. However, we have not established that this obvious truncated form is unprovable in ZFC. We then strengthen the statement by putting an estimate on the magnitudes of the numerators and denominators that are used. It then becomes explicitly Π_1^0, and we can show that the resulting statement is unprovable in ZFC, and in fact provably equivalent to Con(SRP) over EFA.

Section 9 on Computer Investigations is of special note. The discussion suggests, in detail, how exhaustive search algorithms based on truncations of the independent Π_1^0 sentences presented here, to finite initial segments, can either demonstrate inconsistencies in certain large cardinal hypotheses, or arguably, to some extent, confirm the consistency of certain large cardinal hypotheses.

All claims made in this paper that are not proved in Section 10 are to appear in Friedman (forthcoming(b)).

3 Preliminaries

Definition 1. Q is the set of rationals, N is the set of nonnegative integers, Z^+ is the set of positive integers. We use i,j,k,n,m,r,s,t for positive integers, and when indicated, for nonnegative integers. For $x,y \in Q^k$, $x < y \leftrightarrow max(x) < max(y)$, $x \leq y \leftrightarrow x < y \vee x = y$. For $S \subseteq Q^k$ and $p \in Q$, $S_{<p} = \{x \in S : max(x) < p\}$, $S_{\leq p} = \{x \in S : max(x) \leq p\}$. For $S \subseteq Q^k$, $S\#$ is the least $A^k \supseteq S \cup \{0\}^k$.

Definition 2. A rational interval is a $J \subseteq Q$ such that

i. $(\forall a < b < c \text{ from } Q)\ (a,c \in J \to b \in J)$.
ii. $inf(J), sup(J) \in Q \cup \{-\infty, \infty\}$.

We use J exclusively for rational intervals. J can have cardinality 0 or 1.

Definition 3. $Q \leq n$ is the rational interval $Q \cap (-\infty, n)$.

3.1 Invariance

Invariance plays a central role in the development presented here.

Definition 4. *Let R be a binary relation (set of ordered pairs) and W be a set. W is R invariant if and only if for all $(x, y) \in R$, we have $x \in W \leftrightarrow y \in W$.*

We will be using invariance of $W \subseteq X$, where X is viewed as the ambient space for W.

Definition 5. *Let R be a binary relation. $W \subseteq X$ is R invariant if and only if for all $(x, y) \in R \cap X^2$, we have $x \in W \leftrightarrow y \in W$.*

Some authors use "complete invariance" for our notion of invariance.

Theorem 1. *Let R' be the least equivalence relation containing R. The following are equivalent.*

  i. *$W \subseteq X$ is R invariant.*
 ii. *$W \subseteq X$ is R' invariant.*
iii. *W is $R \cap X^2$ invariant.*
 iv. *W is $R' \cap X^2$ invariant.*
  v. *W is the intersection of an R invariant set with X.*
 vi. *W is the union of equivalence classes of the equivalence relation $R' \cap X^2$.*

Definition 6. *Let $S \subseteq Q^k$ and $R \subseteq Q^{2k}$. $R[S] = \{y \in Q^k : (\exists x \in S)(R(x, y))\}$.*

We treat unary functions as sets of ordered pairs. Thus if $R \subseteq Q^{2k}$ is a function from Q^k into Q^k, then $R[S]$ is the same as the forward image of R on S as a function.

Almost all of our statements call for a root or a clique. The root or clique is a subset of an implied ambient space that is clear from the context. This convention will be reiterated when we introduce roots and cliques.

3.2 Order Invariance

Definition 7. *$x, y \in Q^k$ are order equivalent if and only if for all $1 \leq i, j \leq k, x_i < x_j \leftrightarrow y_i < y_j$.*

We also use the following more general definition in Section 10.2.

Definition 8. *Let $(A, <)$ and $(B, <')$ be linear orderings. $x \in A^k$ and $y \in B^k$ are order equivalent if and only for all $1 \leq i, j \leq k, x_i < x_j \leftrightarrow y_i <' y_j$.*

Theorem 2. *For all k, the number of equivalence classes of order equivalence on Q^k is finite.*

Definition 9. *$ot(k)$ is the number of equivalence classes of order equivalence on Q^k.*

Theorem 3. *$ot(k) = \Sigma_{i \leq k} f(k, i)$, where $f(k, i)$ is the number of surjective maps from $\{1, \ldots, k\}$ onto $\{1, \ldots, i\}$. $ot(k)$ is asymptotic to $k!/(2ln^{k+1}2)$. The number of order invariant subsets of Q^k is $2^{ot(k)}$.*

The above asymptotic result is from Gross (1962), which also presents the following table of exact values:

$ot(1) = 1$
$ot(2) = 3$
$ot(3) = 13$
$ot(4) = 75$

$$\mathrm{ot}(5) = 541$$
$$\mathrm{ot}(6) = 4{,}683$$
$$\mathrm{ot}(7) = 47{,}293$$
$$\mathrm{ot}(8) = 545{,}835$$
$$\mathrm{ot}(9) = 7{,}087{,}261$$
$$\mathrm{ot}(10) = 102{,}247{,}563$$
$$\mathrm{ot}(11) = 1{,}622{,}632{,}573$$
$$\mathrm{ot}(12) = 28{,}091{,}567{,}595$$
$$\mathrm{ot}(13) = 526{,}858{,}348{,}381$$
$$\mathrm{ot}(14) = 10{,}641{,}342{,}970{,}443$$

Order invariant sets play a crucial role here.

Definition 10. $W \subseteq Q^k$ *is order invariant if and only if $W \subseteq Q^k$ is R invariant, where R is order equivalence on Q^k.*

Theorem 4. *Let $W \subseteq Q^k$. The following are equivalent.*

   i. W is the union of equivalence classes of order equivalence on Q^k.

   ii. W can be defined as a Boolean combination of inequalities $v_i < v_j, 1 \le i, j \le k$.

   iii. W is 0-definable over $(Q, <)$.

Order invariance extends to ambient spaces via Definition 5.

Definition 11. *Let $X \subseteq Q^k$. $W \subseteq X$ is order invariant if and only if $W \subseteq X$ is R invariant, where R is order equivalence on Q^k.*

Order invariant $W \subseteq J^{2k}$ will play an important role in the theory. By writing $J^{2k} = J^k \times J^k$, we view W as a binary relation on J^k.

Theorem 5. *Let $|J| > 1$. The number of order invariant $W \subseteq J^{2k}$ is $2^{ot(2k)}$. The number of order invariant graphs on J^k is $2^{(ot(k)+ot(2k))/2}$.*

Graphs and order invariant graphs are introduced in Section 3.6.

3.3   N/Order Invariance

We will go beyond order invariance in the following natural way.

Definition 12. *$x, y \in Q^k$ are N/order equivalent if and only if for all $1 \le i, j \le k, (x_i < x_j \leftrightarrow y_i < y_j) \wedge (x_i \in N \leftrightarrow y_i \in N)$. $S \subseteq Q^k$ is N/order invariant if and only if S is R invariant, where R is N/order equivalence on Q^k.*

Theorem 6. *$W \subseteq Q^k$ is N/order invariant if and only if W can be defined as a Boolean combination of statements $v_i < v_j, v_i \in N, 1 \le i, j \le k$. There are finitely many N/order invariant subsets of Q^k.*

N/order invariance of $R \subseteq Q^{2k}$ is used in Section 5.

3.4   Restricted Shifts

The shift function on Q^k is very familiar: just add 1 to all coordinates.

Definition 13. *A restricted shift on Q^k is an $f : Q^k \to Q^k$ such that each $f(x)$ is obtained from x by adding 1 to zero or more coordinates of x. I.e., $f(x) - x : Q^k \to \{0, 1\}^k$.*

Here are some simple examples of restricted shift functions.

Definition 14. *The shift on Q^k adds 1 to all coordinates. The N shift adds 1 to all coordinates in N. The nonnegative shift adds 1 to all nonnegative coordinates.*

We shall see that the last of these restricted shifts play a significant role in the development, beginning with Section 4.4.

An additional restricted shift called the N tail shift is crucial for the theory, and is introduced in Section 4.1.

3.5 $(Q, N, +, <)$

$(Q, N, +, <)$ is a well known tame structure, with quantifier elimination in an appropriately extended language. We use its definable sets to formulate strong forms of some conjectures (see Section 8). We follow the usual convention in model theory that definability allows parameters, and 0-definability does not.

An ultimate goal is to determine exactly which relations on Q^k, definable over $(Q, N, +, <)$ as subsets of Q^{2k}, can be used for the invariance of the roots and cliques throughout the paper. At this point, we do not even know exactly which relations on Q^k, definable over $(Q, <)$ as subsets of Q^{2k}, can be used for many of the results.

3.6 Graphs

All mathematicians are familiar with binary relations, and many are familiar with graphs. Therefore it is advantageous to use binary relations. However, there are some advantages to using graphs rather than relations.

Definition 15. *A graph is a pair $G = (V, E)$, where V is a set, $E \subseteq V^2$, and E is irreflexive and symmetric on V. We say that G is a graph on V. The elements of V are called vertices, and the elements of E are called edges. Two vertices are adjacent if and only if their order pair is an edge.*

Definition 16. *Let $G = (J^k, E)$ be a graph and $p \in Q$. $G_{\leq p}$ is the graph $(J^k_{\leq p}, E_{\leq p})$.*

4 Invariant Maximal Roots and Cliques

Definition 17. *S is a root of $W \subseteq J^{2k}$ if and only if $S \subseteq J^k$ and $S^2 \subseteq W$. S is a maximal root of $W \subseteq J^{2k}$ if and only if S is a root of W and every root $S' \supseteq S$ of $W \subseteq J^{2k}$ is S.*

Definition 18. *S is a clique in the graph G on J^k if and only if $S \subseteq J^k$ and every two distinct elements of S are adjacent in G. S is a maximal clique in G if and only if S is a clique in G and every clique $S' \supseteq S$ in G is S.*

The roots of $W \subseteq J^{2k}$ always have J^k as the implied ambient space. The cliques of a graph on J^k always have J^k as the implied ambient space. Invariance conditions take into account the ambient space as discussed in Section 3.1.

In Definitions 17 and 18, it clearly suffices to use only single point extensions S' of S. Maximal roots and cliques can also be defined by a particularly simple set equation.

Theorem 7. *(RCA_0). Let $W \subseteq J^{2k}$. S is a maximal root of W if and only if $S = \{x \in J^k : \text{for all } y \in S, (x, y), (y, x), (x, x) \in W\}$. Let G be a graph on J^k. S is a maximal clique in G if and only if $S = \{x \in J^k : x \text{ is adjacent in } G \text{ to all } y \in S \backslash \{x\}\}$.*

We use such set equations in Sections 4.4 and 7.

Theorem 8. *(RCA_0). Every $W \subseteq J^{2k}$ has a maximal root. Every graph on J^k has a maximal clique. (ACA_0). Every root of every $W \subseteq J^{2k}$ can be extended to a maximal* *root of W. Every clique in every graph G on J^k can be extended to a maximal clique in G.*

Theorem 9. *The second pair of claims of Theorem 8 are provably equivalent to ACA_0 over RCA_0, even for $J = Q$ and order invariant W, G.*

4.1   Maximal Roots and Cliques in Q^k

We focus on the statements

every order invariant $W \subseteq Q^{2k}$ has an R invariant maximal root
every order invariant graph on Q^k has an R invariant maximal clique

where $R \subseteq Q^{2k}$ is tame. We first try the basic restricted shifts from Definition 14.

Theorem 10. (RCA_0). *Let $k \geq 2$. "Every order invariant $W \subseteq Q^{2k}$ has a shift (N shift) invariant maximal root" fails. "Every order invariant graph on Q^k has a shift (N shift) invariant maximal clique" fails.*

We do not know if Theorem 10 holds for the nonnegative shift. However, see Section 4.4.

This sets the stage for another restricted shift—the N tail shift.

Definition 19. *The N tail of $x \in Q^k$ consists of the x_i such that every $x_j \geq x_i$ lies in N. The N tail shift of $x \in Q^k$ results from x by adding 1 to the N tail of x.*

Theorem. EXAMPLE: *The N tail of (-1, 0, 3, 7/2, 5, 5, 8) consists of both copies of 5 and the single copy of 8. The N tail shift of (-1, 0, 3, 7/2, 5, 5, 8) is (-1, 0, 3, 7/2, 6, 6, 9). The coordinates are in numerical order for the reader's convenience.*

Theorem. INVARIANT MAXIMAL ROOTS *(Ntsf). IMR(Ntsf). For all k, every order invariant $V \subseteq Q^{2k}$ has an N tail shift invariant maximal root.*

Theorem. INVARIANT MAXIMAL CLIQUES *(Ntsf). IMC(Ntsf). For all k, every order invariant graph on Q^k has an N tail shift invariant maximal clique.*

Here Ntsf is read "N tail shift function".

IMR(Ntsf) and IMC(Ntsf) are the simplest statements presented here that we prove from large cardinals, but do not know how to prove in ZFC. However, see IMR$((\forall Q \leq n$, Ntsf), IMC$(\forall Q \leq n$, Ntsf) in Section 4.2.

We now sharpen N tail shift invariance, using the N tail shift relation.

Definition 20. *The N tail shift relation on Q^k is given by $R(x, y) \leftrightarrow (\exists n \geq 0)$ (y results from adding 1 to the part of the N tail of x that is $\geq n$).*

Theorem. INVARIANT MAXIMAL ROOTS *(Ntsr). IMR(Ntsr). For all k, every order invariant $W \subseteq Q^{2k}$ has an N tail shift relation invariant maximal root.*

Theorem. INVARIANT MAXIMAL CLIQUES *(Ntsr). IMC(Ntsr). For all k, every order invariant graph on Q^k has an N tail shift relation invariant maximal clique.*

Here Ntsr is read "N tail shift relation".

Following Theorem 1, we use the least equivalence relation on Q^k containing the N tail shift relation. We call this equivalence relation, N tail equivalence, and the corresponding invariance notion, N tail invariance.

Theorem. INVARIANT MAXIMAL ROOTS *(Nteq). IMR(Nteq). For all k, every order invariant $W \subseteq Q^{2k}$ has an N tail invariant maximal root.*

Theorem. INVARIANT MAXIMAL CLIQUES *(Nteq). IMC(Nteq). For all k, every order invariant graph on Q^k has an N tail invariant maximal clique.*

Here Nteq is read "N tail equivalence".

N tail equivalence has the following simple direct definition.

Theorem 11. *(RCA$_0$). $x, y \in Q^k$ are N tail equivalent if and only if x, y are order equivalent and identical off of their respective N tails.*

Theorem 12. *(RCA$_0$). IMR(Q, Ntsf), IMR(Q, Ntsr), IMR(Q, Nteq), IMC(Q, Ntsf), IMC(Q, Ntsr), IMC(Q, Nteq) are equivalent. They are provable in SRP^+. Furthermore, they are provable in $WKL_0 + Con(SRP)$.*

We do not know if these six statements are provable in ZFC or even in RCA$_0$. The second and third claims are proved here as part of Corollary 1.

In Section 4.2, we will extend these three statements from Q^k to rational intervals, where we claim their equivalence with Con(SRP) over WKL$_0$.

4.2 Maximal Roots and Cliques in J^k

Note that we are not claiming any unprovability results in Section 4.1 when we use $J = Q$. Recall from Definition 2 that J always denotes a rational interval.

Theorem. INVARIANT MAXIMAL ROOTS/CLIQUES *(intervals). IMR/C(intervals). Let $k \geq 2$ and J be given. The following seven statements are equivalent.*

i. For all k, every order invariant $W \subseteq J^{2k}$ has an N tail shift invariant (N tail shift related invariant, N tail invariant) maximal root.

ii. For all k, every order invariant graph on J^k has an N tail shift invariant (N tail shift related invariant, N tail invariant) maximal clique.

iii. It is not the case that: J contains its nonnegative integer left endpoint and the length of J is at least 2.

Theorem 13. *In IMR/C(intervals), RCA$_0$ proves the equivalence of all six forms of i, ii, and the implication from any of these six forms to iii. The implication from iii to any of the six forms of i, ii is provably equivalent to Con(SRP) over WKL$_0$. IMR/C(intervals) is provable in SRP+ but not in any consistent fragment of SRP proving RCA$_0$. Furthermore, it is provably equivalent to Con(SRP) over WKL$_0$.*

Here are six statements corresponding to Con(SRP). They use rational intervals $J = Q \leq n$.

Theorem. INVARIANT MAXIMAL ROOTS *($\forall Q \leq n, Ntsf$), ($\forall Q \leq n, Ntsr$), ($\forall Q \leq n, Nteq$). IMR($\forall Q \leq n, Ntsf$), IMR($\forall Q \leq n, Ntsr$), IMR($\forall Q \leq n, Nteq$). For all k, n, every order invariant $W \subseteq Q_{\leq n}^{2k}$ has an N tail shift invariant (N tail shift related invariant, N tail invariant) maximal root.*

Theorem. INVARIANT MAXIMAL CLIQUES *($\forall Q \leq n, Ntsf$), ($\forall Q \leq n, Ntsr$), ($\forall Q \leq n, Nteq$). IMC($\forall Q \leq n, Ntsf$), IMC($\forall Q \leq n, Ntsr$), IMC($\forall Q \leq n, Nteq$). For all k, n, every order invariant graph on $Q_{\leq n}$ has an N tail shift invariant (N tail shift related invariant, N tail invariant) maximal clique.*

Theorem 14. *IMR($\forall Q \leq n, Ntsf$), IMR($\forall Q \leq n, Ntsr$), IMR($\forall Q \leq n, Nteq$), IMC($\forall Q \leq n, Ntsf$), IMC($\forall Q \leq n, Ntsr$), IMC($\forall Q \leq n, Nteq$) are provable in SRP^+ but not provable in any consistent fragment of SRP that proves RCA$_0$. Furthermore, they are provably equivalent to Con(SRP) over WKL$_0$. These claims remain un-*

changed if we require that the maximal roots (cliques) be recursive in $0'$, resulting in explicitly arithmetic sentences.

$$IMR(\forall Q \leq n, Ntsf), \qquad\qquad IMR(\forall Q \leq n, Ntsr),$$
$$IMR(\forall Q \leq n, Nteq), \qquad\qquad IMC(\forall Q \leq n, Ntsf),$$
$$IMC(\forall Q \leq n, Ntsr), \qquad\qquad IMC(\forall Q \leq n, Nteq)$$

can be put in Π_1^0 form via the Gödel completeness theorem.

The provability claims in Theorem 14 are proved here as part of Corollary 1.

IMR($\forall Q \leq n$, Ntsf), IMC($\forall Q \leq n$, Ntsf) are the simplest statements that we prove here from large cardinals, and know cannot be proved in ZFC (assuming ZFC is consistent). Competitors are ISMR(Ntsf) and ISMC(Ntsf) in Section 4.3, which do have the advantage of using $J = Q$.

4.3 Step Maximal Roots and Cliques in Q^k

Definition 21. *A step maximal root of $W \subseteq J^{2k}$ is an $S \subseteq J^k$ such that for all $n \geq 0$, $S_{\leq n}$ is a maximal root of $W_{\leq n}$.*

Definition 22. *A step maximal clique of the graph G on J^k is an $S \subseteq J^k$ such that for all $n \geq 0$, $S_{\leq n}$ is a maximal clique in $G_{\leq n}$.*

Theorem 15. *(ACA_0). For all k, every $W \subseteq Q^{2k}$ has a step maximal root. For all k, every order invariant graph on Q^k has a step maximal clique.*

Theorem 16. *(RCA_0). Let $k \geq 2$ and $p \in Q$. "Every order invariant $W \subseteq Q^{2k}$ has a shift (N shift) invariant step maximal root" fails. "Every order invariant graph G on Q^k has a shift (N shift) invariant step maximal clique" fails.*

We do not know if Theorem 16 holds for the nonnegative shift. However, see Section 4.4.

Theorem. Invariant Step Maximal Roots ($\forall Q \leq n, Ntsf$), ($\forall Q \leq n, Ntsr$), ($\forall Q \leq n, Nteq$).

$$ISMR(\forall Q \leq n, Ntsf), \qquad\qquad ISMR(\forall Q \leq n, Ntsr),$$
$$ISMR(\forall Q \leq n, Nteq).$$

For all k, n, every order invariant $W \subseteq Q^{2k}_{\leq n}$ has an N tail shift invariant (N tail shift related invariant, N tail invariant) step maximal root.

Theorem. Invariant Step Maximal Cliques ($\forall Q \leq n, Ntsf$), ($\forall Q \leq n, Ntsr$), ($\forall Q \leq n, Nteq$).

$$ISMC(\forall Q \leq n, Ntsf), \qquad\qquad ISMC(\forall Q \leq n, Ntsr),$$
$$ISMC(\forall Q \leq n, Nteq).$$

For all k, n, every order invariant graph on $Q^k_{\leq n}$ has an N tail shift invariant (N tail shift related invariant, N tail invariant) step maximal clique.

Theorem 17. *ISMR(Ntsf), ISMR(Ntsr), ISMR(Nteq), ISMC(Ntsf), ISMC(Ntsr), ISMC(Nteq) are provable in SRP^+ but not provable in any consistent fragment of SRP that proves RCA_0. Furthermore, they are provably equivalent to Con(SRP) over WKL_0. These claims remain unchanged if we require that the step maximal roots and cliques be recursive in $0'$, resulting in explicitly arithmetic sentences. ISMR(Ntsf), ISMR(Ntsr), ISMR(Nteq), ISMC(Ntsf), ISMC(Ntsr), ISMC(Nteq) can be put in Π_1^0 form via the Gödel completeness theorem.*

The following implication is immediate.

Theorem 18. *(RCA^0). ISMR(Nteq) implies all eighteen titled statements in Sections 4.1–4.3, excluding IMR/C(intervals).*

Theorem 18 is also true for IMR/C(intervals). This is proved in Friedman (forthcoming(b)).

We prove ISMR(Nteq) in WKL$_0$ + Con(SRP) in Section 10.2. This establishes the same for all of the eighteen statements, via Theorem 18.

4.4 Inductively Maximal Cliques in Q^k

To simplify the discussion, we now use graphs and cliques rather than relations and roots. The same results hold for relations and roots.

Here we modify step maximality, with the effect of requiring that maximal cliques be built up smoothly (rather than stepwise) from below going up.

We call the most obvious strengthening, naïve inductive maximality, as it is too strong.

Definition 23. *A naive inductively maximal clique in the graph G on J^k is an $S = \{x \in J^k : x$ is adjacent to every $y < x$ from $S\}$.*

Theorem 19. *(RCA$_0$). Let $k \geq 1$ and $\mid J \mid > 1$. There is an order invariant graph on J^{2k} without a naïve inductively maximal clique*

Definition 24. *An inductively maximal clique in the graph G on J^k is an $S = \{x \in J^k \cap S\# : x$ is adjacent to every $y < x$ from $S\}$. The ambient space of S is J^k.*

Note that in an inductively maximal clique, any two elements with different max are adjacent. However, distinct elements with the same max may or may not be adjacent.

Theorem 20. *(RCA$_0$). For all k, every order invariant graph on Q^k has an inductively maximal clique.*

Theorem 20 is trivial since we can simply arrange that the inductively maximal clique be $\varnothing$ or $\{(0,\ldots,0)\}$. However, the imposition of an invariance condition radically changes the situation.

Theorem. INDUCTIVELY MAXIMAL CLIQUES *(Nteq). INMC(Nteq). For all k, every order invariant graph on Q^k has an N tail equivalent inductively maximal clique.*

We also can use a different kind of invariance condition. The nonnegative shift was introduced in Section 3.4. Here is a more powerful operation.

Definition 25. *The N upper shift of $S \subseteq Q^k$ is the union over $n \geq 0$ of the result of adding 1 to all coordinates $\geq n$ of elements of S.*

Note that the N upper shift contains the N tail shift.

Theorem. INDUCTIVELY MAXIMAL CLIQUES *(nsh). INMC(nsh). For all k, every order invariant graph on Q^k has an inductively maximal clique containing its nonnegative shift.*

Theorem. INDUCTIVELY MAXIMAL CLIQUES *(Nush). INMC(Nush). For all k, every order invariant graph on Q^k has an inductively maximal clique containing its N upper shift.*

Theorem. Inductively Maximal Cliques *(Nteq, Nush). INMC(Nteq, Nush). For all k, every order invariant graph on Q^k has an N tail invariant inductively maximal clique containing its N upper shift.*

Theorem 21. *INMC(Nteq), INMC(nsh), INMC(Nush), INMC(Nteq,Nush) are provable in SRP^+ but not provable in any consistent fragment of SRP that proves RCA_0. Furthermore, they are provably equivalent to Con(SRP) over WKL_0. This holds if we add the requirement that the inductively maximal clique be recursive in $0'$, resulting in explicitly arithmetic sentences. INMC(Nteq), INMC(nsh), INMC(Nush), INMC(Nteq, Nush) can be put into Π_1^0 form via the Gödel completeness theorem.*

5   N Tail Equivalence in Maximum

We now discuss the clique statements in Section 4 with R invariance, where $R \subseteq Q^{2k}$ is N/order invariant. Note that N tail equivalence on Q^k is itself an N/order invariant subset of Q^{2k}.

We first note that requiring $R \subseteq Q^{2k}$ to be order invariant leads to a triviality.

Theorem 22. (RCA^0). *Let k, J be given, $\mid J \mid > 1$, and $R \subseteq Q^{2k}$ be order invariant. Suppose every order invariant graph on J^k has an R invariant maximal clique $S \subseteq J^k$. Then $(\forall x, y \in Q^k)(R(x, y) \to x = y)$.*

For step maximality and inductive maximality, we have a complete understanding of the statements that result when we use N/order invariant $R \subseteq Q^{2k}$. For maximality in J^k, we need to impose the following natural condition on $R : R \cap N^{2k}$ is order equivalence on N^k.

Theorem 23. (RCA_0). *Let $k, n \geq 1, R \subseteq Q^{2k}$ be N/order invariant, and $R \cap N^{2k}$ be order equivalence on N^k. Then $i \lor ii \to iii$.*

i. *Every order invariant graph on Q^k has an R invariant maximal clique.*

ii. *Every order invariant graph on $Q^k_{\leq n}$ has an R invariant maximal clique.*

iii. *R is contained in N tail equivalence on Q^k.*

Theorem 24. (RCA_0). *Let $k \geq 1$ and $R \subseteq Q^{2k}$ be N/order invariant. Then $i \lor ii \to iii$.*

i. *Every order invariant graph on Q^k has an R invariant step maximal clique.*

ii. *Every order invariant graph on Q^k has an R invariant inductively maximal clique.*

iii. *R is contained in N tail equivalence on Q^k.*

Nteq Maximum. *Let $k, n \geq 1$. N tail equivalence on Q^k is the maximum $R \subseteq Q^{2k}$ such that*

    i. *R is N/order invariant.*

    ii. *$R \cap N^{2k}$ is order equivalence on N^k.*

    iii. *Every order invariant graph on $Q^k_{\leq n}$ has an R invariant maximal clique.*

Nteq Step Maximum. *Let $k \geq 1$. N tail equivalence on Q^k is the maximum $R \subseteq Q^{2k}$ such that*

    i. *R is N/order invariant.*

    ii. *Every order invariant graph on Q^k has an R invariant step maximal clique.*

Nteq Inductive Maximum. *Let $k \geq 1$. N tail equivalence on Q^k is the maximum $R \subseteq Q^{2k}$ such that*

    i. *R is N/order invariant.*

ii. *Every order invariant graph on Q^k has an R invariant inductively maximal clique.*

Theorem 25. *Nteq Maximum, Nteq Step Maximum, and Nteq Inductive Maximum are provable in SRP^+ but not provable in any consistent fragment of SRP that proves RCA_0. Furthermore, they are provably equivalent to Con(SRP) over WKL_0. These claims remain unchanged if we require that the maximal cliques be recursive in $0'$, resulting in explicitly arithmetic sentences. Nteq Maximum, Nteq Step Maximum, and Nteq Inductive Maximum can be put in Π_1^0 form via the Gödel completeness theorem.*

6 Finite Sequential Cliques

We present an explicitly Π_1^0 form of INMC(nsh). Recall the nonnegative shift, nsh, from Definition 14. The inductively maximal statements have the advantage of a simpler invariance condition involving nsh – merely containing its nonnegative shift. It also has the disadvantage of requiring use of the $\#$ operator from Definition 1. Remarkably, the finite form retains this advantage but does not need the $\#$ operator.

Definition 26. *Let $x, y \in Q^k$. $x \leq y$ if and only if $max(x) < max(y) \vee x = y$.*

Definition 27. *Let $x_1, \ldots, x_r \in Q^k$. $nsh(x_1, \ldots, x_r\}$ is the set of nonnegative shifts of $x_1, \ldots, x_r$.*

Proposition 1. *Let G be an order invariant graph on Q^k, and let $x_1, x_2, \ldots \in Q^k$. There exists $y_1, y_2, \ldots \in Q^k$ such that for all $i \geq 1, y_i \leq x_i$ is not adjacent to x_i and adjacent to each $nsh(y_j), j < i$, other than x_i.*

Theorem 26. *RCA_0 proves that INMC(nsh) and Proposition 1 are equivalent.*

We now truncate Proposition 1 in the obvious way.

Proposition 2. *Let G be an order invariant graph on Q^k, and $x_1, \ldots, x_r \in Q^k$. There exists $y_1, \ldots, y_r$ such that for all $1 \leq i \leq r, y_i \leq x_i$ is not adjacent to x_i, but adjacent to each $nsh(y_j) \neq y_i, j < i$.*

Note that Proposition 2 is explicitly Π_2^0. It can be converted to Π_1^0 form by well known decision procedures, or quantitative estimation. In any case, unfortunately, the only proof that we have of Proposition 2 merely quotes Proposition 1, and so the proof uses large cardinal hypotheses. We do not know if Proposition 2 can be proved in ZFC.

We now sharpen Proposition 2 by what we call "control". We simply put an upper bound on the norm of y_i in terms of the norm of $\{x_1, \ldots, x_i\}$.

For this purpose, a convenient norm $\#(x_1, \ldots, x_r), x_1, \ldots, x_r \in Q^k$, is defined by putting the x's in reduced form, and adding the magnitudes of the resulting $2kr$ integers.

Proposition 3. *Let G be an order invariant graph on Q^k, and $x_1, \ldots, x_r \in Q^k$. There exists $y_1, \ldots, y_r$ such that for all $1 \leq i \leq r, y_i \leq x_i$ is not adjacent to x_i, but adjacent to each $nsh(y_j) \neq y_i, j < i$, where $\#(y_i) \leq \#(x_1, \ldots, x_i)^4$.*

Theorem 27. *Proposition 3 is provably equivalent to Con(SRP) over EFA.*
Proposition 3 is explicitly Σ_2^0. However, c can be specified to be a small universal integer, and Proposition 3 becomes explicitly Π_1^0. Probably exponent 2 works in Proposition 3, as well as much sharper estimates.

7 Inductively Maximal Cliques in $Q^{\leq k}$

Definition 28. *$Q^{\leq k}$ is the set of tuples from Q of nonzero length at most k. Let $S, S' \subseteq Q^{\leq k}$. The 1-sections of $S \subseteq Q^{\leq k}$ are the subsets of Q obtained from S by fixing any k-*

1 coordinates to be specific rational numbers. S 1-contains S' if and only if S contains S', and every 1-section of S' with finite sup is a 1-section of S.

Definition 29. Let G be a graph on $Q^{\leq k}$. An inductively maximal Q^k clique of G is an $S \subseteq Q^{\leq k}$ such that $S \cap Q^k = \{x \in S\# : x$ is adjacent in G to every $y < x$ from $S\}$.

Theorem. Inductively Maximal Cliques $(Q^{\leq k}, 1 \supseteq, nsh)$. $INMC(Q^{\leq k}, 1 \supseteq, nsh)$. For all $k \geq 3$, every order invariant graph on $Q^{\leq k}$ has an inductively maximal Q^k clique that 1-contains its nonnegative shift.

Theorem. Inductively Maximal Cliques $(Q^{\leq k}, 1 \supseteq, Nush)$. $INMC(Q^{\leq k}, 1 \supseteq, Nush)$. For all $k \geq 3$, every order invariant graph on $Q^{\leq k}$ has an inductively maximal Q^k clique that 1-contains its N upper shift.

Theorem 28. $INMC(Q^{\leq k}, 1 \supseteq, nsh)$, $INMC(Q^{\leq k}, 1 \supseteq, Nush)$ are provable in $HUGE^+$, but not in any consistent fragment of $HUGE$ that proves RCA_0. Furthermore, they are provably equivalent to $Con(HUGE)$ over WKL_0. These claims hold if we add the requirement that the maximal clique be recursive in $0'$, resulting in an explicitly arithmetic sentence. $INMC(Q^{\leq k}, 1 \supseteq, nsh)$, $INMC(Q^{\leq k}, 1 \supseteq, Nush)$ can be put into Π^0_1 form via the Gödel completeness theorem.

8 Order Theoretic Invariance Program

Definition 30. $R \subseteq Q^m$ is order theoretic if and only if R is definable without quantifiers over the structure $(Q, <)$, with constants allowed in the quantifier free definition. $Q[0,1] = Q \cap [0,1]$.

Template. Let $R \subseteq Q[0,1]^{2k}$ be order theoretic. Every order invariant graph on $Q[0,1]^k$ has an R invariant maximal clique.

The Order Theoretic Invariance Program seeks to determine for which R the above Template holds.

Conjecture 1. Every instance of the above Template is provable or refutable in SRP.

We know that there are particular instances of the Template that are provable in SRP but not in ZFC. This is the case for IMC (invariant maximal cliques) for a specific choice of small k, r. Instead of using the order theoretic relation Nteq on $Q^k_{\leq n}$, we can use a corresponding order theoretic relation on $Q[0,1]^k$ where the parameters $1/n$, $1/(n-1), \ldots, 1$ in $Q[0,1]$ take the place of the parameters $1, 2, \ldots, n$ in $Q \leq n$. In fact, in this way we see that

Theorem 29. Conjecture 1 does not hold for any consistent SRP[k].

In particular, ZFC is far too weak to carry out the Order Theoretic Invariance Program. The results presented here carry out the program in a limited way.

9 Computer Investigations

In Section 9.1, we present a general infinite length construction associated with a slight modification of the INMC(Nteq, Nush) from Section 4.4, from Q^k to $Q^k_{\leq n}$.

Definition 31. Let $S \subseteq Q^k_{\leq n}$. $Nteq^*[S] = Nteq[S] \cap Q^k_{\leq n}$. $Nush^*[S]$ is the least $S' = Nush[S'] \cap Q^k_{\leq n}$ containing S. $N^*[S]$ is $0, \ldots, n$, together with all $p + i \leq n$, such that p is a coordinate of an element of S, and $i \in N$.

Theorem. Inductively Maximal Cliques ($\forall Q \leq n, Nteq^*, Nush^*$). $INMC(\forall Q \leq n, Nteq^*, Nush^*)$. *For all k, every order invariant graph on $Q_{\leq n}^k$ has an inductively maximal clique $S \subseteq Q_{\leq n}^k$ containing $Nteq^*[S] \cup Nush^*[S]$.*

Theorem 30. *$INMC(\forall Q_{\leq n}, Nteq^*, Nush^*)$ is provable in SRP^+ but not provable in any consistent fragment of SRP that proves RCA_0. Furthermore, it is provably equivalent to Con(SRP) over WKL_0.*

The general infinite length construction is an obvious nondeterministic construction that builds the inductively maximal clique with the required properties.

In Section 9.2, we discuss the exhaustive search for finite initial segments of the infinite construction. This exhaustive search can rapidly require far too much computer resources. These practical considerations are addressed in Section 9.2.

9.1 Infinite and Finite Length Constructions

We fix positive integer parameters $k, n \geq 1$. We also fix an order invariant graph G on $Q_{\leq n}^k$. We nondeterministically build an infinite sequence of finite sets $S_1 \subseteq S_2 \subseteq \cdots \subseteq Q_{\leq n}^k$.

Recall the definition of $\leq$ (Definition 1).

Definition 32. *A G resolution of $x \in Q_{\leq n}^k$ is a $y \leq x$ such that x, y are not adjacent in G.*

We begin by choosing a G resolution of each $x \in \{0, \ldots, n\}^k$, and forming the set S_i of these G resolutions. Suppose S_i has been constructed, $i > 1$. Choose a G resolution of each $x \in N^*[S_i]^k$, and form the set S_{i+1} consisting of these G resolutions together with S_i.

This construction is considered successful if and only if for $S = \cup_i S_i \subseteq Q_{\leq n}^k$, $Nteq^*[S] \cup Nush^*[S]$ is a clique in G.

If we are only carrying out the construction for r steps, then success is for $Nteq^*[S] \cup Nush^*[S]$ to be a clique in G.

Theorem 31. *The assertion "for all k, n and order invariant graphs G on $Q_{\leq n}^k$, this construction can be successfully carried out for infinitely many steps" is provable in SRP^+ but not provable in any consistent fragment of SRP that proves RCA_0. Furthermore, it is provably equivalent to Con(SRP) over WKL_0.*

Theorem 32. *The assertion "for all k, n, r and order invariant graphs G on $Q_{\leq n}^k$, this construction can be successfully carried out for r steps" is provable in SRP^+ but not provable in any consistent fragment of SRP that proves EFA. Furthermore, it is provably equivalent to Con(SRP) over EFA.*

Note that the quoted statement in Theorem 32 is explicitly Π_2^0, and can easily be put into explicitly Π_1^0 form by the remarks made in Section 6.

9.2 Computational Aspects

We know that the construction in Section 9.1 can be successfully carried out for any number of steps—even infinitely many steps—using Con(SRP). However, it is not at all clear how you actually carry out the construction, even for modest k, n, r, in the sense of actually obtaining a computer file—i.e., a certificate residing on your computer.

We will discuss the exhaustive search for such certificates, which will be implementable for certain choices of parameters, based on algorithmic efficiencies and computer resources. There seems to be plenty of opportunity here for clever algorithmic design.

If the search comes up empty, then we know that SRP is inconsistent. In fact, if the search comes up empty, we can obtain an actual inconsistency in SRP that corresponds to the trace of the negative computer search.

Thus, if the computer search turns up a certificate, we have arguably confirmed the consistency of at least ZFC and even SRP[k] for small k.

Of course, the "strength" of the confirmation here corresponds to the extent to which the high powered set theory is actually being "engaged". At this point, we do not know how to confidently judge this, but one test is whether the exhaustive searches show that certificates are extremely rare. And of course, we know that large cardinals are, in fact, fully engaged as the parameters approach infinity, because of Theorem 32.

There are a number of computational issues in order to bring the exhaustive search for certificates down to practice.

(1) The choice of positive integer parameters $k, n, r, s, t_1, \ldots, t_r$.
(2) The choice of order invariant graph G on $Q \leq n^k$.
(3) Managing the rationals (easily handled).
(4) Resolutions of tuples.

The parameters k, n determine the ambient space $Q^k_{\leq n}$. The parameter r is the number of "big stages" in the nondeterministic construction. These big stages result in the finite sets $S_1 \subseteq \cdots \subseteq S_r$, where the goal is for Nteq*[S] $\cup$ Nush*[S] to be a clique in G.

Inside each big stage i, there are t_i resolutions, as discussed below.

The order invariant graph G on $Q^k_{\leq n}$ is given by a randomly generated

$$C \subseteq \{1, \ldots, 2k\}^{2k}$$

of cardinality s, subject to the constraint that no $(x, x) \in C$, and

$$(x, y) \in C \leftrightarrow (y, x) \in C.$$

Here $x, y \in Q^k_{\leq n}$ are considered to be adjacent in G if and only if (x, y) is order equivalent to an element of C.

In our nondeterministic construction of $S_1, \ldots, S_r$, we will arrange that each $S_i = $ Nteq*[S_i] $\cup$ Nush*[S_i] is a clique in G. The set of candidates for resolutions are, respectively, $\{0, \ldots, n\}^k, N^*[S_1]^k, N^*[S^2]^k, \ldots, N^*[S_{r-1}]^k$. However, we do not want to be resolving all of these tuples, as would cause a computational explosion. We will resolve only $t_1, \ldots, t_r$ tuples, respectively, from these r large sets. But precisely which tuples are to be selected for resolution?

We envision community wide agreement on a selection process Ω for picking elements from A, where $A \subseteq Q^k_{\leq n}$ of reasonable size. Ω would give a listing of A without repetition, or at least a reasonable number of elements of A without repetition. The tuples to be resolved would be chosen in order from this list. Thus Ω sets priorities for the choice of tuples to be resolved. Ω should be invariant under order isomorphisms of Q that fix $0, \ldots, n$.

Ω should give some bias toward use of coordinates from $\{0, \ldots, n\}$, and some bias toward use of coordinates among the lesser numbers in $Q \leq n \backslash \{0, \ldots, n\}$. This way, the engagement with Nteq* and Nush* will be intensified. We expect considerable randomness to be built into Ω unless a theory develops that suggests otherwise.

More precisely, in forming S_1, we apply Ω to $\{0, \ldots, n\}^k$, and successively resolve the first t_1 elements of $\Omega(\{0, \ldots, n\}^k)$, throwing the resolutions into S_1. When we insert a new tuple into S_1, we immediately close under Nteq* ∪ Nush*, checking that the S_1 thus far remains a clique in G. If we find otherwise, then we have made an error, and need to try a new resolution.

If we find that we are scheduled to resolve a tuple from the list obtained by applying Ω which already has a resolution in S_1, (because there is already some $y \leq x$ nonadjacent to x in S_1), then we move to the next tuple on the list supplied by Ω. We proceed until we have put t_1 new resolutions into S_1, satisfying the requirement that the closure under Nteq* ∪ Nush* is a clique in G.

After this process has been completed, we have formed a clique $S_1 = $ Nteq$^*[S_1]$ ∪ Nush$^*[S_1]$ in G. Now the set of candidates for resolution is $N^*[S_1]^k$. We proceed as above, to construct a clique $S_2 = $ Nteq$^*[S_2]$ ∪ Nush$^*[S_2]$ in G, after inserting t_2 new resolutions of tuples generated by Ω, as before. We continue in this way constructing $S_1, \ldots, S_r$.

For the exhaustive search, we need to consider all possible resolutions. Let us go back to the formation of S_1 from $\{0, \ldots, n\}^k$. We need to form the set of all possible S_1's arising from the above process. I.e., all possible S_1's up to order isomorphism that fixes $0, \ldots, n$.

At any stage in the formation we manage the rationals that have been used by forming the list of them in strictly increasing order, including $0, \ldots, n$. When making a resolution of a tuple, we usually will be using one or more new rationals, which are to be inserted into this list in the following way. Whenever a new nonnegative rational is inserted, we must immediately also insert the results of adding nonnegative integers, as long as the sum stays $\leq n$.

It is now clear what we mean by constructing all possible S_1's, up to isomorphism fixing $0, \ldots, n$. We have arrived at a list $S[1,1], \ldots, S[1,p_1]$ of all possible S_1's fixing $0, \ldots, n$.

We repeat this process, starting independently with each $S[1, i]$. Each $S[1, i]$ spawns all possible S_2's containing $S[1, i]$, up to isomorphism fixing $0, \ldots, n$. We arrive at a list $S[2, 1], \ldots, S[2, p_2]$ of all possible S_2's up to isomorphism fixing $0, \ldots, n$.

We eventually obtain a complete listing $S[r, 1], \ldots, S[r, p_r]$ of all possible S_r's up to isomorphism. Any one of them provides the sought after certificate. However, if $p_r = 0$ then there is no certificate, and this yields an inconsistency in SRP$[k]$, for small k.

It is a matter for research to see how to best use computational resources under various adjustments of the parameters, and to see how many sets exist at the last stage.

What values of the parameters are reasonable for initial experimentation, where at least ZFC is engaged?

We recommend initially experimenting with $k = n = r = t_1 = t_2 = t_3 = t_4 = 4$, and $s = 100$. There may have to be some experimentation with s in order to obtain traction. We have no way of foretelling that.

We might find that for even modest choices of parameters, it becomes computationally intractable to keep checking for the closure under Nteq* ∪ Nush* being a clique in G. There is the possibility of weakening the closure under Nteq* ∪ Nush* according to some predetermined rule, although this may cause the number of $S[i, j]$'s in the exhaustive search to grow uncontrollably.

Experts in large cardinals have developed a large amount of experience and intuitions that make them comfortable with Con(SRP). How much credence to give this experience and intuition is a deep question that is difficult to address today.

A possible reservation is that these scholars have a good grasp only of humanly intelligible proofs. It could be that the only inconsistencies in large cardinal hypotheses are not humanly intelligible. Certainly any inconsistency found by one of our proposed exhaustive searches coming up negative, will be completely humanly unintelligible. There is the remarkable possibility, not altogether absurd, that the proposed exhaustive searches actually do generate wholly humanly unintelligible inconsistencies. Note that we have arranged for these exhaustive searches to at least plausibly be intensely engaging the large cardinals through their underlying finite combinatorial structure.

There is a real precedent for surprising findings by exhaustive computer searches that turn up completely humanly unintelligible proofs. There is now an approximately 100 terabyte database evaluating all seven piece chess positions via the Lomonosov supercomputer in Moscow, ChessOK (2013). It yields a treasure trove of theorems that seem completely beyond human intelligibility, and have already forced chess grandmasters to revise many previously held evaluations.

10 Proofs of Invariant Maximality

In this section we prove ISMC(Nteq) in WKL$_0$ + Con(SRP). We also show that for all k, SRP proves ISMC(Nteq) for dimension k.

In Section 10.1 we discuss the large cardinal hypotheses that are relevant to this paper (except those relevant to Section 7). We will be using a considerable portion of the notation from Section 10.1 in Section 10.2, but not many of the results. Section 10.1 does provide a lot of useful orienting information about the relevant region of large cardinal hypotheses.

In Section 10.2, we prove ISMC(Nteq) in SRP$^+$ and WKL$_0$ + Con(SRP). We also show that for all k, SRP proves ISMC(Nteq) for dimension k. We focus on ISMC(Nteq) in light of Theorem 18.

10.1 The Stationary Ramsey Property

All results in this section are taken from Friedman (2001). All of these results, with the exception of Theorem 33, $iv \leftrightarrow v \rightarrow vi$, are credited in Friedman (2001) to James Baumgartner. Below, λ always denotes a limit ordinal.

Definition 33. We say that $C \subseteq \lambda$ is unbounded if and only if for all $\alpha < \lambda$ there exists $\beta \in C$ such that $\beta \geq \alpha$.

Definition 34. We say that $C \subseteq \lambda$ is closed if and only if for all limit ordinals $x < \lambda$, if the sup of the elements of C below x is x, then $x \in C$.

Definition 35. We say that $A \subseteq \lambda$ is stationary if and only if it intersects every closed unbounded subset of λ.

Definition 36. For sets A, let $S(A)$ be the set of all subsets of A. For integers $k \geq 1$, let $S_k(A)$ be the set of all k element subsets of A.

Definition 37. Let $k \geq 1$. We say that λ has the k-SRP if and only if for every $f : S_k(\lambda) \rightarrow \{0,1\}$, there exists a stationary $E \subseteq \lambda$ such that f is constant on $S_k(E)$. Here SRP stands for "stationary Ramsey property."

The k-SRP is a particularly simple large cardinal property. To put it in perspective, the existence of an ordinal with the 2-SRP is stronger than the existence of higher order indescribable cardinals, which is stronger than the existence of weakly compact cardinals, which is stronger than the existence of cardinals which are, for all k, strongly k-Mahlo (see Theorem 33 below, and Friedman (2001, Lemmas 1.11)).

Our main results are stated in terms of the stationary Ramsey property. In particular, we use the following extensions of ZFC based on the SRP.

Definition 38. $SRP^+ = ZFC + $ "for all k there exists an ordinal with the k-SRP". $SRP = ZFC + \{there\ exists\ an\ ordinal\ with\ the\ k\text{-}SRP\}_k$. We also use k-SRP for the formal system $ZFC + (\exists\lambda)(\lambda$ has the k-SRP).

For technical reasons, we will need to consider some large cardinal properties that rely on regressive functions.

Definition 39. We say that $f : S_k(\lambda) \to \lambda$ is regressive if and only if for all $A \in S_k(\lambda)$, if $min(A) > 0$ then $f(A) < min(A)$. We say that E is f-homogenous if and only if $E \subseteq \lambda$ and for all $B, C \in S_k(E), f(B) = f(C)$.

Definition 40. We say that $f : S_k(\lambda) \to S(\lambda)$ is regressive if and only if for all $A \in S_k(\lambda)$, $f(A) \subseteq min(A)$. (We take $min(\varnothing) = 0$, and so $f(\varnothing) = \varnothing$). We say that E is f-homogenous if and only if $E \subseteq \lambda$ and for all $B, C \in S_k(E)$, we have $f(B) \cap min(B \cup C) = f(C) \cap min(B \cup C)$.

Definition 41. Let $k \geq 1$. We say that α is purely k-subtle if and only if

i. α is an ordinal;
ii. For all regressive $f : S_k(\alpha) \leftarrow \alpha$, there exists $A \in S_{k+1}(\alpha\backslash\{0,1\})$ such that f is constant on $S_k(A)$.

Definition 42. We say that λ is k-subtle if and only if for all closed unbounded $C \subseteq \lambda$ and regressive $f : S_k(\lambda) \to S(\lambda)$, there exists an f-homogenous $A \in S_{k+1}(C)$.

Definition 43. We say that λ is k-almost ineffable if and only if for all regressive $f : S_k(\lambda) \to S(\lambda)$, there exists an f-homogenous $A \subseteq \lambda$ of cardinality λ.

Definition 44. We say that λ is k-ineffable if and only if for all regressive $f : S_k(\lambda) \to S(\lambda)$, there exists an f-homogenous stationary $A \subseteq \lambda$.

Theorem 33. Let $k \geq 2$. Each of the following implies the next, over ZFC.

i. there exists an ordinal with the k-SRP.
ii. there exists a (k-1)-ineffable ordinal.
iii. there exists a (k-1)-almost ineffable ordinal.
iv. there exists a (k-1)-subtle ordinal.
v. there exists a purely k-subtle ordinal.
vi. there exists an ordinal with the (k-1)-SRP.

Furthermore, i, ii are equivalent, and iv, v are equivalent. There are no other equivalences. ZFC proves that the least ordinal with properties i–vi (whichever exist) form a decreasing ($\geq$) sequence of uncountable cardinals, with equality between i, ii, equality between iv, v, and strict inequality for the remaining consecutive pairs.

Proof. $i \leftrightarrow ii$ is from Friedman (2001, Theorem 1.28), $iv \leftrightarrow v$ is from Friedman (2001, Corollary 2.17). The strict implications $ii \to iii \to iv \to vi$ are from Friedman (2001, Theorem 1.28). Same references apply for comparing the least ordinals. $\square$

Definition 45. We follow the convention that for integers $p \leq 0$, a p-subtle, p-almost ineffable, p-ineffable ordinal is a limit ordinal, and that the ordinals that are 0-subtle,

0-almost ineffable, 0-ineffable, or have the 0-SRP, are exactly the limit ordinals. An ordinal is called subtle, almost ineffable, ineffable, if and only if it is 1-subtle, 1-almost ineffable, 1-ineffable.

Definition 46. *SRP⁺ is the formal system ZFC + (∀k)(there exists an ordinal with the k-SRP). SRP is the formal system ZFC + {there exists an ordinal with the k-SRP}ₖ. For each k, we write SRP[k] for the formal system ZFC + "there exists an ordinal with the k-SRP".*

10.2 Invariant Step Maximal Roots (Nteq)

In this section, we prove ISMR(Nteq) in SRP⁺ and WKL₀ + Con(SRP).

Definition 47. *ISMR(Nteq, k) is ISMR(Nteq) for order invariant $W \subseteq Q^{2k}$ only. Obviously ISMR(Nteq) and $\forall k(ISMR(Nteq,k))$ are provably equivalent over RCA₀. We show that for all k, ISMR(Nteq, k) is provable in SRP.*

Lemma 1. *RCA₀ proves ISMR(Nteq, 1).*

Proof. Let $W \subseteq Q^2$ be order invariant. There are three principal cases.
case 1. W has no (p, p). Use $S = \varnothing$.
case 2. W has all (p, p) but not all elements of Q^2. Use $S = \{-1\}$.
case 3. $W = Q^2$. Use $S = Q$. □

We now fix n ≥ k ≥ 2 and λ to be the least (k-1)-subtle ordinal. We derive ISMR(Nteq,k), using λ, within ZFC. We make no claims for the optimality of λ here.

Note that by Theorem 33, λ is strictly between the least ordinal with the (k-1)-SRP and the least ordinal with the k-SRP (assuming the latter exists). In any case, λ is available to us in SRP.

We first make some definitions in general linear orderings.

For any set U, we think of U^r as the set of functions from $\{1,\ldots,r\}$ into U. We extend this as follows.

Definition 48. *U^{r-} is the set of all nonempty partial functions from $\{1,\ldots,r\}$ into U.*

Definition 49. *Let $(A, <^*)$ be a linear ordering, $w \in A^{r-}, X \subseteq A^{r-}, S^* \subseteq A^k, W^* \subseteq A^{2k}, E \subseteq A, x \in A$. $max(w)$ is the largest value of w under $<^*$. $X[<^* x] = \{y \in X : max(y) <^* x\}$. $X[\leq^* x] = \{y \in X : max(y) \leq^* x\}$. S^* is a root of W^* if and only if $S^{*2} \subseteq W$. S^* is a maximal root of $W^* \subseteq A^{2k}$ if and only if S^* is a root of W^* and every root $S^{*\prime} \supseteq S^*$ of W^* is S^*. The E tail of $z \in A^k$ consists of the z_i such that every $z_j \geq^* z_i$ lies in E. $z, w \in A^k$ are E tail equivalent if and only if z, w order equivalent (under $<^*$) and identical off of their E tails. S^* is E tail invariant if and only if for all E tail equivalent $z, w \in S^*, z \in X \leftrightarrow w \in X$.*

Definition 50. *Q[0,1) is the rational interval [0,1). $(\lambda \times Q[0,1), <_\lambda)$ is the linear ordering where $(\alpha,p) <_\lambda (\beta,q) \leftrightarrow \alpha < \beta \vee (\alpha = \beta \wedge p < q)$. $x \leq_\lambda y \leftrightarrow x <_\lambda y \vee x = y$.*

It is obvious that $(\lambda \times Q[0,1), <_\lambda)$ is a dense linear ordering with the left endpoint $(0,0)$. Since we are going to transfer from the transfinite to Q, and Q has no left endpoint, it is convenient to use a slightly different linear ordering.

Definition 51. *$(\lambda \times' Q[0,1), <'_\lambda)$ is the same linear ordering as $(\lambda \times Q[0,1), <_\lambda)$, but with the left endpoint $(0,0)$ removed. $x \leq'_\lambda y \leftrightarrow x <'_\lambda y \vee x = y$. For $x \in (\lambda \times' Q[0,1))^{r-}, | x |$ is the largest of the first coordinates of values of x. We also use $\alpha \times' Q[0,1)$ for $(\alpha \times Q[0,1))\setminus\{(0,0)\}$.*

Definition 52. *Let* $f : S_{k-1}(\lambda) \to S((\lambda \times' Q[0,1))^{k-})$. *$f$ is regressive if and only if for all* $x \in S_{k-1}(\lambda), y \in f(x)$, *we have* $\mid y \mid < min(x)$. *$E \subseteq \lambda$ is f-homogenous if for all* $x, y \in S_{k-1}(E)$ *and* $z \in S((\lambda \times' Q[0,1))^{k-})$ *with* $\mid z \mid < min(x \cup y)$, *we have* $z \in f(x) \leftrightarrow z \in f(y)$.

Lemma 2. *λ is a strongly inaccessible cardinal. Let $C \subseteq \lambda$ be closed and unbounded, and let* $f_1, \ldots, f_m : S_{k-1}(\lambda) \to S((\lambda \times' Q[0,1))^{k-})$ *be regressive. There exists $E \subseteq C$ of order type ω, where for all i, E is f_i-homogenous.*

Proof. λ is a strongly inaccessible cardinal by Friedman (2001, Lemma 1.10). Let C, $f_1, \ldots, f_m$ be as given. Set C to be the set of uncountable cardinals $< \lambda$, which is closed and unbounded in λ by the first claim. Define $f : S_{k-1}(\lambda) \to S(\lambda)$ by $f(x) = < f_1(x), \ldots, f_m(x) >$. Here we can use any convenient one-one map from $S((\lambda \times' Q[0,1))^{k-})^m$ into $S(\lambda)$ which, for each uncountable cardinal $\kappa < \lambda$, maps $S((\kappa \times' Q[0,1))^{k-})^m$ into $S(\kappa)$. Now f may not be regressive, but clearly f is regressive on $S_{k-1}(C)$. So we take the values of f to be the empty set, off of $S_{k-1}(C)$. According to Friedman (2001, Lemma 1.10), there exists f-homogenous $E \subseteq C$ of order type ω. Then for all i, E is f_i-homogenous. □

Lemma 3. *Let $X \subseteq (\lambda \times' Q[0,1))^k$. There exists $E \subseteq \lambda$ of order type ω such that X is $E \times \{0\}$ tail invariant.*

Proof. Let X be as given. We apply Lemma 2. We define finitely many functions $f : S_{k-1}(\lambda) \to S((\lambda \times' Q[0,1))^{k-})$ as follows. The $w \in (\lambda \times' \{0\})^{k-}$ with domain not all of $\{1, \ldots, k\}$ (see Definition 49), fall naturally into finitely many kinds. The kind of w is determined first by its domain (a nonempty subset of $\{1, \ldots, k\}$), and second by the order type of the ordinal components (first terms) of its values, listed from left to right. Write these kinds as $\sigma_1, \ldots, \sigma_m$, without repetition.

We define $f_1, \ldots, f_m : S_{k-1}(\lambda) \to S((\lambda \times' Q[0,1))^{k-})$ as follows. Let $x \in S_{k-1}(\lambda)$. To evaluate $f_i(x)$, let $y \in \lambda^{k-}$ be unique with kind σ_i, where the values of y form an initial segment of the elements of x. Set $f_i(x) = \{z \in (min(x) \times' Q[0,1))^{k-} : dom(z) = \{1, \ldots, k\} \backslash dom(y) \wedge y \times \{0\} \cup z \in X\}$.

By Lemma 2, let $E \subseteq \lambda$ be of order type ω and f_i-homogeneous for all $1 \le i \le m$. We can assume that $0 \notin E$. Let $u, v \in (\lambda \times' Q[0,1))^k$ be $E \times \{0\}$ tail equivalent, where $u, v \notin (E \times \{0\})^k$. We claim that $u \in X \leftrightarrow v \in X$.

Let $u' \in \lambda^{k-}$ be the restriction of u to its coordinates in $E \times \{0\}$, where all coordinates higher in $<'_\lambda$, are in $E \times \{0\}$. Let $v' \in \lambda^{k-}$ be the restriction of v to its coordinates in $E \times \{0\}$, where all coordinates higher in $<'_\lambda$, are in $E \times \{0\}$. Then u', v' result in the evaluation of some $f_i(x), f_i(y)$, respectively, where $x, y \in S_{k-1}(E)$. Hence $u \in X \leftrightarrow v \in X$.

It remains to show that for order equivalent $u, v \in (E \times \{0\})^k, u \in X \leftrightarrow v \in X$. However, this may not be the case. But we can use the ordinary infinite Ramsey theorem to replace E by a suitable subset of E of order type ω for which this is the case. □

Lemma 4. *Let $W^* \subseteq (\lambda \times' Q[0,1))^{2k}$. There exists $S^* \subseteq (\lambda \times' Q[0,1))^k$ such that for all $\alpha < \lambda, S^*[\le'_\lambda (\alpha, 0)]$ is a maximal root of $W^*[\le'_\lambda (\alpha, 0)]$.*

Proof. Let W^* be as given. We build S^* by transfinite recursion along λ. Let S_0^* be the maximal root of $W^*[\le'_\lambda (0,0)]$, which is $\{(0,0), \ldots, (0,0)\}$ or $\varnothing$. Suppose S_α^* is a maximal root of $W^*[\le'_\lambda (\alpha, 0)]$. Take $S_{\alpha+1}^*$ to be a maximal root of $W^*[\le'_\lambda (\alpha+1, 0)]$ extending $S^{\alpha*}$. Now suppose that for all $\alpha < \gamma < \lambda, S_\alpha^*$ is a maximal root of $W^*[\le'_\lambda (\alpha, 0)]$, where

γ is a limit ordinal, and we have for all $\beta < \delta < \gamma$, $S_\alpha^* \subseteq S_\beta^*$. Let $S_\gamma^{**} = \cup_{\alpha < \lambda} S_\alpha^*$. Then S_γ^{**} is a maximal root of $W^*[<_\lambda' \ (\gamma, 0)]$. Let S_γ^* be a maximal root of $W^*[\leq_\lambda' \ (\gamma, 0)]$ extending S_γ^{**}. Finally, define $S = \cup_\alpha S_\alpha$. $\qquad\square$

Lemma 5. *Let* $W^* \subseteq (\lambda \times' Q[0,1))^{2k}$. *There exists* $E \subseteq \lambda$ *of order type* ω *and* $S^* \subseteq (sup(E) \times' Q[0,1))^k$, *where for all* $\alpha \in E$, $S^*[\leq_\lambda' \ (\alpha, 0)]$ *is a maximal root of* $W^*[\leq_\lambda' \ (\alpha, 0)]$, *and where* S^* *is* $E \times \{0\}$ *tail invariant.*

Proof. Let W^* be as given. Let S^* be as given by Lemma 4. By Lemma 3, let $E \subseteq \lambda$ be of order type ω, where S^* is $E \times \{0\}$ tail invariant. $\qquad\square$

Lemma 6. *Let* $(A, <^*)$ *be a dense linear ordering with no endpoints, and* $x_1, x_2, \ldots$ *be an infinite strictly increasing sequence from* A *with no upper bound. Let* $W' \subseteq Q^{2k}$ *and* $S' \subseteq A^k$, *where for all* n, $S'[\leq^* x_n]$ *is a maximal root of* $W'[\leq^* x_n]$, *and* S' *is* $\{x_1, x_2, \ldots\}$ *tail invariant. There exists* $B \subseteq A$, *where* $(B, <^* \cap B^2)$ *is a countable dense linear ordering with no endpoints,* $x_1, x_2, \ldots \in B$, *for all* n, $(S' \cap B^k)[(\leq^* \cap B^2)x_n]$ *is a maximal root of* $(W' \cap B^k)[(\leq^* \cap B^2)x_n]$, *and where* $S' \cap B^2$ *is* $\{x_1, x_2, \ldots\}$ *tail invariant.*

Proof. This can be proved either from the basic model theoretic theorem that every structure has a countable elementary substructure (in a finite language), or directly by an easy sequential construction. The former is preferable because the model theoretic setup facilitates the proof of Lemma 7. We work with the structure $(A, <^*, W', S', x_1, x_2, \ldots)$. The hypothesized properties are all first order. Now simply take any countable elementary substructure $(B, <^* \cap B^2, W' \cap B^{2k}, S' \cap B^k, x_1, x_2, \ldots)$. The first order properties still hold, and they have the proper meaning for the desired conclusion. $\qquad\square$

Lemma 7. *ISMR(Nteq,k).*

Proof. Let $W \subseteq Q^{2k}$ be order invariant. Let $W^* = \{x \in (\lambda \times' Q[0,1))^{2k} : (\exists y \in W) (x, y \text{ are order equivalent})\}$. Let $S^* \subseteq (\lambda \times' Q[0,1))^k$ and $E = \{x_1 < x_2 < \ldots\}$ be given by Lemma 5. Let $sup(E) = (\gamma, 0)$, where γ is a limit ordinal. Let $A, <^*, W', S'$ be the restrictions of $\lambda \times' Q\{0,1), <_\lambda', W^*, S^*$ to $\gamma \times' Q[0,1)$. Then the hypotheses of Lemma 6 hold. Note that $W' = \{x \in (\gamma \times' Q[0,1))^{2k} : (\exists y \in W) (x, y \text{ are order equivalent})\}$.

Now apply Lemma 6. We obtain $(B, <^{**}, W'', S'', x_1, x_2, \ldots)$, where $(B, <^{**})$ is a countable dense linear ordering without endpoints, $x_1 <^{**} x_2 <^{**} \ldots$ has no upper bound in $<^{**}$, each $S''[<^{**} x_i]$ is a maximal root of $W''[<^{**} x_i]$, and S'' is $\{x_1, x_2, \ldots\}$ tail invariant. Once again, $W'' = \{x \in B^{2k} : (\exists y \in W) (x, y \text{ are order equivalent})\}$.

Let h be any isomorphism from $(B, <^{**})$ onto $(Q, <)$ mapping $x_1, x_2, \ldots$ onto $0, 1, \ldots$. Then h is an isomorphism from $(B, <^{**}, W'', S'', x_1, x_2, \ldots)$ onto $(Q, <, h[S''], h[W''], 0, 1, \ldots)$. It is clear that $h[W''] = W$, because the definition of W'' as the set of all tuples with certain order types must be preserved under h. It is immediate that $S = h[S'']$ is an N tail invariant step maximal root of W. $\qquad\square$

Theorem 34. *(ZFC). For all* $k \geq 1$, *if there exists a* $(k-1)$-*subtle ordinal then ISMR(Nteq,k).*

Proof. Note that the argument from Lemmas 2–7 used only that $k \geq 2$, and λ is the least $(k-1)$-subtle ordinal. Lemma 1 takes care of the trivial case $k = 1$. $\qquad\square$

Theorem 35. *RCA_0 proves that for all k, SRP proves ISMR(Nteq,k). SRP^+ proves ISMR(Nteq).*

Proof. The first claim is immediate from the proof of Theorem 34. It is also clear that the entire argument can be conducted using k as a variable. Therefore SRP^+ proves ISMR(Nteq). □

It remains to prove ISMR(Nteq) in $WKL_0 + $ Con(SRP). We work in $WKL_0 + $ Con(SRP).

Lemma 8. *There is a countable M with satisfaction relation which satisfies SRP.*

Proof. By the formalized Gödel completeness theorem, which is provable in WKL_0. □

We now fix M as given by Lemma 8.

Lemma 9. *Let k and order invariant $W \subseteq Q^{2k}$ be given. Let K be the finite set of order types of elements of W. There exists $(Q^*, <^*, W^*, N^*, S^*)$ internal to M, where, according to M,*

  i. *$(Q^*, <^*)$ is the rationals with its usual linear ordering.*
  ii. *$W^* \subseteq Q^{*k}$ consists of all $x \in Q^{*k}$ with order type in K.*
  iii. *N^* is the set of all nonnegative integers.*
  iv. *S^* is a step maximal root of W^*.*
  v. *S^* is N^* tail invariant.*

Proof. Here k, W, K are standard. There may be nonstandard rationals in Q^* and nonstandard integers in N^*. The steps in iv are given by elements of N^*. The tail invariance uses N^* and not N. We simply carry out the proof of Theorem 34 in the model M. □

Theorem 36. *$WKL_0 + $ Con(SRP) proves ISMR(Nteq).*

Proof. We work in $WKL_0 + $ Con(SRP), with M provided by Lemma 8. Let $W \subseteq Q^{2k}$ be order invariant. Let $Q^*, <^*, W^*, N^*, S^*$ be as given by Lemma 9. Obviously $(Q^*, <^*)$ is a countable dense linear ordering without endpoints, let $x_1 <^* x_2 <^* \ldots$ be elements of N^* that have no upper bound in $(Q^*, <^*)$. Let h be any isomorphism from $(Q^*, <^* , W^*, S^*, x_1, x_2, \ldots)$ onto $(Q, <, W, S, 0, 1, \ldots)$, $S \subseteq Q^{2k}$. Note that $h[W^*] = W$, and we have set $h[S^*] = S$. Then S is an N tail equivalent step maximal root of W. □

Corollary 1. *SRP^+ and $WKL_0 + $ Con(SRP) prove all eighteen titled statements in Sections 4.1–4.3, excluding IMR/C(intervals).*

Proof. By Theorems 35, 36, and 18. □

In Friedman (forthcoming(b)), we derive Con(SRP) from $RCA_0 + ISMR$(Nteq). This together with Theorem 36 establishes that ISMR(Nteq) is provably equivalent to Con(SRP) over WKL_0. In Friedman (forthcoming(b)), we also show that IMR/ C(intervals) is provably equivalent to Con(SRP) over WKL_0.

11 Appendix: Formal systems used

EFA Exponential function arithmetic. Based on exponentiation and bounded induction. Same as $I\Sigma_0(exp)$, Hájek and Pudlák (1993, p. 37, 405).

RCA_0 Recursive comprehension axiom naught. Our base theory for Reverse Mathematics. Simpson (1999, 2009).

WKL_0 Weak Konig's Lemma naught. Our second level theory for Reverse Mathematics. Simpson (1999, 2009).

ACA_0 Arithmetic comprehension axiom naught. Our third level theory for Reverse Mathematics. Simpson (1999, 2009).

ZF(C) Zermelo set theory (with the axiom of choice). ZFC is the official theoretical gold standard for mathematical proofs. Jech (2006).

SRP[k] ZFC + $(\exists\lambda)(\lambda$ has the $k-$SRP), for fixed k. Section 10.1.

SRP ZFC + $(\exists\lambda)(\lambda$ has the $k-$SRP), as a scheme in k. Section 10.1.

SRP^+ ZFC + $(\forall k)(\exists\lambda)(\lambda$ has the $k-$SRP). Section 10.1.

HUGE[k] ZFC + $(\exists\lambda)(\lambda$ is $k-$HUGE), for fixed k.

HUGE ZFC + $(\exists\lambda)(\lambda$ is $k-$HUGE), as a scheme in k.

HUGE+ ZFC + $(\forall k)(\exists\lambda)(\lambda$ is $k-$HUGE).

λ is $k-huge$ if and only if there exists an elementary embedding $j : V(\alpha) \to V(\beta)$ with critical point λ such that $\alpha = j^{(k)}(\lambda)$. (This hierarchy differs in inessential ways from the more standard hierarchies in terms of global elementary embeddings). For more about huge cardinals, see Kanamori (1994, p. 331).

12 Acknowledgement

This research was partially supported by an Ohio State University Presidential Research Grant and by the John Templeton Foundation grant ID #36297. The opinions expressed here are those of the author and do not necessarily reflect the views of the John Templeton Foundation.

References

ChessOK. 2013. Lomonosov supercomputer announcement. URL http://chessok.com/?page_id=27966.

Cohen, P. J. 1963. The independence of the continuum hypothesis. *Proceedings of the National Academy of Sciences of the United States of America* 50(6):1143–1148.

—. 1964. The independence of the continuum hypothesis. *Proceedings of the National Academy of Sciences of the United States of America* 51(1):105–110.

Ferrario, R. and V. Schiaffonati. 2012. *Formal Methods and Empirical Practices: Conversations with Patrick Suppes.* Stanford, CA: CSLI Publications.

Friedman, H. M. 2001. Subtle cardinals and linear orderings. *Annals of Pure and Applied Logic* 107(1-3):1–34. doi:10.1016/S0168-0072(00)00019-1.

—. 2011. My forty years on his shoulders. In M. Baaz, C. H. Papadimitrou, H. W. Putnam, D. S. Scott, and J. C. L. Harper, eds., *Kurt Gödel and the Foundations of Mathematics, Horizons of Truth,* pages 399–432. Cambridge: Cambridge University Press. URL http://u.osu.edu/friedman.8/files/2014/01/Fortyyears111909-1taw9j8.pdf.

—. forthcoming(a). Boolean Relation Theory and Incompleteness. Lecture Notes in Logic, ASL Publications, https://u.osu.edu/friedman.8/foundational-adventures/boolean-relation-theory-book/.

—. forthcoming(b). Maximality and incompleteness. Ohio State University.

Gödel, K. 1986. On formally undecidable propositions of principia mathematica and related systems I. In S. Feferman, J. W. Dawson, S. C. Kleene, G. H. Moore, R. M. Solovay, and J. V. Heijenoort, eds., *Kurt Gödel: Collected Works*, vol. I, pages 145–195. Oxford University Press, 1931.

—. 1986-2003. *Kurt Gödel: Collected Works*, vol. I-V. New York and Oxford: Oxford University Press. S. Feferman, J. W. Dawson, Jr., W. Goldfarb, C. Parsons, and R. M. Solovay (eds).

—. 1989. The consistency of the axiom of choice and the generalized continuum hypothesis with the axioms of set theory. In S. Feferman, J. W. Dawson, S. C. Kleene, G. H. Moore, R. M. Solovay, and J. V. Heijenoort, eds., *Kurt Gödel: Collected Works*, vol. II, pages 33–101. Oxford University Press, 1940.

Gross, O. A. 1962. Preferential arrangements. *American Mathematical Monthly* 69(1):4–8. doi:10.2307/2312725.

Hájek, P. and P. Pudlák. 1993. *Metamathematics of First-Order Arithmetic, Perspectives in Mathematical Logic*. Berlin: Springer-Verlag.

Jech, T. J. 2006. *Set Theory: The Third Millennium Edition, revised and expanded*. New York: Springer-Verlag Berlin Heidelberg.

Kanamori, A. 1994. *The Higher Infinite. Large Cardinals in Set Theory from their Beginnings, Perspectives in Mathematical Logic*. Berlin: Springer-Verlag.

Simpson, S. G. 1999. *Subsystems of second order arithmetic*. Berlin: Springer.

—. 2009. *Subsystems of Second Order Arithmetic: Perspectives in Logic*. New York: ASL and Cambridge University Press, 2nd edn.

Suppes, P. 1992. Patrick Suppes, Lucie Stern Professor of Philosophy, Emeritus. URL `http://web.stanford.edu/~psuppes/`.

—. 1999. Suppes Brain Lab, Stanford University. URL `http://suppes-brain-lab.stanford.edu`.

—. 2002. Collected works of Patrick Suppes. URL `http://suppes-corpus.stanford.edu`.

Part II

Philosophy of Science

4

The Structure, the Whole Structure, but not Nothing but the Structure

Thomas Ryckman

My thanks to Paul Humphreys for helpful comments on the version presented in March, 2012, and my apologies to Stathis (2006) for almost cribbing his title.

> "One of the besetting sins of philosophers of science is to oversimplify the structure of science"
> (Suppes 1962)

The Stichworte for the following remarks are to be found in this revealing epigram; I offer them to Patrick Suppes with respect and humility.

1 The Structure of Science

Philosophers of science now rather tend to shy away from using the term. It has the stale odor of an earlier era when global pronouncements about the nature and character of science were widespread, when references to "the scientific method" appeared to connote something so well-understood that they occurred literally everywhere in the culture, and not, as today, largely in the expert testimony of philosophers of science before courts of law, deciding whether creationist doctrine should be accorded equal time in high school biology classes. Yet even contemporary philosophers cannot forget the title of Ernest Nagel's book of (1961), a work that in fact dialectically succeeded in broadly characterizing general canons of scientific method. Dialectically indeed, since each chapter of Nagel's book is presented as a clash of the often one-sided merits of conflicting viewpoints, regarding theoretical reduction, the meaning of probabilistic laws, of explanation, causality and indeterminism, and perhaps most famously, realism and instrumentalism ('The Cognitive Status of Theories'). Even as Nagel dared package his *chef d'oeuvre* using the very term Kuhn would soon diachronically transmute, at the same time he gave many convincing illustrations of the shallow and unwarranted character of sweeping philosophical generalizations fashioned independently of detailed investigations of the practice of science.

The continuing impact of Nagel's instruction throughout Suppes' long career is not difficult to detect, even though, like another influential philosopher of science (Carnap), His youthful desire to write on the axiomatization of physical theories was frustrated by

Foundations and Methods from Mathematics to Neuroscience.
Colleen E. Crangle, Adolfo García de la Sienra and Helen Longino.
Copyright © 2014, CSLI Publications.

his dissertation supervisor. It is a great pity that Nagel did not live to witness Suppes' election to the National Academy of Science, joining Nagel as one of the few philosophers ever so honored. Justly emphasizing Nagel's prominence in the intellectual life of Columbia University and of the city of New York for over forty years, the principal theme of Suppes' moving 1994 obituary of Nagel in the NAS's *Biographical Memoirs* is "Nagel's characteristic skepticism of philosophers who propose simple and general theories for complex matters . . .". *Ye fruit doesn't fall far from ye tree!* One recalls Nagel's critical attitude toward the then prevalent syntactic conception of scientific theories, his refusal to engage in grand philosophical pronouncement, and his insistence that formal analysis is an effective tool that should be used wherever possible, particularly in that area of scientific methodology Nagel characterized by a "general fuzziness", the application of theory to an empirical subject matter. Of course, in the development of formal methods for stating theories and characterizing their models through representation theorems, as well as in studying the applications of theory through the theory of measurement, data analysis and statistical test of hypotheses, Suppes has taken Nagel's baton and far surpassed his (or for that matter, anyone else's) achievements. For now, I only observe that in the above epigram to this paper, the intent of the term *the structure of science*, taken in context, is precisely the Nagelian sense, as referring to "the logic of scientific inquiry and the logical structure of its intellectual products", Nagel (1961, p. viii). And I can think of no more fitting accolade for Patrick Suppes than to say of him in 2012 what he wrote of Nagel in 1994:

> "To many generations of students he [is] the outstanding spokesman of what philosophy could offer in terms of analysis of the scientific method, as it is practiced in many different sciences, and in the relation between science and perennial problems of philosophy . . ."

Suppes' immense body of work is both a grand tribute to, and a hardly imaginable development of, the Nagelian philosophical temperament, combining characteristic skepticism and detailed investigation.

2 Besetting Sins

Here we are reminded of the implacable moral ferocity of an Old Testament prophet, an association not completely unsuited in view of Suppes' frequent complaints about the lack of any "serious discussion" in the philosophy of science literature of many issues arising in the methodology of experimentation and testing of scientific theories. While only few philosophers of science have risen to his challenge to produce interesting descriptive analyses of actual experiments in science, a larger number have followed his lead in thinking about the formalization or representation of scientific theories. Whether he sees himself this way or not, many contemporary philosophers regard Suppes as the lofty Elijah of structuralism in philosophy of science, foretelling the triumph of semantic view of theories while enjoining against idolatry of the syntactic (or Standard) view disseminated with such effect by the logical empiricists and their imitators. Alas, as we shall see, some among his erstwhile structuralist followers have strayed into philistinism, sacrificing to Baal.

3 Oversimplifying

A recurring theme in Suppes' work (and conversation!) is how very distant, abstract and inaccurate are most philosophers of science' conceptions of science, and of scientific theories, from the practice of science, i.e., science familiar to the working scientist. In a forceful attempt to rectify this disciplinary shortcoming, he once managed to publish a paper that begins, implausibly, with several pages of longish quotations from the scientific literature where the term 'model' is employed Suppes (1960, pp. 287–88). Despite the various, seemingly quite distinct, senses of the term in fields as far apart as economics, psychology and physics, Suppes argued that all these uses could be subsumed under Tarski's sense of 'model'. A few years later, in understated Nagelian fashion, he suggested that even radical instrumentalist conceptions of theories must also "yield the ordinary talk about models and theories as a first approximation" (1967, p. 66). Here a model or structure is a non-linguistic object of the appropriate type in which a theory is realized or satisfied, a characterization sufficiently broad to encompass both the mathematician's sense of model and the great diversity of uses of the term 'model' in the empirical sciences. In brief compass, Suppes stated that the set-theoretic conception is the "basic and fundamental concept of model needed for an exact statement of any branch of empirical science; indeed it is "more fundamental" than the (various) concepts of model employed in the empirical sciences in the sense that one can use it to give an exact statement of the theory or use in the exact analysis of the data under investigation. In his conception, a theory is presented by the specification of a 'set-theoretical predicate', essentially, an ordered n-tuple of set-theoretically defined objects. The theory can then be viewed as the collection of structures picked out by the set-theoretical predicate.

4 Set-theoretic Structures

It would appear that considering theories in the empirical sciences as set-theoretic structures is a reflection not only of Tarski's model theory but also of the significant impact of the Bourbaki mathematics texts in the immediate post-WWII period, an influence both pedagogical and broad enough to be considered cultural. Thus (an observation Suppes has frequently made) the mathematician's statement of a theory, say, the theory of abelian groups, is quite distinct from the syntactical account of an axiomatized theory. It is the mathematician's sense of axiomatization that underlies Suppes' papers on axiomatization of classical particle mechanics (with McKinsey et al. (1953)) and of relativistic mechanics (with Rubin and Suppes (1954)). A methodological theme of these papers, and in his logic book of 1957, is that formalization of theories under a restriction to first order logic is like boxing with one hand. One might possibly do it but the result can be neither pretty, nor pleasing, nor efficient. The language of set-theory (which can of course be given a first order formulation) is required. In the 1950s climate of American philosophy of science, this was something of a heresy. After all, it was only in the 1950s that a passing familiarity with first-order logic became a requirement of philosophical education. Moreover, there was a widespread prejudice, among other influences promoted by Goodman and Quine (1947), to the effect that the responsible (i.e., anti-metaphysical) philosopher of science must attempt to eschew abstract objects, of which sets or classes are the exemplars. That Quine quickly but *sub rosa* backtracked from nominalism was not generally apparent till much later. Tarski, whose model theory was a principal stimulus to the conception of models of empirical theories that Suppes pioneered, certainly also had strong nominalist leanings. Moreover, Sol Feferman has pointed out that set theory was, for Tarski, also the paradigmatic instrument for purposes of conceptual analysis. And Tarski early on submerged any nominalist metaphysical qualms by appeal to something like the pragmatic indispensability argument for the use of classical mathematics in empirical science that we associate with the later Quine. Suppes, so far as I can see, has also never allowed metaphysical qualms to impede conceptual analysis. The upshot is that scientific theories are to be presented using mathematics, i.e., set theory, not first order logic. Lest Platonists take comfort in the Suppesian profligacy with sets, let us state for the record his conviction that in the applications of mathematics in physics, "one can go the entire distance, or certainly almost the entire distance, in a purely finitistic way" Suppes (2002, p. 305).

This salutary message to philosophers of science was in any case fully in step with the developing practice of how theories are represented for certain purposes in physics. A striking example is Stephen Hawking and George Ellis' (1973) text on General Relativity. To my knowledge, this is the first (if not the first, the most influential) explicit presentation of any modern physical theory in structural form. It initiated what is now called the "space-time theories" approach: broadly speaking, space-time theories are presented within a model-theoretical format, i.e., by specification of a set-theoretic predicate. Space-time theories are then possible realizations of a structure, an ordered pair whose first member is a differentiable space-time manifold, followed by the metric tensor, a geometrical object field with dynamical content. By the mid-1970s, the model theoretical format had extended further into mathematical physics with the geometric representation of renormalizable ('gauge') quantum field theories, utilizing the fiber bundle formalism of modern differential geometry (e.g., Choquet-Bruhat et al. (1977)), a mathematical development that again only emerged in the immediate post-WWII period.

5 The So-called Semantic View of Theories

Though Suppes is widely cited as an originator, if not the *fons et origo* of this view, he will surely exclaim, as did Newton to Bentley, "Pray, do not ascribe that notion to me!" One readily sees the reason for the attribution: by encompassing (but generalizing upon) the sense of the term 'model' in the empirical sciences under the Tarskian conception, the term semantic view appears altogether natural. But Suppes' understanding is not The Semantic View as is widely understood, for that notion commonly and crucially relies either on the idea of a truth-making structure or a truth-making map. In the latter case, models are not non-linguistic objects, for they invoke an interpretation function whose domain is the non-logical vocabulary of a particular language. As for truth-making structure, let us recall a fact that philosophers claim to know well but often manage to forget: *set-theoretic relationships hold only between mathematical structures*, not between such structures and 'the world'. The idea of a truth—making structure, in the realist sense of a correspondence to 'the world', has no clear meaning. One productive use of the structural view of theories concerns inter-theoretic relations, an issue to which Nagel devoted much attention. It is fair to say that the some of the best current thinking about reduction of one theory to another (I think of Bob Batterman's work) is not in the Nagelian but in the structural tradition pioneered by Suppes.

So in the Suppesian sense, a model of an empirical theory is a mathematical model, a set-theoretic structure used to represent a system or kind of system and its behavior. The theory itself is illuminatingly thought of as a collection of such representative mathematical models. To say then that a representative mathematical model realizes an empirical theory is to say that the model satisfies the description given by the theory's characteristic set-theoretic predicate, i.e., fits the description of a certain kind of mathematical structure. We shall follow Suppes (2011) (web-based course(s)) in calling his view "the informal structural view" to distinguish it from the Semantic View with its (often implicit) reference to truth-making structures or truth-making maps.

6 Data Models

The Semantic View (as commonly understood) is thought to be an answer to the question, *What is a scientific theory?* (doxological chorus, in unison: *a collection of models*). Suppes, on the other hand, has, I think, maintained that this is not a definition but a perspective on theories more in accord with scientific practice. And he has long held the view that "too much philosophy of science is 'theory-talk', it is as if science were only about theories" and his advocacy of the informal structural view of theories should be understood in this spirit. His "Models of Data" paper of (1962) showed us that if we attempt a precise analysis of the relation between a model of a theory and the "complete experimental experience" we should have to consider a whole hierarchy of models interposed between that experience and the model of the theory we are attempting to test. At each level of the hierarchy is a theory, to be given an empirical interpretation at the next lower level. The complexity of this hierarchical view of testing a theory reflects the methodology of modern mathematical statistics, where distinctions between systematic and measurement error, parameter estimation, goodness of fit and other issues are prominent. Nonetheless, we can still hope to use formal, set-theoretic tools to display the relations between models in this hierarchy to bring illumination to that area of scientific methodology that Nagel characterized as a "general fuzziness", the application of theory to an empirical subject matter.

7 The Sins of the Children: Structural Realism

In the past decade and a half, structuralist approaches to physics among philosophers of physics has been a rocketing industry with near exponential growth. Surely it is a good time to get out of the market! For what the most prolific faction of these folks claim is that physics tells us not about the objects of which the world is made, but about the structure of the world. Advocates of its most extreme form, so-called "ontic structural realism", hold that science itself is concerned with only structural claims about the world; non-structural claims are either false or meaningless because there are no non-structural facts. Out with objects, they are just epiphenomena: the world, in and of itself, is structure. This is intended not only as a radical revision of fundamental ontology but also (in Oxford at least) perhaps a necessary consequence of "taking quantum mechanics seriously", i.e., adopting the Everett or Many Worlds interpretation of quantum mechanics. For present purposes, the point is that the originators of so-called "Structural Realism" claim descent from two sources. First, an observation going back to Poincaré and Duhem (and even earlier) that experimental laws are preserved across theory change This is needed to preserve both Realism and its opponents from what was once termed "Kuhn losses", the discontinuities of paradigm shifts. The second source of Structural Realism (in the virulent "ontic" version) is the Semantic View of Theories (SVT), that theories just are set-theoretic structures latching on to ontological furniture of the World. Patrick Suppes, the alleged patriarch of the SVT, is accordingly recognized as their prophet, foreshadowing the claim that everything, ultimately, is structure. I hope to have made clear that this idolatry is misplaced, that Suppes' Nagelian proclivities absolutely forbid such over-reaching blasphemy, and that the Structural Realists are sacrificing their ontological calf at altogether the wrong altar.

References

Choquet-Bruhat, Y., C. DeWitt-Morette, and M. D. Bleick. 1977. *Analysis, Manifolds and Physics*. Amsterdam: North Holland Pub. Co.

Goodman, N. and W. V. Quine. 1947. Steps toward a constructive nominalism. *Journal of Symbolic Logic* 12(4):105–122. doi:10.2307/2266485.

Hawking, S. and G. F. R. Ellis. 1973. *The Large Scale Structure of Space-Time*. Cambridge: Cambridge University Press.

McKinsey, J., A. Sugar, and P. Suppes. 1953. Axiomatic foundations of classical particle mechanics. *Journal of Rational Mechanics and Analysis* 2(1):253–272.

Nagel, E. 1961. *The Structure of Science: Problems in the Logic of Scientific Explanation*. New York: Harcourt, Brace, and World, Inc.

Psillos, S. 2006. The structure, the whole structure, and nothing but the structure? *Philosophy of Science* 73(5):560–570.

Rubin, H. and P. Suppes. 1954. Transformations of systems of relativistic particle mechanics. *Pacific Journal of Mathematics* 4(4):563–601. doi:10.2140/pjm.1954.4.563.

Suppes, P. 1957. *Introduction to Logic and Axiomatic Set Theory*. New York: Van Nostrand Reinhold Co. Reprinted Dover, New York, 1999.

—. 1960. A comparison of the meaning and uses of models in mathematics and the empirical sciences. *Synthese* 12(2/3):287–301.

—. 1962. Models of data. In E. Nagel, P. Suppes, and A. Tarski, eds., *Logic, Methodology, and Philosophy of Science: Proceedings of the 1960 International Congress*, pages 252–261. Stanford: Stanford University Press.

—. 1967. What is a scientific theory? In S. Morgenbesser, ed., *Philosophy of Science Today*, pages 55–67. New York: Basic Books, Inc.

—. 1994. Ernest Nagel. In *Biographical Memoirs*, vol. 65, pages 257–272. Washington, D.C.: National Academy of Sciences.

—. 2002. *Representation and Invariance of Scientific Structures*. Stanford, CA: CSLI Publications.

—. 2011. Future development of scientific structures closer to experiments: Response to F.A. Muller. *Synthese* 183(1):115–126.

5

Models of Data and Inverse Methods

PAUL HUMPHREYS

Fifty years after its publication, Patrick Suppes' paper 'Models of Data' (1962), stands as a remarkable achievement in the philosophy of science. I shall briefly lay out some of the central features of his paper and then use the basic idea of a hierarchy of models to explore the relation between data and inference in a very different setting than the one that Suppes used. Let me emphasis at the outset that I am not attempting an exegesis of Suppes' paper and that I am using many of the ideas outside their original domain of application. The theoretical contexts to which Suppes applied his hierarchy of models differ in important ways from the contexts that I shall discuss, the most important of these being that Suppes was interested in situations where there exists a well established parametric theory of the top level domain and statistical estimation of the parameters from experimental data is the goal. The present project is to see how inferences from data are constrained when the quantity to be estimated is not part of an explicit theory.

Here are some of the main points of Suppes' paper.

A. A hierarchy of models exists between theory and data when the latter are connected with the former. This account of how empirical content is injected into formal theories is significantly more sophisticated and detailed than was the earlier logical empiricists' use of coordinating definitions. One virtue, consistent with the semantic account's de-emphasis on particular syntactic formulations of theories, is that Suppes' approach is largely unconcerned with linguistic meaning and focuses instead on how quantitative estimates are provided for parameters that occur in abstractly formulated theories.

B. The models in the hierarchy are of different logical type. There is a variety of reasons for this. One is that there are concepts in the theory that have no observable analogues in the experimental data, a point that anticipates in certain respects the well-known distinction between phenomena and data made in Bogen and Woodward (1988). A second reason is that some levels in the hierarchy will contain models with continuous variables or infinite data sequences while others will have models using discrete variables and finite data sets only. This distinction between continuous and discrete versions of models was later critical to understanding how the models that drive computer simulations differ from the continuous mathematical models back of the simulations. These moves are mathematically non-trivial and often require ad-

Foundations and Methods from Mathematics to Neuroscience.
Colleen E. Crangle, Adolfo García de la Sienra and Helen Longino.
Copyright © 2014, CSLI Publications.

ditional correction techniques to avoid errors and artifacts arising from the discrete approximations.

C. In order for there to be models of the data there has to be a theory of the generating conditions for the data. Put another way, the models of data with which Suppes is concerned are perhaps better called 'models of data from correctly designed experiments'. For example, within the learning model that Suppes uses as a running example, there are possible realizations of the data that fail to satisfy the stationarity condition that applies to the reinforcement schedule in a model of the experiment. Although such realizations count as possible data, they are inappropriate for estimating the parameter θ of the linear learning model. The need for a theory of the experiment makes the position radically different from most empiricist accounts of data in which minimal reference to theories is desirable and the generating conditions of the data are frequently not a part of the empiricist analysis.

D. One of the themes of Suppes' article is that many details of an experimental arrangement cannot be included in the hierarchy of models either because they cannot be couched in terms of the language of the theory to be tested or because they involve heuristics that cannot easily be formalized. This orientation is part of a more general methodological claim that '... the only systematic results possible in the theory of scientific methodology are purely formal...' (p. 261). This position is appealing, at least in the sense of seeing how far a purely formal analysis of data processing can be taken.

1 Computerized Tomography and Inverse Methods

My running example will be data generated by imaging devices that use computed tomography (CT). These instruments take data that originate with physical sources, such as X-rays, and use computationally intensive processing of that data to produce the final image. Because of space constraints, I shall provide only the basic features of these instruments; further details can be found in Humphreys (2013b, Forthcoming).

One of the core features of CT instruments is that they construct solutions to inverse problems, a kind of problem that occurs in many other contexts such as geophysics and radioastronomy. Although an inverse problem is sometimes taken to be one that involves an inference from data to model parameters, and hence is directly related to Suppes' framework, there is not a uniform use of the term and generically an indirect problem involves an inference from the data to the generating conditions of that data. One of the morals of this paper is that philosophers can benefit from methods that have been developed to solve inverse problems since they are directly relevant to a number of topics in the philosophy of science, including scientific realism and the application of mathematics. More on that later.

In two dimensional computerized tomography instruments, M parallel X-ray beams, collimated to lie in a plane, traverse the object to be imaged and impinge on M detectors on the far side of the object. The energy of the X-rays is attenuated by traveling through the object and the degree of attenuation depends upon the varying densities of the materials through which the X-rays are traveling. From the detector measurements and the initial intensities, the total attenuation along each ray is easily calculated. The sources and detectors are then rotated by an angle π/N and the process is iterated. (Here, the value of N determines at how many points around the half-circle the X-ray beams are triggered.) Possible realizations of the data thus consist of an $M \times N$ matrix of rational numbers.

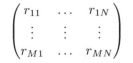

$$\begin{pmatrix} r_{11} & \cdots & r_{1N} \\ \vdots & \vdots & \vdots \\ r_{M1} & \cdots & r_{MN} \end{pmatrix}$$

where $r_{ij} \in [0, I_{max}]$. I note that from the perspective of the algorithms involved, it is of no importance whether such possible realizations come from actual measurements, from simulations, or are simply numerical arrays.

The inverse problem is then to construct a two dimensional image of the target in the plane of the beams from that data about total attenuation. The image can be represented by a function μ on the Cartesian plane, where $\mu(x, y)$ is the value of the X-ray attenuation coefficient at the point (x, y). The attenuation coefficient values in a given region are strongly associated with the density of the material in that region and the former can therefore be taken to represent the latter. The most common construction method is filtered backprojection.

In broad outline, filtered backprojection contains these steps:

Step 1 Calculate the total attenuation along a given ray between a source and its detector by integrating the values of μ along that ray. This gives a representation of the data values in terms of μ.

Step 2 Convolve these spatial projections with a filter. The filter compensates for distortions introduced into the representations by coordinate transformations in discrete models.

Step 3 Fourier transform these convolutions into the frequency domain.

Step 4 Compute the convolutions as products in the frequency domain.

Step 5 Inverse Fourier transform the results back to the spatial domain. Steps 3 through 5 are primarily to accommodate computational load constraints, something that does not appear in traditional analyses of models but that is of central concern in the real-life application of this methods.

Step 6 Compute the inverse Radon transforms in the spatial domain to arrive at values of $\mu(x, y)$ for the desired points (x, y) within the target frame.[1] I note here that these inverse transforms introduce a severely non-local aspect to the relation between data and image, in that a given image pixel is reconstructed from multiple backprojections taken at different values of θ_i and a given datum contributes to the reconstruction along the whole ray associated with that datum. This requires a very different attitude towards data correction than does the usual compositional approach to which philosophical discussions are largely directed.

Now, instead of the target frame, consider a detector frame which is oriented at an angle θ to the target frame. Each ray can be represented mathematically by the line parameterized by r and θ: $L_\theta(r) = \{(x, y) : r = x \cos(\theta) + y \sin(\theta)\}$ where r is the radial coordinate. The total attenuation along the line L is given by $\int_L \mu(x, y) dL$. This represents projected values of $\mu(x, y)$ along the ray orthogonal to the r axis of the detector frame when it is oriented at angle θ to the target frame.

---

[1]The inverse Radon transform $f(x, y)$ can be represented by $\frac{1}{2\pi} \int_{-\infty}^{\infty} \frac{d}{dy} H[R(r, y - rx)] dr$, where H and R are Hilbert and Radon transforms, respectively. Thanks to Tom Ryckman for pointing this out.

To represent this value using the target frame coordinates we have:

$$\L_\theta(r) = \int_{y=-\infty}^{\infty} \int_{x=-\infty}^{\infty} \mu(x,y)\delta(xcos(\theta) + ysin(\theta) - r)dxdy \qquad (1)$$

which is the Radon transform of μ over L. It is here that we have the first connection between the data values and a formal representation of them. The Radon transform is continuous, as is the theoretical Fourier transform of steps 3 and 5, but the computational implementations of these are inevitably discrete and this move shows that we must step down to logically different type of models even in the absence of an explicit theory of the phenomena.

Perhaps the most important difference between the application discussed here and Suppes' original account is that the values of μ occur as stand alone values rather than as parameters of a broader theory. Data analysis in the absence of theory or hypothesis testing has become increasingly important in recent years because of the enormous increase in data that is available in high energy physics, astrophysics, climate modeling, financial markets, and other areas and there have been interesting suggestions that non-theoretical approaches to data may be the most appropriate methods in certain areas. (See e.g. Napoletani et al. (2011); Humphreys (2013a)) Rather than theories and their associated models, our focus is thus methods that operate on data. Despite the absence of a parametric top level theory, the goal of accurately estimating values of $\mu(x,y)$ from the data fits the general motivation behind Suppes' restriction that 'The central idea... is to restrict models of the data to those aspects of the experiment which have a parametric analogue in the theory.' (258). In the present case, models act not as possible realizations of a theory but provide constraint conditions on the methods used to transform data.

2   Models of the Instrument

Next are models of the instrument. Suppes' focus was data from laboratory experiments that are used to estimate a parameter included in a linear learning theory. In our case we must have a theory of the instrument rather than a theory of the experiment and models of the data will then be possible realizations of the data that satisfy the theory of the instrument. Suppes is right to emphasize that in his example the learning theory itself is assumed to be correct and is not being tested. With imaging devices of the kind considered here, we are also not testing the theories that lie behind the design of the instrument and the interpretation of the inverse inferences. They are sufficiently well established that they are taken to be true; nobody seriously questions the existence of X-rays and their experimentally established properties. We do have a theory of the instrument that determines in generic terms what kind of data should be produced. Here the physical theory behind the instrument design is, in its basic form, extremely simple: it simply asserts that X-rays are transmitted from a discrete array of sources, are transmitted through the target object with varying degrees of absorption depending upon the material composition of the target, and the intensity of the arriving X-rays is recorded by a discrete array of detectors along a line parallel to the set of sources. This apparent simplicity is misleading, for complications such as Compton scattering, photoelectric absorption, detector noise, partial volume artifacts, and many others have to be taken into consideration. In addition, the beams must be properly collimated, the intensity of a given source must be constant across all indices i and j, corrections must be made for detector inefficiencies, and so on. I list these not simply to note that the path from data to image is thoroughly infused with models but that many, if not most, of these factors not only can be formally represented but must be in order for the image construction algorithms to make appropriate corrections. The theory of the instrument is then captured by a complex and interlocking set of models that represent both physical and computational processes. Just as in the case of experiments, we have to provide these models of the instrument in order to ensure that a possible realization of the data counts as a model of the data but there is one aspect that requires specific discussion and which raises an important issue connected with Suppes' point C above.

Whether it is desirable or even acceptable to alter and adjust data from instruments and experiments is a delicate methodological issue. Although adjusting a data set so that it is a model of the data will usually leave the data matrix within the class of possible realizations of the data, both the process and the motivation in the case of data correction differ from situations in which the model of the data was arrived at as raw data from an experiment. In the case of an experiment, because a model of the data is a sequence of possible data points from a well conducted experiment, we are entitled to exclude as a model of the data a sequence of data points that violates what we know about the generating conditions that constitute the experiment. Analogously, in the case of instruments we are entitled to adjust data that are generated by the instruments in the light of theories and models that apply to those generating conditions. One of the things that is of particular interest in the CT instruments is to avoid artifacts of the instrument and this is only possible by adjusting data sets using the networks of models of the instrument mentioned above. These artifacts can arise either from physical processes in the instrument or from computational processes in the image production. A simple example of this involves beam hardening which results from the complete absorption by the target of all X-rays that fall below a threshold energy level, resulting in errors in the image construction. Corrections for beam hardening can be made by either using physical filters or by software correction algorithms, both of which can be formally modeled.

Such models do not fit into a neat hierarchy. The order in which they are deployed can vary and their use may require iterated cycles of application. That said, the use of models of the instrument stands in sharp contrast to an empiricist tradition that views theoretical transformations of, and changes to, so-called 'raw data' as epistemologically counterproductive. Yet for CT instruments and elsewhere, such data manipulation is not merely desirable but necessary in many cases to arrive at accurate outputs. This of course has its dangers and an explicit acknowledgement of such manipulations is required. This point must be distinguished from a better known issue that I shall now discuss.

One of the challenges of dealing with the kind of data that comes from instruments is to relate the languages of all the different theories that come into play in the operation of the instrument and the generation of the data. Because the entire inference and reconstruction process from data to image is automated, Suppes' goal, stated in point D, of restricting the analysis of models to formal methods can be satisfied in a completely general way. Although it is undeniable that some aspects of experimental procedure and of instrument use require heuristic tricks of the trade to successfully produce reliable data, one should not underestimate the extent to which many such adjustments can be given formal representations.

Let me relate this to point A above. The view that empirical content diffuses through the entire theoretical apparatus of science, affecting even the most mathematical parts, has long dominated the philosophy of science and this broadly Quinean view has itself become something of a dogma. In a previous publication (Humphreys (2008, Section 4)) I argued that in the case of probability theory one can preserve the purely formal character of the measure-theoretic formulation of probability theory and the statistical models that are used to apply it, by restricting the empirical input to a mapping between the last statistical model and the data. Suppes' hierarchy of formal models gives support to this position in the case where a general background theory exists, especially since he notes (p. 252) that for present purposes, it is unimportant whether the formalization takes place within the semantic or the syntactic account.

3 Inverse Problems

In a more general setting, moves from observations to unobservable entities can be considered as inverse inferences from the data to their source. For philosophers, the problem of scientific realism can then be seen to consist in providing a solution to a particular type of inverse problem. Once we view matters in this way many familiar philosophical issues take on a different cast. An inverse inference problem is said to be ill-posed just in case there is either not a unique solution to the problem or the solution does not depend continuously on the data. Thus, if we can show that the representation of an inverse problem is well-posed, this gives us a response to certain types of anti-realism, albeit at the price of shifting the inductive problem to justifying the use of that representation for the case at hand. It is regularly said, and correctly, that such inferences from the observable to the unobservable are always under-determined. Yet there are well developed techniques within the area of inverse methods to deal with such under-determination. For example, given a finite data set, the inverse Radon transform based on those data is not unique. Yet by using results such as the Nyquist sampling theorem, which determines how often a given continuous function must be sampled for the sampled signal to contain the same amount of information as the continuous signal, errors in the reconstructed image can be drastically reduced. (See Buzug 2008, pp. 135 ff.)

Finally, there is one point made by Suppes that I believe needs interpretation. He says that 'From a conceptual standpoint the distinction between pure and applied mathematics is spurious—both deal with set-theoretic entities and the same is true of theory and experiment.' (p. 260) Interestingly, this view is maintained in his comprehensive treatise (Suppes (2002, p. 33)) while explicitly addressing the finitistic character of much applied mathematics (ibid. pp. 303–311). Viewed from an abstract perspective, this is correct but that abstraction disguises some philosophically important differences. Put in the form of an aphorism, we can say that applied mathematics is not always the application of pure mathematics. The reason is this: in some areas of applied mathematics, and optimization methods are a good example, theorems do not exist that guarantee the success of a method when applied to particular situations. Rather than the deductive procedures that are usually discussed in the philosophy of mathematics literature, heuristically justified trial and error procedures are often used that lend an inductive aspect to applied mathematics. Nonconstructive proofs have been much discussed in the philosophy of mathematics, but pure existence proofs in applied mathematics raise issues that are important to the philosophy of science, not the least because existence results for optimization procedures do not always provide an effective method for finding the optimum. Take one example of a standard numerical methods procedure, attempting to find the global minimum of a function. Given an objective function f, an optimization method will arrive at either a local or a global minimum but with many complicated functions, there will be no proof that the minimum reached is global rather than local. Consider this example:

A set $S \subseteq R^n$ is convex if it contains the line segment between any two of its points, i.e., $\{\alpha x + (1 - \alpha)y : 0 \leq \alpha \leq 1\} \subseteq S$ for all $x, y \in S$. A function $f : S \subseteq R^n \to R$ is strictly convex on a convex set S if its graph along any line segment in S lies on or below the chord connecting the function values at the endpoints of the segment, i.e., if $f(\alpha x + (1 - \alpha)y) < \alpha f(x) + (1 - \alpha)f(y)$ for all $\alpha \in (0, 1)$ and all $x \neq y \in S$. Then any local minimum of a strictly convex function f on a convex set $S \subseteq R^n$ is the unique global minimum of f on S. But it is often impossible to determine for a given objective function f whether it satisfies the convexity conditions needed for the theorem. Methods such as steepest descent and conjugate gradient can be used to converge on a solution if it exists but cannot be guaranteed to succeed.

The general point is straightforward. The distinction between pure and applied mathematics is not always sharp, but assuming some such distinction can be made, results in pure mathematics that are used in science come with a set of conditions that must be satisfied in order for those results to correctly be applied. This is as true for arithmetical results as it is for martingales used on time series data from financial markets. While the mathematical results themselves can be assessed a priori, the truth of the application conditions for a given situation usually cannot and it is in the absence of such a guarantee that heuristic methods must often be applied. Many of those methods, which are a legitimate part of applied mathematics, can be formally represented but the standards of rigor expected of pure mathematics must be relaxed. So I do not disagree with Suppes' claim concerning the conceptual equivalence of pure and applied mathematics when applied mathematics is taken as a self-contained subject, but more broadly construed it can support a different epistemological attitude while retaining its formal qualities.

4   Conclusion

'Models of Data' is a canonical element in Suppes' set-theoretical approach to theories. As data become increasingly available, often in vast quantities, their relation to theory is evolving, especially in situations in which the role of explicit theory is small. I have tried to show, albeit in a specialized domain, that explicit models of the data, of the instrument, and of the transformations made on the data, are consistent with Suppes' overall message that formal methods can carry us a long way towards a philosophical appreciation of techniques that are rapidly advancing in importance yet do not fit the standard picture of a connection between theory and data.

References

Bogen, J. and J. Woodward. 1988. Saving the phenomena. *The Philosophical Review* 97(3):303–352. doi:10.2307/2185445.

Buzug, T. M. 2008. *Computed Tomography: From Photon Statistics to Modern Cone-Beam CT*. Berlin: Springer.

Humphreys, P. 2008. Probability theory and its models. In D. Nolan and T. P. Speed, eds., *Probability and Statistics: Essays in Honor of David A. Freedman*, pages 1–11. Beachwood, Ohio: Institute of Mathematical Statistics.

—. 2013a. Data analysis: Models or techniques? *Foundations of Science* 18(3):579–581. doi: 10.1007/s10699-012-9317-4.

—. 2013b. What are data about? In E. Arnold and J. Duran, eds., *Computer Simulations and the Changing Face of Experimentation*. Cambridge: Cambridge Scholars Publishing.

—. Forthcoming. X-ray data and empirical content. LMPS XIV: Proceedings of the 14th Logic, Philosophy, and Methodology of Science Congress, P. Bour et al. (eds). London: College Publications.

Napoletani, D., M. Panza, and D. C. Struppa. 2011. Agnostic science: Towards a philosophy of data analysis. *Foundations of Science* 16(1):1–20.

Suppes, P. 1962. Models of data. In E. Nagel, P. Suppes, and A. Tarski, eds., *Logic, Methodology, and Philosophy of Science: Proceedings of the 1960 International Congress*, pages 252–261. Stanford: Stanford University Press.

—. 2002. *Representation and Invariance of Scientific Structures*. Stanford, CA: CSLI Publications.

6

Representational Measurement in Economics

Adolfo García de la Sienra

To Patrick Suppes in his ninetieth anniversary, with affection, admiration and gratitude for his teachings.

My aim in the present paper is to discuss the role of the representational theory of measurement (RTM from now on) in economics. According to Luce and Suppes (2002),

> Representational measurement is, on the one hand, an attempt to understand the nature of empirical observations that can be usefully recoded, in some reasonably unique fashion, in terms of familiar mathematical structures. [...] representational measurement goes well beyond the mere construction of numerical representations to a careful examination of how such representations relate to one another in substantive scientific theories, such as in physics, psychophysics, and utility theory. These may be thought of as applications of measurement concepts for representing various kinds of empirical relations among variables.[1]

Suppes' beautiful example of measuring weights by means of a balance, quoted by Morgan (2007, p. 106), as represented by a suitable set-theoretical structure, suggests that RTM involves a claim about measuring procedures (perhaps a form of instrumentalism), and that every measuring procedure consists of constructing some homomorphism. Curiously enough, other authors have got the impression that the set-theoretical structures RTM makes use of, far from representing empirical instrumentalist measurement procedures, are not even about the "natural world". Michell (2007, p. 36), for one, writes that RTM

> is based upon an inconsistent triad: first, there is the idea that mathematical structures, including numerical ones, are about abstract entities and not about the natural world; second, there is the idea that representation requires at least a partial identity of structure between the system represented and the system representing it; and third, there is the idea that measurement is the numerical representation of natural systems. The second and third ideas imply that natural systems instantiate mathematical structures and when the natural system involves an unbounded, continuous quantity, it provides an instance of the system of positive real numbers. Thus the second two refute the first idea, the principal *raison d'être* for the representational theory.

[1]By far, the most important presentation of RTM is found in Krantz et al. (1971); Suppes et al. (1989); Luce et al. (1990)

Foundations and Methods from Mathematics to Neuroscience.
Colleen E. Crangle, Adolfo García de la Sienra and Helen Longino.
Copyright © 2014, CSLI Publications.

Let me start by distinguishing between metrization and measuring. By 'metrization of magnitude M' I understand a demonstration, out of empirically meaningful conditions, of the fact that M is measurable. RTM accomplishes the metrization of M by building suitable set-theoretical structures and proving the existence of a certain homomorphism, which is a mathematical representation of M. By 'measuring procedure' I understand a procedure for actually finding the value of M for given objects, with respect to a unit of reference. It seems to me that RTM *does not* claim that every measuring procedure consists of explicitly constructing some homomorphism. What it *does* claim is that any (putative) magnitude M must satisfy certain conditions in order for it to be metrizable (Díez (2000, p. 20, n. 5)). The task of the theory of fundamental metrization is to probe into the conditions that M must satisfy in order to guarantee the existence of such representation when the same does not presuppose any other previous metrizations. The fundamental measuring procedures determine specific empirical procedures for qualitative comparison of the specific property involved, and chooses a standard with which the assignment begins (ibid).

It is certain that RTM has never maintained the claim that "mathematical structures, including numerical ones, are about abstract entities and not about the natural world". Clearly, mathematical structures *are* abstract entities, insofar as they are set-theoretical structures. And they *are* "about" the natural (and the "social") world, but the meaning of 'about' in this context needs philosophical clarification. I will try to clarify the idea that natural systems "instantiate" mathematical structures. It will be seen that the proof of the existence of important metrizations in economic theory requires the adoption of rather severe restrictions.

1   Prolegomena

Let me start by making some distinctions. I distinguish target (concrete, real) systems, given in pre-theoretical experience, from model systems on one hand, and set-theoretical structures, on the other. I take for granted that scientific theories can be identified with certain ordered classes of set-theoretical structures (cf. Suppes (2002); Balzer et al. (1987)). Moreover, I assume that theories can be formulated by means of the definition of set-theoretical predicates (cf. Suppes (2002)).

The relationship between target systems, model systems and set-theoretical structures is roughly the following: a model system is an idealized replica of a certain class of target systems, a replica that can be described by means of axioms defining a set-theoretical predicate. Model systems are not set-theoretical structures but rather (imagined) physical, economic or geometric objects that represent their corresponding target systems. Thus models systems can be seen

> as imagined physical systems, i.e. as hypothetical entities that, as a matter of fact, do not exist spatio-temporally but are nevertheless not purely mathematical or structural in that they would be physical things if they were real. If the Newtonian model system of sun and earth were real, it would consist of two spherical bodies with mass and other concrete properties such as hardness and colour, properties that structures do not have; likewise, the populations in the Lotka-Volterra model would consist of flesh-and-blood animals if they were real, and the agents in Edgeworth's economic model would be rational human beings. (Frigg 2010, p. 253)

Hence, it is important to stress that the representation of a target system by means of a model system cannot be identified with any kind of homomorphism, since homomorphisms are mappings from one set-theoretical structure into another. My version and defense of RTM is grounded upon these distinctions.

2 Empirical Structures?

When the question arises whether the property of components of a given type of concrete system is measurable, RTM proceeds to build a typical model system to represent that class and, based upon the properties of such model, builds a class of set-theoretical structures by means of the definition of a set-theoretical predicate. It is in this sense that mathematical structures *are* about the natural world: they are true of (more or less) idealized model systems that purport to *represent* some type of concrete target system in the world. These structures, insofar as they represent target systems, may be called "empirical structures".

Suppes' famous example of measuring weights by means of a balance is a classical instance of an "empirical structure", but it can be misleading in two ways. It may suggest both that measurement structures are always empirical, in the sense that they are not describing idealized model systems, and that RTM is a general theory of concrete empirical measuring procedures. The main criticisms against RTM in the field of economic methodology[2] begin by misconstruing it as a theory of measuring procedures, which it is not except for some cases: It is a theory about the conditions that properties of certain classes of target systems must satisfy in order to be measurable. It functions, thus, as a sort of ideal control for empirical measuring procedures (that can also be suggestive of empirical measuring procedures).

A natural or social system (a target system) "instantiates" a mathematical structure if, and only if, it is exactly described by the axioms defining the structure (under a certain empirical interpretation) without the mediation of a model system. In such a case, the structure may be called an "empirical structure". For instance, a structure $\mathfrak{A} = \langle X, \succsim \rangle$ would be an empirical structure faithfully recording the preferences of a given consumer if X is interpreted as the collection of consumption menus over which he has to make a choice and his pairwise preferences are correctly recorded by $\succsim$. But then nothing guarantees *a priori* that $\mathfrak{A}$ will be a weak order (i.e. a connected and transitive relation).

In general, it is not target systems that are exactly described by mathematical structures, but rather model systems representing the former. Hence, many mathematical structures that we find in economics are not instantiated by target systems, but only by model systems obtained out of idealized conditions.

It is usual, in the process of elaboration of economic theories, to find the construction of ideal objects out of concepts that, even though they originate in experience and correspond to real properties of real beings, have suffered a sort of transformation. For example, there is no doubt that real economic agents are human beings who handle certain information, who can remember events, and who can also perform arithmetic operations. That is why the predicates

$$\text{`}x\text{ possesses information'} \tag{1}$$

$$\text{`}x\text{ has memory'} \tag{2}$$

---

[2] Cf. Boumans (2007, pp. 7–8). Boumans clearly thinks that RTM claims something like that it is impossible to measure some empirical magnitude if an axiomatic system of the proper sort is not provided, see also Boumans (2008).

and

$$\text{`}x \text{ is able to perform calculations'} \qquad (3)$$

are true of human beings who effect economic transactions anywhere. Nevertheless, even though economic theory has to refer somehow to the properties they denote, the use of the mathematical method requires the use of deformed versions of such predicates, or of the concepts they express thereby. Some specializations of economic theory require, for instance, the economic agents to possess perfect information, perfect memory and unlimited computational powers. Thus, for instance, predicates (1), (2) and (3) are deformed in order to obtain

$$\text{`}x \text{ possesses perfect information'} \qquad (4)$$
$$\text{`}x \text{ has perfect recall'} \qquad (5)$$

and

$$\text{`}x \text{ has unlimited computational powers'.} \qquad (6)$$

The conjunction of predicates

$$\text{`}x \text{ possesses perfect information} \wedge x \text{ has perfect recall} \wedge$$
$$\wedge \, x \text{ has unlimited computational powers'} \qquad (7)$$

defines a type or ideal object—what we have called a model system—nonexistent in reality but required to channel mathematical reasoning. As Walras (1954, p.71) put it:

> the mathematical method is not an *experimental* method; it is a *rational* method, . . . the pure science of economics should then abstract and define ideal-type concepts in terms of which it carries its reasoning. The return to reality should not take place until the science is completed and then only with a view to practical applications.

Let us call 'idealized concepts' (or, briefly, idealizations) the concepts obtained by deformation out of concepts which are truly predicable of real beings if, in spite of this deformation, the same are meaningful—even if they are false—of these real objects, and keep an intension akin to those. We have called 'ideal objects' those model systems built by means of conjunctions of predicates that express idealizations.

The axioms of a consistent economic theory define directly a set-theoretical predicate (as '$\mathfrak{A}$ is a regular preference structure') but they are also obliquely true, at least under a typical interpretation of the primitive terms, of ideal objects. For instance, the just mentioned predicate can be defined canonically as follows: $\mathfrak{A}$ is a *regular preference structure* iff $\mathfrak{A}$ is a weak order. It is debatable whether the preference relation (partially) exhibited by a human agent at a given moment is connected and transitive. Nevertheless, the systematic development of the theory can go on as if it were so.

Utility theory has proceeded through the construction of ideal objects endowed with idealized properties. It has assumed, for instance, that the "consumer" (an ideal object) defines his preference relation over a space of consumption menus which is representable as the nonnegative orthant Ω of $\mathbb{R}^L$. Other idealizations introduced over and above these are those of a continuous or non-satiated preference relation. All these idealizations are needed in order to prove the existence of utility functions possessing certain characteristics required by economic theory (like continuity or monotonicity).

Hence, the use of RTM to obtain such functions cannot be seen as a case of empirical measurement. The application of RTM methods to obtain representations of idealized magnitudes is usual in economics. For instance, Beaver and Demski (1979) used such methods in trying to fundamentally measure income. They found that income can be fundamentally measured only in a world of complete and perfect markets, but not necessarily otherwise. They concluded that "at a fundamental level the central feature of financial reporting cannot be income measurement".

3 Models of Data in Measuring Procedures

The view of theories as systems of set-theoretical structures naturally leads to a particular view of theory-application and testing. Typically, to conceptualize a certain phenomenon means to "apply" the conceptual apparatus of the theory fleshing out the terms with a particular concrete meaning. For instance, in order to conceptualize a target system c, a real human consumer, demand theory (T) claims that c be represented by a certain model system m, which is a consumer with unlimited memory that knows beforehand what would be his choice in confronting any budget set $\{\mathbf{x} \in \mathbf{X} \mid \mathbf{px} \leq \mathbf{w}\}$, where $\mathbf{p} \in P$ is a system of prices and $w \in W$ is his income. The information of this ideal consumer is represented in a set-theoretical structure by means of a demand function $\eta \colon \mathscr{B} \to X$. If the Slutsky matrix corresponding to η is symmetric and negative-semidefinite, m knows that η coincides with a Walrasian demand function μ derived from a utility function that represents his preferences; moreover, since m has unlimited computational capabilities, he can determine instantaneously the preference relation from which μ (which turns out to be η) derives.

If the theory represents c with m, the scientist may try to apply T to c. Whereas whether T is applicable to m is a matter of the definition of m, whether T is applicable or not to c is an entirely empirical matter. To say that it is applicable is a *claim* that has to be held on the basis of empirical data. Varian (1983) has stressed that there are two approaches to this problem. One, the approach based upon calculus, originated in the work of Antonelli (1986) and Slutsky (1915), derives necessary and sufficient conditions on the derivatives of the demand function η. The second, originated in the work of Samuelson (1938, 1947, 1948) is called 'non-parametric' because it assumes no specific form whatsoever for η.

> The distinction between the two approaches is very important in empirical work. The calculus approach assumes the entire demand function [our η] is available for analysis, while the algebraic approach assumes only a finite number of observations on consumer behavior is available. Since all existing data on consumer behavior does consist of finite number of observations, the latter assumption is much more realistic.
>
> (Varian 1983, p. 99)

Certainly, virtually all empirical studies in all fields are based on a finite set of data. The difference between the "calculus" and the "algebraic" approach lies in the way the data are used. In the calculus approach they are used to build a continuous demand function, which is then used to integrate the utility function. In the algebraic approach they are used to build a utility function directly.

We can explain the non-parametric method by means of the concept of data structure (or, as Suppes prefers to call it, model of data). Observing the behavior of the agent we get the structure of data:

$$\hat{\mathfrak{D}} = \langle X, F, \hat{\eta} \rangle,$$

where $\hat{\eta}$ is a function defined over a finite subset F of $P \times W$, precisely that of pairs of systems of prices and income levels under which the behavior of the consumer has been observed. The empirical (actually observed) demand function $\hat{\eta}$ is obviously finite and discrete, and that is why structure $\mathfrak{D}$ can be depicted by means of Table 1.

TABLE 1

Argument	Value of $\hat{\eta}$
$(\mathbf{p}^1, w^1)$	$\hat{\mathbf{x}}^1 = \hat{\eta}(\mathbf{p}^1, w^1)$
$(\mathbf{p}^2, w^1)$	$\hat{\mathbf{x}}^2 = \hat{\eta}(\mathbf{p}^2, w^2)$
$\vdots$	$\vdots$
$(\mathbf{p}^k, w^k)$	$\hat{\mathbf{x}}^n = \hat{\eta}(\mathbf{p}^k, w^k)$
$\vdots$	$\vdots$
$(\mathbf{p}^N, w^N)$	$\hat{\mathbf{x}}^N = \hat{\eta}(\mathbf{p}^N, w^N)$

The non-parametric method does not try to build directly the complete demand function η, but rather uses an algorithm to build directly a utility function that "rationalizes" the data, provided that structure $\mathfrak{D}$ satisfies the conditions of the following theorem.

Theorem 1. (AFRIAT) *The following conditions are equivalent:*

(1) *There exists a non-satiated utility function that rationalizes the data.*

(2) *The data satisfy the general axiom of revealed preference.*

(3) *There are numbers U^i, $\lambda^i > 0$ $(i = 1, \ldots, N)$ that satisfy Afriat's inequalities:*

$$U^i \le U^j + \lambda^j \mathbf{p}^j (\mathbf{x}^i - \mathbf{x}^j)$$

for $i, j = 1, \ldots, N$.

(4) *There is a concave, monotonic, continuous and non-satiated utility function that rationalizes the data.*[3]

Afriat (1967) and Varian (1983) provided algorithms by means of which it is possible to build a utility function u that rationalizes the data of structure $\mathfrak{D}$. This is "jumping", so to say, from a model of data to the theoretic function u. Then it is possible to embed $\mathfrak{D}$ in the partial structure consisting of the Walrasian demand function μ induced by u. That is to say, μ restricted to F coincides with $\hat{\eta}$. The parametric method, on the other hand, tries first to determine the non-theoretic demand function η without involving u and afterward tries to recover u out of η.

The upshot of this example is that there are some applications of economic theories that require methods other than those of RTM in order to flesh out the terms of the theory, to actually measure the magnitudes referred to by the terms of the theory. It is frequently impracticable to measure theoretical magnitudes in a direct way, as these may be non-directly observable or require the use of theoretical systematizations in order to be calculated. For an example taken from the natural sciences, consider the measurement of the mass of the sun. Clearly, the sun cannot be placed in the pan of a balance in order to be compared, as to its weight, with other objects. Rather, an essential use of the laws of mechanics is required to do the calculation, assuming that the sun and the earth satisfy the law of universal gravitation.

Suppes himself used methods more similar to this (fleshing out the magnitudes out of models of data) in his classical experiments on learning theory.[4] At least for the case of utility measurement, Suppes sounds very skeptical about the possibility of measuring utility of given agents by showing that their preference structures satisfy certain axioms. He relies instead on experimental methods that might, applying response theory, go

---

[3]For a proof see Varian (1983).

[4]For a description of these see García de la Sienra (2011).

beyond "the individual preference orderings to the environmental and constitutional conditions that produced them" (Suppes and Atkinson (1960, p. 233)).

Hence, more than a method to flesh out the terms of a theory in empirical application or testing, in economics RTM seems to be rather a methodology to establish the existence of metrizations, sometimes even for idealized magnitudes, and especially for these.

4 Acknowledgements

The present paper was produced with the support of CONACYT Project 127380, Filosofía de la Economía. I appreciate the comments and suggestions by professors Russell Hardin, Kenneth J. Arrow and Helen Longino, upon a previous version of this paper.

References

Afriat, S. N. 1967. The construction of utility functions from expenditure data. *International Economic Review* 8(1):67–77. doi:10.2307/2525382.

Antonelli, G. 1986. Sulla teoria matematica della economia politica. In J. S. Chipman, L. Hurwicz, M. K. Richter, and H. F. Sonnenschein, eds., *Preferences, Utility, and Demand*, pages 333–364. Harcourt Brace & Jovanovich.

Balzer, W., C. M. Ulises, and J. D. Sneed. 1987. *An Architectonic for Science. The Structuralist Program*. Dordrecht: D. Reidel Publishing Company.

Beaver, W. H. and J. S. Demski. 1979. The nature of income measurement. *The Accounting Review* 54(1):38–46.

Boumans, M. 2007. Introduction. In *Measurement in Economics. A Handbook*, pages 3–18. Amsterdam: Elsevier.

—. 2008. Measurement. In D. S. N. and L. E. Blume, eds., *The New Palgrave Dictionary of Economics*. London: Palgrave Macmillan, 2nd edn.

Díez, J. A. 2000. Structuralist analysis of theories of fundamental measurement. In W. Balzer, C. U. Moulines, and J. Sneed, eds., *Structuralist Knowledge Representation. Paradigmatic Examples (Poznań Studies in the Methodology of the Sciences and the Humanities)*, vol. 75, pages 19–49. Amsterdam: Rodopi.

Frigg, R. 2010. Models and fiction. *Synthese* 172(2):251–268. doi:10.1007/s11229-009-9505-0.

García de la Sienra, A. 2011. Suppes' methodology of economics. *Theoria* 26(3):347–366. doi:10.1387/theoria.2559.

Krantz, D. H., R. D. Luce, P. Suppes, and A. Tversky. 1971. *Foundations of measurement, Volume I: Additive and polynomial representations*. New York: Academic Press.

Luce, R. D., D. Krantz, P. Suppes, and A. Tversky. 1990. *Foundations of Measurement, Vol. III: Representation, Axiomatization, and Invariance*. New York: Academic Press.

Luce, R. D. and P. Suppes. 2002. Representational measurement theory. In J. Wixted and H. Pashler, eds., *Stevens' Handbook of Experimental Psychology*, vol. 4, pages 1–41. New York: John Wiley and Sons, 3rd edn.

Michell, J. 2007. Representational theory of measurement. In M. Boumans, ed., *Measurement in Economics. A Handbook*, pages 19–39. Amsterdam: Elseiver.

Morgan, M. S. 2007. An analytical history of measuring practices: The case of velocities of money. In *Measurement in Economics. A Handbook*, pages 105–132. Amsterdam: Elsevier.

Slutsky, E. 1915. Sulla teoria del bilancio del consumatore. *Giornali degli Economisti e Rivista di statistica* 3(51):1–26.

Suppes, P. 2002. *Representation and Invariance of Scientific Structures*. Stanford, CA: CSLI Publications.

Suppes, P. and R. C. Atkinson. 1960. *Markov Learning Models for Multiperson Interactions*. Stanford, CA: Stanford University Press.

Suppes, P., D. Krantz, R. D. Luce, and A. Tversky. 1989. *Foundations of Measurement, Vol. II: Geometrical, Threshold, and Probabilistic Representations.* New York: Academic Press.

Varian, D. 1983. *The inventor and the pilot: Russell and Sigurd Varian.* Palo Alto, CA: Pacific Books.

Walras, L. 1954. *Elements of Pure Economics.* London: Allen and Unwin.

7

Imprecise Probabilities in Quantum Mechanics

STEPHAN HARTMANN

It is a pleasure to thank Patrick Suppes for his great support and for many years of stimulating discussions about a wide range of topics of mutual interest. The present project grew out of these discussions, and I look forward to work with him on it and other projects for many more years.

1 Introduction

In his entry on "Quantum Logic and Probability Theory" in the *Stanford Encyclopedia of Philosophy*, Alexander Wilce (2012) writes that "it is uncontroversial (though remarkable) that the formal apparatus of quantum mechanics reduces neatly to a generalization of *classical probability* in which the role played by a Boolean algebra of events in the latter is taken over by the *'quantum logic'* of projection operators on a Hilbert space." For a long time, Patrick Suppes has opposed this view (see, for example, the papers collected in Suppes and Zanotti (1996). Instead of changing the logic and moving from a Boolean algebra to a non-Boolean algebra, one can also 'save the phenomena' by weakening the axioms of probability theory and work instead with *upper and lower probabilities*. However, it is fair to say that despite Suppes' efforts upper and lower probabilities are not particularly popular in physics as well as in the foundations of physics, at least so far. Instead, quantum logics is booming again, especially since quantum information and computation became hot topics. Interestingly, however, imprecise probabilities are becoming more and more popular in formal epistemology as recent work by authors such as James Joyce (2010) and Roger White (2010) demonstrates.

In this essay I would like to give *one more reason* for the use of upper and lower probabilities in quantum mechanics and outline the research program that they inspire. The remainder of this essay is organized as follows. Sec. 2 introduces upper and lower probabilities. Sec. 3 turns to quantum mechanics and presents the CHSH inequality. We show that there is not always a joint probability distribution that reproduces observed quantum correlations. Sec. 4 argues that imprecise probabilities can be defined in these cases, and Sec. 5 concludes with a number of open questions.

Foundations and Methods from Mathematics to Neuroscience.
Colleen E. Crangle, Adolfo García de la Sierra and Helen Longino.

2 Imprecise Probabilities

Imprecise probabilities are well known from the theory of uncertain reasoning, Halpern (2005); Walley (1991). The starting point of the formal developments is the question of how to represent one's ignorance about a probability value. One way to do this is to introduce a lower probability measure P_* and an upper probability measure P^*, where the difference between the two is an agent's measure of her uncertainty about a probability assignment. To illustrate this, consider a coin tossing experiment and start with $P_*(\text{Heads}) = 0$ and $P^*(\text{Heads}) = 1$, which means that the agent is in a state of full uncertainty about the outcomes of the coin tossings. Then collect evidence and update $P_*(\text{Heads})$ and $P^*(\text{Heads})$ accordingly. If the coin is fair, then both measures will eventually converge to $1/2$, i.e. the probability of a fair coin to land heads. Note that the use of uppers and lowers is compatible with the existence of a probability value. The uppers and lowers only express our uncertainty about the probability value.

Upper and lower probability measures are defined as follows Suppes and Zanotti (1996).

Definition 1 (Upper Probability). *Let Ω be a nonempty set, $\mathcal{B}$ a Boolean algebra on Ω, and P^* a real-valued function on $\mathcal{B}$. Then $\Omega = (\Omega, F, P^*)$ is an upper probability space if and only if for every A and B in $\mathcal{B}$, (i) $0 \leq P^*(A) \leq 1$, (ii) $P^*(\emptyset) = 0$ and $P^*(\Omega) = 1$, (iii) if $A \cap B = \emptyset$, then $P^*(A \cup B) \leq P^*(A) + P^*(B)$.*

Definition 2 (Lower Probability). *Let Ω be a nonempty set, $\mathcal{B}$ a Boolean algebra on Ω, and P_* a real-valued function on $\mathcal{B}$. Then $\Omega = (\Omega, F, P_*)$ is a lower probability space if and only if for every A and B in $\mathcal{B}$, (i) $0 \leq P^*(A) \leq 1$, (ii) $P_*(\emptyset) = 0$ and $P_*(\Omega) = 1$, (iii) if $A \cap B = \emptyset$, then $P_*(A \cup B) \geq P_*(A) + P_*(B)$.*

We also note the following definition:

Definition 3 (Upper-Lower Pair). *We call a pair (P_*, P^*) an upper-lower probability pair Ω, if for every A in $\mathcal{B}$ we have $P_*(A) \leq P^*(A)$.*

Note that lower probabilities are super-additive and upper probabilities are sub-additive, which has several consequences: First, the sum over all atoms of the algebra may lead to a value greater than 1 for uppers and smaller than 1 for lowers. Second, while for a probability measure $P(A) = \sum_{A',B,B'} P(A, A', B, B')$ holds, the following inequalities hold for uppers and lowers:

$$P^*(A) \rightarrow \sum_{A',B,B'} P^*(A, A', B, B')$$

$$P_*(A) \rightarrow \sum_{A',B,B'} P_*(A, A', B, B').$$

Interestingly, if *monotonicity* holds, then uppers and lowers are related in the following way: $P_*(A) = 1 - P^*(\overline{A})$, where $\overline{A}$ is the complement of A in $\mathcal{B}$. (We will see later that this relation does not hold in quantum mechanics.) For an interpretation of upper and lower probabilities in terms of betting odds, see Walley (1991).

3 Quantum Mechanics and the CHSH Inequality

Let us consider four binary random variables A, A', B and B' that can take the values $a_i, a'_i, b_i, b'_i = \pm 1$ for $i = 1, 2$. We assume *symmetry*, i.e. we only consider situations where $E(A) = E(A') = E(B) = E(B') = 0$ with the expectation value E defined in the usual way, i.e. $E(A) := \sum_{i=1}^{2} a_i\, p(a_i)$. Next, we define the quantity

$$\mathcal{F} := |E(AB) + E(AB') + E(A'B) - E(A'B')|, \qquad (1)$$

where the expectation value

$$E(AB) := \sum_{i,k=1}^{2} a_i\, b_k\, P(a_i, b_k) = \sum_{i,j,k,l=1}^{2} a_i\, b_k\, P(a_i, a_j', b_k, b_l') \qquad (2)$$

measures the *correlation* between the random variables A and B. P is a probability measure. Note that $E(AB)$ takes values in the interval $[-1, 1]$ and that these correlations can be measured. Generalizing Bell's theorem, Clauser et al. (1969) effectively showed the following.

Theorem 1. *If there is a joint probability distribution* $P(A, A', B, B')$*, then* $\mathcal{F} \leq 2$ ("CHSH inequality").

The proof is in the appendix.

As is generally known, the CHSH inequality does not always hold. There are experimental setups that exhibit (quantum) correlations which violate the CHSH inequality. In experiments with correlated photons, for example, one can measure values of $\mathcal{F}$ up to $2\sqrt{2}$. These experiments starts with an EPR state of correlated photons, i.e. with the state $|\text{EPR} >= 1/\sqrt{2} \cdot (|10> - |01>)$ where $|0>$ and $|1>$ represent the photon polarizations of the two subsystems **A** and **B**. One can then find measurement angles α and α' (at **A**) and β and β' (at **B**) such that the CHSH inequality is violated. Hence, there is not always a joint probability distribution over A, A', B and B' that reproduces the expectation values $E(AB)$ etc. Note that these expectation values can be calculated from quantum mechanics and that the experiments confirm the theory.

Let us now study the CHSH inequality for atoms. Experiments similar to the just-mentioned photon experiments can be performed with an EPR state of two 2-level atoms that are trapped in a cavity. Here $|0>$ and $|1>$ represent the states of a single 2-level atom being in the ground state or the excited state, respectively. Let $A := X_1$, $A' := Z_1$, $B := X_2 + Z_2$ and $B' := X_2 - Z_2$, where X_1 denotes the Pauli matrix σ_x applied to the state of subsystem 1. Z_1, X_2 etc. are defined accordingly. Note that *symmetry* holds i.e. $E(A) = E(A') = E(B) = E(B') = 0$. Next, we calculate $E(AB) = E(AB') = E(A'B) = -1/2\sqrt{2}$ and $E(A'B') = 1/2\sqrt{2}$. Hence $\mathcal{F} = 2\sqrt{2}$, i.e. the CHSH inequality is maximally violated.

Next, we examine what happens if the quantum state under consideration decays under the influence of decoherence Schlosshauer (2007). Clearly, how fast the state decays will depend on the experimental context. It is known, for example, that the decay is slower in a cavity than in free space. What is important to us is that if the EPR state decoheres, then the correlations in the system also decay and the CHSH inequality will eventually be satisfied after some time τ_0. Once the CHSH inequality is satisfied, the correlations can be explained classically, i.e. by a non-contextual local hidden variables model. Moreover, these correlations can then be accounted for by a joint probability distribution.

Let us now calculate the time τ_0 when this is the case. One way of modeling decoherence is by coupling the quantum system to a reservoir. One can then write down the Schrödinger equation for the system plus the reservoir (environment), make the Born-Markov approximation, trace out the environment and obtain a quantum master equation for the reduced state ρ of the system. ρ then satisfies the following quantum master equation, which is of the Lindblad form Breuer and Petruccione (2002):

$$\frac{d\rho}{dt} = -\frac{B}{2} \sum_{i=1}^{2} [\sigma_+^{(i)} \sigma_-^{(i)} \rho + \rho \sigma_+^{(i)} \sigma_-^{(i)} - 2\sigma_-^{(i)} \rho \sigma_+^{(i)}], \quad (3)$$

with the decay constant B. Using the theory of Generalized Dicke States, Hartmann (Forthcoming), this equation can be solved analytically. We then obtain for the time evolution of the initial state $\rho(0) = |\text{EPR} >< \text{EPR}|$:

$$\rho(\tau) = e^{-\tau} \rho(0) + (1 - e^{-\tau}) |00 >< 00|, \quad (4)$$

with $\tau := B\,t$.

Next, we calculate the expectation values of A, A', B and B' as defined above for a system in the state $\rho(\tau)$ and obtain:

$$< A >= 0, \quad < A' >=< B >= - < B' >= e^{-\tau} - 1 \quad (5)$$

To make sure that *symmetry* holds for all times τ, we replace $A \to \tilde{A} := A- < A >$ etc. Clearly, we then have $E(\tilde{A}) = E(\tilde{A}') = E(\tilde{B}) = E(\tilde{B}') = 0$. For the correlations, we obtain:

$$< \tilde{A}\tilde{B} >=< AB > \quad , \quad < \tilde{A}'\tilde{B} >=< A'B > -(e^{-\tau} - 1)^2$$
$$< \tilde{A}\tilde{B}' >=< AB' > \quad , \quad < \tilde{A}'\tilde{B}' >=< A'B' > +(e^{-\tau} - 1)^2$$

Next, we calculate $\tilde{\mathcal{F}}$ as a function of τ (see Eq. (1)). It is easy to see that a joint probability distribution over $\tilde{A}, \tilde{A}', \tilde{B}$ and $\tilde{B}'$ exists if $\tau > \tau_0 := 245$, i.e. after a relatively short period of time after the quantum state starts to decay (in units of the inverse decay constant B). Figure 1 shows $\tilde{\mathcal{F}}$ and, for comparison, also $\mathcal{F}$ as a function of τ, where $\mathcal{F}$ is calculated using the original operators A, A', B and B'.

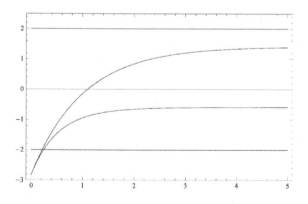

FIGURE 1 $\mathcal{F}$ (upper) and $\tilde{\mathcal{F}}$ (lower) as a function of τ

4 Imprecise Probabilities in Quantum Mechanics

We have seen that there is a joint probability distribution P for $\tau \geq \tau_0$ that reproduces the experimentally measurable correlations in the decaying EPR state. But how can we account for the correlations before that time? Hartmann and Suppes (2010) have explicitly constructed an upper probability distribution P^* that accounts for the correlations of a decaying EPR state at all times, i.e. before, at, and after τ_0. We therefore have unified account, which allows us to stick to a Boolean algebra of events. It is not necessary to work with a non-Boolean algebra in the quantum domain and a Boolean algebra in the classical domain, as quantum logicians do. All correlations can be accounted for by an upper probability distribution. This measure is explicitly sub-additive for times $\tau < \tau_0$ and turns into an additive probability measure for $\tau \geq \tau_0$. I take this to be a main advantage of the proposed approach to work with imprecise probabilities in quantum mechanics compared to the alternative quantum logical account, which do not allow for such a unified treatment.

It is interesting to note that the situation discussed here is similar to the learning situation discussed in Sec. 2. In the learning case, the upper probability distribution approximates the proper joint probability distribution more and more as the number of coin tosses increases. They coincide in the limit of an infinite number of coin tosses. In the quantum mechanical case, the upper probability distribution approximates the proper joint probability distribution more and more as the state decays. It coincides with the joint probability distribution once the CHSH inequality is satisfied (after a finite decay time). The joint probability distribution emerges from the interaction of the quantum state with its environment.

For the decaying EPR state, there is also a lower probability measure. This measure also converges into a probability measure which is defined for times $\tau \geq \tau_0$. However, the lower and the upper probability distributions are not related via $P_*(A) - 1 \quad P^*(\overline{A})$, i.e. they do not form an upper-lower pair. This is in line with the fact that there is no joint distribution for times $\tau < \tau_0$. Consequently, the *monotonicity* condition is violated in quantum mechanics, and upper and lower probability distributions have to be calculated independently by fitting them to the quantum mechanical expectation values. It is interesting to further explore the implications of the failure of monotonicity in quantum mechanics.

5 Open Questions

In future work, we plan to address the following four questions. *First*, how do our results generalize? Is it always possible, i.e. for all quantum states and corresponding sets of measurement operators, to fit an upper and a lower probability distribution? It would be nice to have a general proof that this is always possible, or a counter example showing that it is not. Our evidence so far is only episodic as we focused on the EPR state. *Second*, what is the proper interpretation of upper and lower probabilities in quantum mechanics? To address this question, the failure of *monotonicity* in quantum mechanics has to be understood. It will also be interesting to relate the discussion of upper and lower probabilities in quantum mechanics to the recent work on Quantum Bayesianism, Caves et al. (2007), which may shed some light on interpretational questions regarding upper and lower probabilities in quantum mechanics. *Third*, to further explore the relation between logic and probability in quantum mechanics, Gleason's Theorem has to be analyzed Hughes (1989). Here special attention has to be paid to the additivity assumption, which shows up in the proof of the theorem. We ask: What follows if one allows for sub- and super additive measures? *Fourth and finally*, what is the advantage of upper and lower probabilities compared to negative probabilities for which our decoherence story can be told as well? Negative probabilities were famously discussed by Feynman (1987) and have recently attracted the interest of Patrick Suppes. It will be worth to compare negative probabilities with imprecise probabilities.

Appendix: Proof of Theorem 1

To prove Theorem 1, we first simplify the notation and denote the value -1 by 0. Next, we introduce the following abbreviations:

$$P(1111) = P(0000) := x_1 \quad , \quad P(1110) = P(0001) = x_2$$
$$P(1101) = P(0010) := x_3 \quad , \quad P(1100) = P(0011) = x_4$$
$$P(1011) = P(0100) := x_5 \quad , \quad P(1010) = P(0101) = x_6$$
$$P(1001) = P(0110) := x_7 \quad , \quad P(1000) = P(0111) = x_8,$$

where we have made use of the *symmetry* requirement. Note that $0 \leq x_i \leq 1$ for $i = 1, \ldots, 8$ and that $\sum_{i=1}^{8} x_i = 1/2$. We then obtain by using Eq. (2) and similar equations for the other expectation values:

$$
\begin{aligned}
\mathcal{F} &= 4|x_1 + x_2 - x_3 - x_4 + x_5 - x_6 + x_7 - x_8| \\
&\leq 4|x_1 + x_2 + x_3 + x_4 + x_5 + x_6 + x_7 + x_8| \\
&\leq 4 \times 1/2 = 2,
\end{aligned}
$$

which completes the proof.

References

Breuer, H. P. and F. Petruccione. 2002. *The Theory of Open Quantum Systems*. Oxford: Oxford University Press.

Caves, C. M., C. A. Fuchs, and R. Schack. 2007. Subjective probability and quantum certainty. *Studies in History and Philosophy of Modern Physics* 38(2):255–274. doi:10.1016/j.shpsb.2006.10.007.

Clauser, J. F., M. A. Horne, A. Shimony, and R. A. Holt. 1969. Proposed experiment to test local hidden-variable theories. *Physics Review Letters* 23(15):880–884.

Feynman, R. P. 1987. Negative probability. In B. J. Hiley and F. D. Peat, eds., *Quantum Implications : Essays in Honour of David Bohm*, pages 235–248. London: Routledge and Kegan Paul.

Halpern, J. 2005. *Reasoning about Uncertainty*. Cambridge, Mass.: MIT Press.

Hartmann, S. Forthcoming. Generalized Dicke states. ArXiv: quant-ph 1201.1732.

Hartmann, S. and P. Suppes. 2010. Entanglement, upper probabilities and decoherence in quantum mechanics. In M. Suarez, M. Dorato, and M. Redei, eds., *EPSA Philosophical Issues in the Sciences: Launch of the European Philosophy of Science Association*, vol. 2, pages 93–103. Berlin: Springer.

Hughes, R. I. G. 1989. *The Structure and Interpretation of Quantum Mechanics*. Cambridge, Mass.: Harvard University Press.

Joyce, J. M. 2010. A defense of imprecise credences in inference and decision making. *Philosophical Perspectives* 24(1):281–323. doi:10.1111/j.1520-8583.2010.00194.x.

Schlosshauer, M. A. 2007. *Decoherence and the Quantum-to-Classical Transition*. Berlin: Springer.

Suppes, P. and M. Zanotti. 1996. *Foundations of Probability with Applications*. Cambridge: Cambridge University Press.

Walley, P. 1991. *Statistical Reasoning with Imprecise Probabilities*. London: Chapman and Hall.

White, R. 2010. Evidential symmetry and mushy credences. In T. S. Gendler and J. Hawthorne, eds., *Oxford Studies in Epistemology*, vol. 3, pages 161–186. Oxford University Press.

Wilce, A. 2012. Quantum logic and probability theory. In *The Stanford Encyclopedia of Philosophy*. Fall 2012 edn. E. Zalta (ed.), URL http://plato.stanford.edu/archives/fall2012/entries/qt-quantlog/.

Part III

Mathematical Representation (across disciplines)

8

The Incompleteness Of Hölder's Theorem During Most of The 20th Century

R. Duncan Luce

This report is based on my presentation at the Stanford University celebration of Patrick Suppes 90th birthday—actually a week before St. Patrick's Day, 2012, his actual 90th birthday. Pat and I have known each other for 57 years and have collaborated extensively on the representational theory of measurement, both on original research and on an exposition that resulted in the 3 volume "Foundations of Measurement". What is reported here was only noticed after Volume III was published in 1990 and so can be thought of as an addendum to it.

Abstract

In 1901, O. Hölder formulated a set of axioms for extensive measurement from which he showed there is a ratio scale, additive measure into the additive real numbers. Neither he nor anyone else apparently noticed until 1991 that had he considered mappings into the additive and multiplicative real numbers, then there are two other types of representation. It turns out that for the physical cases that he had in mind, they are ruled out. This, however, is not true for many attributes of subjective intensity in economics and psychology. This note summarizes what we currently know about such representations including a complex of cross-modal matching predictions.

1 Hölder's (1901) Theorem

1.1 The Classic Formulation

One very important measurement result involved structures $\langle X, \succsim, \odot \rangle$ where X is a set of objects; $\succsim$ is a weak ordering according to some attribute of the objects; and $\odot$ is a binary concatenation of X that is closed under $\odot$, i.e., if $x, y \in X$, then $x \odot y \in X$. A concrete example: X is a set homogeneous masses, which are ordered by using a equal-arm pan balance to determine which of two masses is heavier, and the concatenation operation $\odot$ simply means placing two masses on the same pan of the balance.

Hölder's (1901) theorem, formalizing some of Helmholtz's earlier ideas, stated axioms about $\langle X, \succsim, \odot \rangle$ such that when mapped via an order preserving function φ into $\langle \mathbb{R}^+, \geq$

Foundations and Methods from Mathematics to Neuroscience.
Colleen E. Crangle, Adolfo García de la Sienra and Helen Longino.
Copyright © 2014, CSLI Publications.

, $+\rangle$, then

$$\varphi(x \odot y) = \varphi(x) + \varphi(y). \tag{1}$$

Moreover, this class of additive representation form a ratio scale. This class of models has been called *extensive* measurement. For a detailed critique of Hölder's formulation and a discussion of related but more satisfactory axiomatizations, see pp. 53–55 of Krantz, Luce, Suppes, & Tversky (1971).

1.2 What Hölder and the Rest of Us Missed for Most of 20th Century

Typically, one does not invoke Hölder's theorem in isolation. For example, in mass measurement one also considers the multiplicative conjoint representation of mass being equal to volume times density. And in order to force a common measure φ of mass for both the concatenation and conjoint scales, a linking property must be discovered between the two structures. In this case, two quite standard distribution conditions serve as that link.

But, given that this involved both addition and multiplication, why did Hölder map $\langle X, \succsim, \odot \rangle$ just into $\langle \mathbb{R}^+, \geq, + \rangle$ rather than into the full real numbers $\langle \mathbb{R}^+, \geq, +, \times \rangle$?

In fact, if we do map into $\langle \mathbb{R}^+, \geq, +, \times \rangle$, then his axioms about $\langle X, \succsim, \odot \rangle$ lead to three possible, very distinct, representations, namely,

$$\varphi(x \odot y) = \varphi(x) + \varphi(y) + \delta\varphi(x)\varphi(y), \qquad \delta = -1, 0, 1. \tag{2}$$

This representation is called *polynomial-additive* (for short *p-additive*) because it is the only polynomial that can be transformed into an additive representation, Aczél (1966). That (2) can be so transformed is easily shown (see Section 2.2); that it is the only such polynomial is more subtle. It first appeared in Luce (1991, 2000), for $\delta = 0, 1$ in the context of utility theory where there is additional gambling structure well beyond the Hölder axioms. So it was not until the work described in Section 2 that we fully realized the impact of the incompleteness of Hölder's theorem.

Clearly, p-additivity (2) when $\delta = 0$ is, in fact, additive. The distribution laws of the physical applications, see Luce et al. (1990, pp. 125–126) force $\delta = 0$. So the cases $\delta = -1$ and 1 do not matter for physics.

But they very much do matter for the behavioral sciences, as I outline next.

2 Testable Predictions From Luce's (2004) Global Psychophysical Theory

2.1 Binary Senses

Primitives for Binary Receptors

The ears, eyes, and arms are binary receptors that function as cooperative pairs of organs. Of course, as was long ago remarked (at least by me) were we the result of a grand designer, rather than evolution, we would undoubtedly have 3 arms and hands–2 to hold and the middle one to manipulate. But, in reality, we need only consider binary (2-D) and unary (1-D) senses (Section 2.2).

For any intensity attribute, let X denote the set of physical intensities (not, e.g., dB) less their respective threshold intensities. Thus, for any non-intensive aspect of the signal, such as frequency, then intensity $x = 0$ denote the threshold level of that signal. If $x, u \in X$, then the stimulus presented is $(x, u) \in X \times X$. The subjective intensity ordering $\succsim$ over $X \times X$ is assumed to be a weak order.

Suppose that a respondent selects z such that (z, z) matches (x, u). Then, define the operator $\oplus$ by

$$x \oplus u := z. \tag{3}$$

Thus,

$$(x, u) \sim (z, z) = (x \oplus u, x \oplus u).$$

The operator $\oplus$ characterizes the intensity trade-off between the binary receptors.

Suppose also that the experimenter presents signal x and a number $p > 0$, and assume there is a reference signal $\rho < x$ either given by the experimenter or generated by the respondent. The restriction $\rho < x$ is a strong one that means, to a degree, ρ cannot remain fixed when x is varied sufficiently. Note that p can be either less than, equal to, or greater than, 1. The respondent is asked to report the signal y such that the "interval from the reference signal ρ to y" is perceived as p "times" as intense as the "interval from ρ to x". It is convenient to think of y as an operator: $y = x \circ_p \rho$.

When $\rho = 0$, this is nothing but S. S. Stevens' (1975) method of magnitude production.

Linking $\oplus$ and $\circ_p$

Luce (2002, 2004, 2008) formulated behavioral (i.e., testable) axioms that assert invariances among the primitives. Included were two linking properties between matching and production somewhat analogous to the distribution properties in physics. These allow us to use the same psychophysical function for both matching and production.

Representations of Binary Intensities

These assumed properties imply the following numerical representation: A $p-$ *additive* order preserving psychophysical function ψ:

$$\psi(x \oplus y) = \psi(x \oplus 0) + \psi(0 \oplus y) + \delta\psi(x \oplus 0)\psi(0 \oplus y), \qquad \delta = -1, 0, 1, \tag{4}$$

with

$$\psi(x \oplus 0) = \gamma\psi(0 \oplus x). \tag{5}$$

And a weighting function W over positive numbers such that

$$W(p) = \frac{\psi(x \circ_p \rho) - \psi(\rho)}{\psi(x) - \psi(\rho)}. \tag{6}$$

For loudness estimates using individual respondents, Steingrimsson and Luce (2005a, 2005b, 2006, 2007) strongly supported the behavioral axioms. Steingrimsson (2009, 2011, forthcoming(a), forthcoming(b), forthcoming(c)) has and is running a parallel series for brightness and for "perceived" contrast, and the axioms are equally strongly supported.

Form of the Psychophysical Function ψ

A simple behavioral invariance implies ψ is a power function

$$\psi(x) = \alpha x^\beta. \tag{7}$$

The invariance has been empirically supported for loudness and brightness in the third article of above sequences.

In the 1960s and 1970s this was thought to be sustained for all intensity (prothetic) attributes (summarized in Stevens 1975). Using geometric averaging, the log-log plots of binary attributes were "plausibly" linear. Caution: fitted lines do guide one's eyes.

Several Familiar Operator Properties

Commutativity: $\qquad\qquad\qquad\qquad x \oplus u \sim u \oplus x. \tag{8}$

This property was rejected for loudness, brightness, perceived contrast with individuals analyzed separately (Steingrimsson and Luce (2005a); Steingrimsson (2011, forthcoming(b))).

Associativity: $$(x \oplus u) \oplus v \sim x \oplus (u \oplus v). \tag{9}$$

So far no data have been reported. However, we will shortly see why it is expected to fail.

Bisymmetry: $$(x \oplus y) \oplus (u \oplus v) \sim (x \oplus u) \oplus (y \oplus v). \tag{10}$$

Note that (10) entails the interchange of the two "interior" intensities y and u. Bisymmetry has been empirically accepted for loudness, brightness, and contrast with individuals separately evaluated (Steingrimsson and Luce (2005b); Steingrimsson (2011, forthcoming(c),f)).

None of these properties have yet been tested for two-arm weight lifting.

Predictions of the Binary Theory

Suppose the p-additive representation holds. Then Luce (2012) has proved that in the present context:

- For $\delta = 0$, bisymmetry is satisfied.
- For $\delta \neq 0$, bisymmetry is satisfied iff commutativity is satisfied. Conclusion: The data—Yes to bisymmetry and No to commutativity—imply $\delta = 0$, i.e. simple additivity.

This argument corrects Luce's (2004) unconditional claim that $\delta = 0$. Dr. C. T. Ng pointed out the error to me (Luce 2008).

- If bisymmetry holds, then associativity cannot hold.

2.2 Unary Theory

Many intensity senses are unary, not binary. Examples: taste, electric shock, vibration, force, linear extent, preference for money, and others.

Consider those unary attributes for which a physical signal concatenation $\odot$ exists that has an additive physical ratio scale representation. So $\odot$ must satisfy both commutativity and associativity.

Unary Magnitude Production

Exactly as with the binary theory, we assume magnitude production and a linking axiom between the $\circ_p$ production structure and the $\odot$ structure. And that means we definitely need $\langle \mathbb{R}^+, \geq, +, \times \rangle$, so we have to allow the p-additive form. And, quite unlike the binary case, $\odot$ satisfies all of: bisymmetry, commutativity, and associativity.

I do not know of any principled argument that forces the $\delta = 0$ case for unary attributes.

p-Additive Scale Types

The following observations were first made in connection with utility theory in Luce (2010, 2011).

For the additive case ($\delta = 0$), φ is well known to be a ratio scale.

For the non-additive cases ($\delta = -1, 1$), p=additivity is equivalent to

$$1 + \delta\varphi(x \odot y) = [1 + \delta\varphi(x)][1 + \delta\varphi(y)] = 1 + \delta\varphi(y \odot x)$$

Clearly, ln of this is an additive representation. Because we have $1 + \delta\varphi(x)$ and $\delta = -1, 1, \varphi$ has to be dimensionless and so is an *absolute*, not a ratio, scale.

2.3 Form of the Psychophysical Function φ

Unlike the binary case, there are 3 types corresponding to the value of δ, and Luce (2012) shows that for $x \in \mathbb{R}^+$

$$\varphi_0(x) = \eta x (\eta > 0) \qquad \text{if } \delta = 0 \qquad (11)$$

$$\varphi_1(x) = e^{\lambda x} - 1 (\lambda > 0) \qquad \text{if } \delta = 1 \qquad (12)$$

$$\varphi_{-1}(x) = 1 - e^{\kappa x} (\kappa > 0) \qquad \text{if } \delta = -1. \qquad (13)$$

For $\delta = 0$, this is a special case of a power function. For $\delta \neq 0$, these two exponential function clearly are not power functions.

2.4 But the Empirical Claim is Just Power Functions

The empirical literature seemed to defend power functions; what gives?

For example, Stevens (1959) reported averaged cross-modal matches between loudness of noise, vibration, and shock each measured in dB:

- loudness versus vibration seemed to be a power function—but fitted lines can deceive.
- shock versus loudness and versus vibration were equally not power functions, although Stevens tried—not very convincingly—to explain that fact away.

3 Predictions of Cross-Modal Matches

From the binary and unary theories, it is fairly apparent that predictions of cross-modal matches—matching signal z of attribute b to the presented signal x of attribute a—should follow. They do. They are moderately complicated because of the unary case's 3 representations $\delta = -1, 0, 1$. Because we know of no $\delta = -1$ attribute other than utility of money for some people, that attribute is omitted from the columns in the following already complex table, reprinted with permission from Luce (2012).

TABLE 1 Predictions of the Two Theories for Cross-Modal Matches of Modality b to Modality a

Theory		2-D	1-D
2-D			
$\delta_a = 0$	$\delta_b = 0$	$\delta_b = 0$	$\delta_b = 1$
x	power	power	$\left[\frac{1}{\lambda_b}\ln(1 + \eta_a x_a^\beta)\right]^{\frac{1}{\beta_b}}$
1-D			
$\delta_a = 0$	power	proportion	$\left[\frac{1}{\lambda_b}\ln(1 + \eta_a x)\right]^{\frac{1}{\beta_b}}$
$\delta_a = 1$	$\left[\frac{1}{\alpha_b}(e^{\lambda_a x} - 1)\right]^{\frac{1}{\beta_b}}$	$\frac{1}{\eta_b}(e^{\lambda_a x} - 1)$	proportion
$\delta_a = -1$	$\left[\frac{1}{\alpha_b}(1 - e^{-\kappa_a x})\right]^{\frac{1}{\beta_b}}$	$\frac{1}{\eta_b}(1 - e^{-\kappa_a x})$	$\frac{1}{\lambda_b}\ln(2 - e^{-\kappa_a x})$

Note. Because there are three representations possible for one-dimensional (1-D) modalities, the predictions are more complex than normally recognized. 2-D = two-dimensional.

- The unary predictions differ from a power function only for most cases when $\delta \neq 0$.

- I suspect that further research will confirm that for shock, pain, and vibration $\delta \neq 0$.

- Utility of money, which is unusual because the domain includes losses and well as gains, appears to be a case where all 3 can occur and correspond to risk seeking, risk neutral, and risk averse types.

4 Closing Remarks

The overlooked solutions to Hölder's axiomatizations did not matter at all for physics, but they certainly appear to matter greatly for the behavioral and economic sciences. Much experimentation is needed to check these predictions. But it is most important to realize that such experiments *must be* analyzed for each individual respondent; averaging respondents is clearly inappropriate especially when they have different values of δ.

5 Acknowledgements

This research was supported in part by the Air Force Office of Research grant FA9550-08-1-0468–any opinion, finding, and conclusions or recommendations expressed in this material are those of the author and do not necessarily reflect the views of the Air Force Office of Research.

I also thank Drs. A. A. J. Marley and Ragnar Steingrimsson for suggestions on the exposition.

References

Aczél, J. 1966. *Lectures on Functional Equations and Their Applications.* New York: Academic Press.

Hölder, O. 1901. Die axiome der quantität und die lehre vom mass. *Ber. Verh. Kgl. Sächsis. Ges. Wiss. Leipzig, Math.-Phys.Classe* 53:1–64.

Krantz, D. H., R. D. Luce, P. Suppes, and A. Tversky. 1971. *Foundations of measurement, Volume I: Additive and polynomial representations.* New York: Academic Press.

Luce, R. D. 1991. Rank- and sign-dependent linear utility models for binary gambles. *Journal of Economic Theory* 53(1):75–100.

—. 2000. *Utility of Gains and Losses: Measurement - Theoretical and Experimental Approaches.* Mahwah, NJ: Erlbaum.

—. 2002. A psychophysical theory of intensity proportions, joint presentations, and matches. *Psychological Review* 109(3):520–532.

—. 2004. Symmetric and asymmetric matching of joint presentations. *Psychological Review* 111(2):446–454.

—. 2008. Symmetric and asymmetric matching of joint presentations: Correction to Luce (2004). *Psychological Review* 115(3):601.

—. 2010. Interpersonal comparisons of utility for 2 of 3 types of people. *Theory and Decision* 68(1-2):5–24.

—. 2011. Inherent individual differences in utility. *Frontiers in Psychology* 2(297).

—. 2012. Predictions about bisymmetry and cross-modal matches from global theories of subjective intensities. *Psychological Review* 119(2):373–387.

Luce, R. D., D. Krantz, P. Suppes, and A. Tversky. 1990. *Foundations of Measurement, Vol. III: Representation, Axiomatization, and Invariance.* New York: Academic Press.

Steingrimsson, R. 2009. Evaluating a model of global psychophysical judgments for brightness: I. Behavioral properties of summations and productions. *Attention, Perception, & Psychophysics* 71(8):1916–1930.

—. 2011. Evaluating a model of global psychophysical judgments for brightness: II. Behavioral properties linking summations and productions. *Attention, Perception, & Psychophysics* 73(3):872–885. doi:10.3758/s13414-010-0067-5.

—. forthcoming(a). Evaluating a model of global psychophysical judgments for brightness: III. Forms for the psychophysical and the weighting function.

—. forthcoming(b). Evaluating a model of global psychophysical judgments for perceived contrast II: Behavioral properties linking summations and productions.

—. forthcoming(c). Evaluating a model of global psychophysical judgments of perceived contrast: I. Behavioral properties of summation and production.

Steingrimsson, R. and R. D. Luce. 2005a. Evaluating a model of global psychophysical judgments: I. Behavioral properties of summations and productions. *Journal of Mathematical Psychology* 49(4):290–307.

—. 2005b. Evaluating a model of global psychophysical judgments: II. Behavioral properties linking summations and productions. *Journal of Mathematical Psychology* 49(4):308–319.

—. 2006. Empirical evaluation of a model of global psychophysical judgments: III. A form for the psychophysical and perceptual filtering. *Journal of Mathematical Psychology* 50(1):15–29.

—. 2007. Empirical evaluation of a model of global psychophysical judgments: IV. Forms for the weighting function. *Journal of Mathematical Psychology* 51(1):29–44.

Stevens, S. S. 1959. Cross-modality validation of subjective scales for loudness, vibration, and electric shock. *Journal of Experimental Psychology* 57(4):201–209.

—. 1975. *Psychophysics: Introduction to its perceptual, neural, and social prospects.* New York: Wiley.

9

The Economic System as Trade in Information

KENNETH J. ARROW

Patrick Suppes has been an intellectual and personal colleague since his arrival at Stanford in 1950. The attributes which make him such a good friend are also those that have made him such an outstanding figure in the conduct of scholarship: breadth, exemplified by the variety of papers presented in this volume; rigor; relevance, empirical and otherwise.

In this tribute to Suppes, I want to give emphasis to some aspects of economic behavior which the standard model fails to express. Specifically, I want to give more systematic expression to the key role (more precisely, roles) of information and belief in the conduct of economic life and to the behavior observed regularly, sometimes pleasantly, sometimes very unpleasantly.

Some natural scientists have expressed disdain for economics because of its failures to predict. Suppes and I have had our fling with natural science, having both served as meteorologists in the United States Air Force during World War II. We did not find predictability to be an outstanding characteristic of this application of physics and chemistry, despite a genuinely deep understanding of thermodynamics, the energy consequences of phase transitions, and the implications of the Earth's rotation.

Economics has made a real contribution to the understanding of its field. But there are many aspects not understood, as current conditions and debates show. The market under capitalism is understood as a coordinating device, which enables a reasonably efficient allocation of resources to quite remote uses without any central planning, a self-organizing system. But this system has broken down at frequent, though irregular, intervals since the early 19th century, and there are periods, as now, when resources stand idle though they could be put to use.

1 Economics as Trade in Goods

The typical view of the world in the central version of economic theory is that the economy is the mode in which goods are traded for each other. This trade may be quite indirect. One essential element is that some goods, e.g., labor or capital goods, are transformed through production into other goods, e.g., manufactured goods, which are then ultimately bought by the laborers or capital owners.

Foundations and Methods from Mathematics to Neuroscience.
Colleen E. Crangle, Adolfo García de la Sienra and Helen Longino.
Copyright © 2014, CSLI Publications.

The great insight of mainstream theory is that the system of markets coordinates all these individual economic choices. Suppose that for each commodity there is a price at which individuals and firms can buy or sell. Then the individual can choose how much of his assets (labor or other) he or she wishes to supply and how much of various goods he or she wishes to use, subject to the condition that income from sales of assets equals expenditure on goods purchased. At the same time, each firm can decide how much of each input it wants and how much of each product it will wish to sell. The individual presumably makes his or her choice so as to maximize his or her satisfaction, or, *utility*, to use the economist's term, while the firm seeks to maximize its profit, defined as the difference between the total value of its sales and the total value of its inputs, both calculated at the given prices.

What prices will in fact prevail? It stands to reason that the price for each good will be such that the supplies offered and the amounts demanded are equal (with some qualifications about goods where supply exceeds demand even at zero price; these goods, like air, are free).

If these *equilibrium* prices do prevail, then a remarkably complex system is brought into coordination. Each individual can choose his or her behavior, knowing only what it is supposed to know, its marketable assets and the satisfaction he or she gets from any amounts supplied or demanded, and the prices it faces, Similarly, each firm need know only its production techniques, that is the amounts of its products obtainable from any given levels of inputs, and the prices it faces.

Not only is the system smoothly running but the outcome can be shown to be efficient in a strong sense: there is no other allocation of the resources available within the system which will make everyone better off (according to his or her definition of satisfaction) than the equilibrium outcome just described.

Even within the discussion so far, there are some unresolved theoretical issues. One is how the equilibrium prices are arrived at. But I wish to enlarge the discussion, to introduce the limitations imposed by considerations of time and uncertainty.

What can one say empirically? Goods really are distributed through very remote channels. There are not normally huge surpluses or deficits of supply relative to demand. When oil supply is interrupted or drought reduces the supply of corn, prices really do rise to curb demand. New methods of extracting natural gas are followed by a reduction in price.

But the economic record also shows sharp contradictions to the hypothesis of smoothly-functioning markets. From the early nineteenth century on, there have been recurring periods in which outputs go down, workers are unemployed, and productive capital is underutilized. These events are certainly evidence of inefficiency (clearly. more goods could be produced) and of disequilibrium (supply of labor exceeds demand). They do not last forever; every recession is followed by a period of prosperity, but every period of prosperity is followed by a recession. This phenomenon was recognized early and already written about in the middle of the 19th century (see, e.g., Mill and Ashley (1909, originally published in 1848, pp. 644–649), Juglar (1862)). The problem of explaining what are sometimes called, "business cycles," has several aspects, but I will stress one, the deficiencies in the market representation of reality and their implications for the role of information and belief.

2   Relative Roles of Goods and Information in the Economy

Information already plays a role in the standard model of the economy sketched above. Individuals are presumed to know how their sales and purchases affect their utilities. Firms are presumed to know their production possibilities, how different sets of inputs can give rise to different sets of outputs. But the information demanded in the model so far is only what might be termed, *private information*. What is not required is information about the private information of others. The entire effect of others and their information is conveyed in the prices faced by the individual.

But this supposes that there are markets for all possible goods. What makes it problematic that markets arise? The major reasons are the elements of time and uncertainty. The very term for our modern economic system, *capitalism*, stresses the role of time. What is meant by capital, after all, is a product, durable in time, which is capable of helping to give rise to products over a period of time. Capital is exemplified by durable machines or by buildings.

A decision to invest in capital, then, is motivated by an intention to produce goods in the future. Therefore, to use the standard model sketched above, we would need a market now to sell goods for delivery in the future. Although the significance of capital and the lag between input and output was recognized very early, I think it fair to say that a clear statement along the lines above is only to be found in the 1930s, with the works of Lindahl and Fernholm (1939, Part III, originally published in Swedish in 1929) and Hicks (1939).

But these markets for future goods rarely exist. We do have some markets for agricultural and mineral commodities for a limited period ahead. Much more consequential are markets for payments of money in the future. Bonds are purchased today and promise to pay money in the future, in the form of interest for a period followed by a repayment of the principal. They therefore permit purchasing goods for consumption in the future, but with uncertainty about the prices to be paid for those goods. Stocks are also promises but promises contingent on future events, specifically on the performance of the issuing firm.

Securities therefore permit taking account the future in making present decisions. Ideally, the Lindahl-Hicks program of dated commodities can be extended to take account of uncertainty, by a system of markets for bets on all possible outcomes. This complete set of markets does not, of course, exist, and therefore individuals are guided by beliefs which vary from one individual to another. A belief in general recognizes uncertainty. It might be represented as a probability distribution held by an individual, though some have felt that beliefs cannot be regarded as satisfying all the assumptions of probability theory.

3   The Socialist Controversy

There was an interesting episode in the history of economic thought, now concluded, which nevertheless illuminated sharply the informational content of markets. This was a discussion of the possibility of operating a socialist economic system. This discussion led to a deeper understanding of the role of information in the economy.

In the second half of the nineteenth century, socialist parties entered democratic politics and achieved significant parliamentary representation, especially in Germany, France, and Italy. The possibility that socialism might be established as the economic system, replacing capitalism and private property became real. The great Italian economist, Vilfredo Pareto, first posed this as a research question: how would a socialist system actually allocate resources and, in particular, how could it do so efficiently? Pareto was himself very much opposed to socialism, but he regarded the question as intellectually interesting.

A student of his, Enrico Barone (1908) gave an answer. The socialist state should imitate the workings of the ideal capitalist system. The central authority should announce a set of prices. Firms, though publicly owned, are asked to compute the inputs and outputs which would maximize profits at those prices, and households are asked how much they would buy and sell at those prices. If supplies and demands balance, the allocation so determined should be carried out. If they do not balance, then prices should be adjusted until equilibrium is attained.

The prices in this view were explicitly messages which conveyed information, not facts which compelled responses.

After the end of World War I, there was a period in which the Socialist party formed the government in Austria, a country with a distinguished set of economists, many of whom argued about the feasibility and efficiency of socialism. Ludwig von Mises held that prices were essential to an allocation and that socialism was impossible. Others rediscovered the ideas of Barone. Friedrich von Hayek took an antagonistic view which was, however, somewhat more nuanced than von Mises's. Prices, in Hayek's view, had to be real to create the necessary incentives; mere exchange of information was not compelling enough.

The papers on both sides of the controversy were collected in a useful compendium by von Hayek et al. (1935). The most complete presentation of the market socialist viewpoint is that of Oskar Lange (1936, 1937). Still later came the remarkable paper of Leonid Hurwicz (1960), which, for the first time, spelled out completely the general theme of an information system to guide the economy and the special role of prices as messages. He showed that, under certain circumstances, the price system requires the least amount of information among all communication systems for achieving an efficient allocation of resources.

4   The Infinite Regress in Beliefs About Others

Once it is understood that most conceivable markets for goods conditional on time and uncertainties do not exist, it is clear that savings and investment are governed by expectations.

One bold hypothesis is that the expectations are always correct. In effect, the prices that occur are those that would occur in a complete system of markets. This hypothesis is called, "perfect foresight." A weaker version, called, "rational expectations," is that the expectations are correct on the average. The latter has some but not much empirical support in some contexts.

The economist Oskar Morgenstern was director of a business-cycle research institute in Vienna during the 1930s. He began to consider what would be meant by successful forecasting. He noted that forecasting by any one entrepreneur would be about general business conditions and, in particular, the investment decisions of other entrepreneurs. But those decisions would be based on the forecasts made by those entrepreneurs. Of course, by the same token, each of the latter forecasts would be based in part on the original entrepreneur's forecasts.

Morgenstern (1935) pointed this apparently paradoxical situation out and suggested that the concept of perfect foresight was impossible. He gave as an illustration an example drawn from a Sherlock Holmes story, "The Final Problem." Holmes is on a train from London to Dover, in order to escape the villain, Moriarty. Moriarty is on another train. Either can stop at Canterbury, an intermediate station, or go on to Dover. If they make the same choice, then Moriarty will be able to kill Holmes. Clearly, if Holmes chooses Dover, it is optimal for Moriarty to go to Dover, but then, given Moriarty's choice, it would have been optimal for Holmes to choose Canterbury. But if Holmes had chosen Canterbury and Moriarty responds optimally for him, he should have chosen Dover.

Later, as is well-known, Morgenstern joined with John von Neumann in the classic work on game theory, von Neumann and Morgenstern (1944). They evolved an interesting point of view, carried forward later by Nash (1950). Suppose each of the opponents chooses a random device for making the choice among alternative points to leave the train. Then there can be an equilibrium. That is, each one's randomized strategy, even if known to the other, leaves the other best off by continuing with his randomized strategy. There is an equilibrium in the set of mutual beliefs.

Prices constitute an analogous kind of equilibrium. But there is one important difference. Prices are real, not just beliefs. There is no way of knowing that the parties have in fact coordinated on beliefs.

A somewhat similar point was made about the same time by Keynes (1936, p. 156). He cited an advertising campaign in the United States by a brewery. It posted the faces of six models. Anyone could choose one model and would get a prize if that model had the plurality of votes. Clearly, any chooser would opt not for the one he or she prefers but for the one the chooser thought was most popular with the others. But if everyone acted that way, it would pay an individual to guess that model others guessed was most popular, and so forth. Keynes did not observe that this process does have an equilibrium, in fact, six; everyone agrees on any one candidate. But equally it is true that it is not easy to see how the individuals would converge on any one.

5 The Acquisition of Information

We have seen that economic actors are making forecasts as a basis for action. We have stressed so far the problem of information about the behavior of other economic actors. But, in general, information, about others, about technological change, or about political events, is a commodity: it is valuable, it can be acquired, it is costly to produce, it may be costly to transmit, though less costly than to produce. Above all, it has a unique property. It is not destroyed by use or transmission.

This last property implies that the production and sale of information cannot be handled through the usual market system. Let me remark on some implications.

One is the prevalence of specialization. Since information, once acquired, can be used over and over again, it pays some individuals to acquire the information to be used repeatedly. Professions, such as medicine or law, exemplify this economic principle. An individual has a medical problem could, in theory, go to medical school and learn what is needed. It is obviously more economic for the individual and for society to go instead to a trained physician who can use this same information to treat many people.

However, this specialization has a further implication. Individuals trading with each other have different information, and they know they are dealing with others with different information. Clearly, this is going to create difficulties. Problems of this kind have long been known in the insurance industry. "Moral hazard" refers to a situation in which the insured may take measures or fail to take measures making payment of the insurance policy more likely. "Adverse selection" is a situation in which the insured has a better knowledge of his or her risk situation than the insurer does. It is, for example, well known that those taking out annuities live longer than the average; the individuals know their longevity prospects better than the insurance company even after medical tests.

This situation of *asymmetric information* has become widely recognized and plays an important role wherever there is shared risk-bearing. Typically, markets needed for complete coverage fail to exist. Instead, non-market relations become more important. The implications of asymmetric information has been drawn for the markets for medical care and insurance and to financial markets. For an outstanding survey, see Laffont and Martimort (2002).

References

Barone, E. 1908. Il ministro della produzione nello stato collettivista. *Giornale degli Economisti* 37(19):267–293.

Hicks, J. 1939. *Value and Capital: An Inquiry Into Some Fundamental Principles of Economic Theory.* Oxford: Clarendon Press.

Hurwicz, L. 1960. Optimality and informational efficiency in resource allocation processes. In K. J. Arrow, S. Karlin, and P. Suppes, eds., *Mathematical Methods in the Social Sciences 1959*, chap. 3, pages 27–46. Stanford, CA: Stanford University Press.

Juglar, C. 1862. *Des Crises Commerciales et de Leur Retour Périodique en France, en Angleterre, et Aux États-Unis.* Paris: Guillaumin.

Keynes, J. M. 1936. *The General Theory of Employment, Interest and Money.* New York: Harcourt, Brace.

Laffont, J. J. and D. Martimort. 2002. *The Theory of Incentives: The Principal-agent Model.* Princeton, NJ: Princeton University Press.

Lange, O. 1936. On the economic theory of socialism: Part one. *The Review of Economic Studies* 4(1):53–71.

———. 1937. On the economic theory of socialism: Part two. *The Review of Economic Studies* 4(2):123–142.

Lindahl, E. R. and T. Fernholm. 1939. *Studies In the Theory of Money and Capital.* London: G. Allen & Unwin.

Mill, J. S. and W. J. Ashley. 1909. *Principles of Political Economy: With Some of Their Applications to Social Philosophy.* London: Longmans, Green, and Co.

Morgenstern, O. 1935. Vollkommene voraussicht und wirtschaftliches gleichgewicht. *Zeitschrift für Nationalökonomie* 6(3):337–357.

Nash, J. F. 1950. Equilibrium in n-person games. *Proceedings of the National Academy of Sciences* 36(1):48–49.

von Hayek, F. A., N. G. Pierson, L. von Mises, G. N. Halm, and E. Barone. 1935. *Collectivist Economic Planning: Critical Studies On the Possibilities of Socialism.* London: G. Routledge.

von Neumann, J. and O. Morgenstern. 1944. *Theory of Games and Economic Behavior.* Princeton, N. J.: Princeton University Press.

10

On a Class of Meaningful Permutable Laws

Jean-Claude Falmagne

This paper is dedicated to Patrick Suppes, whose work and counsel have shaped much of my scientific life. I am also grateful to Duncan Luce for his many useful remarks, and to Chris Doble and Louis Narens for their reactions to an earlier version of this paper.

Abstract

The permutability equation $F(F(y,r),t) = F(F(y,t),r)$ is satisfied by many scientific and geometric laws. A few examples among many are: The Lorentz-FitzGerald Contraction, Beer's Law, the Pythagorean Theorem, and the formula for computing the volume of a cylinder. We show here that if we required that a permutable law be meaningful, the possible forms of a law are considerably restricted.

The mathematical expression of a scientific law typically does not depend on the units of measurement of its variables. The most important rationale for this convention is that measurement units do not appear in nature[1]. Thus, any mathematical model or law whose form would be fundamentally altered by a change of units would be a poor representation of the empirical world. As far as I know, however, there is no agreed upon formalization of this type of invariance of the form of scientific laws, even though there has been some proposals (see Falmagne and Narens 1983; Narens 2002; Falmagne 2004). The concept of 'meaningfulness' was discussed by Suppes and Zinnes (1963, pp. 64–74). This paper was the inspiration of our work in this area Suppes (2002, see also pp. 110–112).

As a first step toward a formalization, expanding on the just cited references, I discuss here a condition of 'meaningfulness' constraining a priori the form of certain functions describing scientific or geometric laws. In this paper, I only deal with functions of two ratio scale variables, such as such as mass, length, or time, satisfying some conditions. This meaningfulness condition is defined in the second section of this paper. In this definition, the units of the variables are explicitly specified by the notation, as opposed to being implicitly embedded in the concepts of 'quantities' and 'dimensions' of dimensional analysis (cf. for example Sedov 1959).

---

[1]The only exception is the counting measure, as in the case of the Avogadro number.

Foundations and Methods from Mathematics to Neuroscience.
Colleen E. Crangle, Adolfo García de la Sienra and Helen Longino.
Copyright © 2014, CSLI Publications.

The interest of such a meaningfulness condition from a philosophy of science standpoint is that, in its context, abstract constraints on the function, formalizing 'gedanken experiments', may conceivably yield the short list of exact possible forms of a law, up to some real valued parameters.

An example of such an abstract constraint is the condition below, which applies to a real, positive valued function F of two real non negative measurement variables. It is formalized by the equation

$$F(F(y,r),t) = F(F(y,t),r), \tag{1}$$

where F is strictly monotonic and continuous in both real variables. An interpretation of $F(y,r)$ in Equation (1) is that the second variable r modifies the state of the first variable y, creating an effect evaluated by $F(y,r)$ in the same measurement variable as y. The left hand side of (1) represents a one-step iteration of this phenomenon, in that $F(y,r)$ is then modified by t, resulting in the effect $F(F(y,r),t)$. Equation (1), which is referred to as the 'permutability' equation by Aczél (1966), formalizes the concept that the order of the two modifiers r and t is does not matter.

Many, and various, scientific laws are 'permutable' in the sense of Equation (1). Some examples of permutable laws are the Lorentz-FitzGerald Contraction, Beer's law, the formula for computing the volume of a cylinder, and the Pythagorean theorem. For the Lorentz-FitzGerald Contraction, for example, written in the form

$$L(\ell,v) = \ell\sqrt{1 - \left(\frac{v}{c}\right)^2} \tag{2}$$

in which c is the speed of light, we have

$$L(L(\ell,v),s) = L(\ell,v)\sqrt{1 - \left(\frac{s}{c}\right)^2} = \ell\sqrt{1 - \left(\frac{v}{c}\right)^2}\sqrt{1 - \left(\frac{s}{c}\right)^2} = L(L(\ell,s),v). \tag{3}$$

Not all scientific laws are permutable. Van der Walls Equation, for instance is not: see the Counterexample 1(e).

Under fairly general conditions of continuity and solvability making empirical sense, the permutability Equation (1) implies the existence of a representation

$$F(y,r) = f^{-1}(f(y) + g(r)), \tag{4}$$

where f and g are some real valued, strictly monotonic continuous functions. This is stated precisely in Lemma 2, which is due to Hosszú (1962a,b,c) (cf. Aczél 1966). It is easily shown that the representation (4) implies the permutability condition (1): we have

$$\begin{aligned}
F(F(y,r),t) &= f^{-1}(f(F(y,r)) + g(t)) && \text{(by (4))} \\
&= f^{-1}(f(f^{-1}(f(y) + g(r))) + g(t)) && \text{(by (4) again)} \\
&= f^{-1}(f(y) + g(r) + g(t)) && \text{(simplifying)} \\
&= f^{-1}(f(y) + g(t) + g(r)) && \text{(by commutativity)} \\
&= F(F(y,t),r) && \text{(by symmetry)}.
\end{aligned}$$

We will also use a more general condition, called 'quasi-permutability', which is defined by the equation

$$F(G(y,r),t) = F(G(y,t),r) \tag{5}$$

and lead to the representation

$$F(y,r) = m((f(y) + g(r)))$$ (6)

(see also Lemma 2).

Our contention is that the combined consequences of meaningfulness and permutability or quasi-permutability are capable of delineating a *priori* the possible forms of physical laws. For example, suppose that the function F is symmetric and quasi-permutable and also satisfies reasonable solvability and monotonicity conditions. We will show in Theorem 1 that, under the relevant meaningfulness condition, there are then only two possible forms for the function F, which are:

1. $$F(y,r) = \theta yr \qquad\qquad (\theta > 0)$$ (7)

and if F is homogeneous: $F(\alpha y, \alpha r) = \alpha F(y,r)$,

2. $$F(y,r) = \left(y^\theta + r^\theta\right)^{\frac{1}{\theta}} \qquad\qquad (\theta > 0).$$ (8)

With $\theta = 1$, the first equation is the formula for the area of a rectangle. With $\theta = 2$, the second one is the Pythagorean Theorem, up to the exponent.

In our first section, we state basic definitions and describe a few examples of laws, taken from physics and geometry, in which the permutability condition applies. We also give one example, van der Waals Equation, which is not permutable. The second section is devoted to meaningfulness and ancillary concepts. The third section recalls a representation theorem for permutable and quasi-permutable functions. This section also contains a new result stating that, under the meaningfulness condition, some properties satisfied by a function in a meaningfulness family are automatically transferred to all the functions in the same family. The last section contains an exemplary result, combining quasi-permutability with meaningfulness, which supports our contention.

Some Basic Concepts and Examples

Definition 1. *We write $\mathbb{R}_{++}$ for the positive reals. Let J, J', and H be three real non negative intervals of positive length. A (numerical) code is a function*

$$M : J \times J' \longrightarrow H$$ (9)

that is continuous in its two arguments, strictly increasing in the first and strictly monotonic in the second.

A code M is *solvable* if it satisfies the following two conditions.

[S1] If $M(x,t) < p \in H$, there exists $w \in J$ such that $M(w,t) = p$.

[S2] The function M is *1-point right solvable*, that is, there exists a point $x_0 \in J$ such that for every $p \in H$, there is $v \in J'$ satisfying $M(x_0, v) = p$. In such a case, we may say that M is x_0-*solvable*.

By the strict monotonicity of M, the points w and v of [S1] and [S2] are unique.

Two functions $M : J \times J' \to H$ and $G : J \times J' \to H'$ are *comonotonic* if

$$M(x,s) \le M(y,t) \iff G(x,s) \le G(y,t), \qquad (x,y \in J; s,t \in J').$$ (10)

In such a case, the equation

$$F(M(x,s)) = G(x,s) \qquad\qquad (x \in J; s \in J')$$ (11)

defines a strictly increasing continuous function $F : H \xrightarrow{\text{onto}} H'$. We may say then that G is F-*comonotonic* with M.

We turn to one of the two key conditions of this paper.

Definition 2. *A function $M : J \times J' \longrightarrow H$ is* quasi-permutable *if there exists a function $G : J \times J' \to J$ co-monotonic with M such that*

$$M(G(x,s),t) = M(G(x,t),s) \qquad (x,y \in J; s,t \in J'). \tag{12}$$

We say in such a case that M is *permutable with respect to G*, or *G-permutable* for short. When M is permutable with respect to itself, we simply say that M is *permutable*, a terminology consistent with Aczél (1966, Chapter 6, p. 270).

Lemma 1. *A function $M : J \times J' \to H$ is G-permutable only if G is permutable.*

Proof. Suppose that G is F-comonotonic with M. For any $x \in J$ and $s, t \in J'$, we get
$G(G(x,s),t) = F(M(G(x,s),t)) = F(M(G(x,t),s)) = G(G(x,t),s).$ □

Many scientific laws embody permutable codes, and hence can be written in the form of Equation (4). We give four examples below. In each case, we derive the forms of the functions f and g in the representation Equation (4).

1 Four Examples of Permutable Functions and One Counterexample

(a) The Lorentz-FitzGerald Contraction. This term denotes a phenomenon in special relativity, according to which the apparent length of a rod measured by an observer moving, along a line parallel to the rod at the speed v with respect to that rod, is a decreasing function of v, vanishing as v approaches the speed of light. This function is specified by the formula

$$L(\ell, v) = \ell \sqrt{1 - \left(\frac{v}{c}\right)^2}, \tag{13}$$

in which $c > 0$ denotes the speed of light, ℓ is the actual length of the rod (for an observer at rest with respect to the rod), and $L : \mathbb{R}_+ \times [0, c[\xrightarrow{\text{onto}} \mathbb{R}_+$ is the length of the rod measured by the moving observer.

The function L satisfies the strict monotonicity and continuity requirements of a code, and we have shown in our introduction that it was permutable, that is, Equation (1) hold (with $L = F$). Accordingly, we have for some functions f and g (see Lemma 2(ii)),

$$L(\ell, v) = \ell \sqrt{1 - \left(\frac{v}{c}\right)^2} = f^{-1}(f(\ell) + g(v)). \tag{14}$$

Solving the functional equation of the second equality leads to the Pexider equation (cf. Aczél 1966, pp. 141–165)

$$f(\ell y) = f(\ell) + k(y) \tag{15}$$

with

$$y = \sqrt{1 - \left(\frac{v}{c}\right)^2} \quad \text{and} \quad k(y) = g\left(c\sqrt{1 - y^2}\right).$$

As the background conditions (monotonicity and domains of the functions[2]) are satisfied, the unique forms of f and k in (15) are determined. They are: with $\xi > 0$,

$$f(\ell) = \xi \ln \ell + \theta \tag{16}$$
$$k(y) = \xi \ln y.$$

So, we get for the function g in (14):

$$g(v) = \xi \ln \left(\sqrt{1 - \left(\frac{v}{c}\right)^2} \right). \tag{17}$$

(b) BEER'S LAW. This law applies in a class of empirical situations where an incident radiation traverses some absorbing medium, so that only a fraction of the radiation goes through. In our notation, the expression of the law is

$$I(x, y) = x\, e^{-\frac{y}{c}}, \qquad (x, y \in \mathbb{R}_+,\ c \in \mathbb{R}_{++} \text{ constant}) \tag{18}$$

in which x denotes the intensity of the incident light, y is the concentration of the absorbing medium, c is a reference level measured with the same unit as y, and $I(x,y)$ is the intensity of the transmitted radiation. The form of this law is similar to that of the Lorentz-FitzGerald Contraction and the same arguments apply. Thus, the function $I: \mathbb{R}_+ \times \mathbb{R}_+ \xrightarrow{\text{onto}} \mathbb{R}_+$ is also a permutable code. The solution of the functional equation

$$x\, e^{-\frac{y}{c}} = f^{-1}(f(x) + g(y))$$

follows a pattern similar to that of Equation (14) for the Lorentz-FitzGerald Contraction. The only difference lies in the definition of the function g, which is here

$$g(y) = -\xi\, \frac{y}{c}.$$

The definition of f is the same, namely (16). So, we get

$$I(x, y) = f^{-1}(f(x) + g(y)) = \exp\left(\frac{1}{\xi}(\xi \ln x + \theta - \xi \frac{y}{c} - \theta \right) = x\, e^{-\frac{y}{c}}.$$

(c) THE MONOMIAL LAWS. These are functions of the form

$$G(x, y) = \phi x y^\theta. \tag{19}$$

An example is the formula for computing the volume $V(r, h)$ of a cylinder of radius r and height h, that is

$$V(r, h) = \pi h\, r^2.$$

(In this case $\phi = \pi$ is constant.) The function G is permutable. We have

$$G(G(x, y), z) = G(\phi x y^\theta, z) = \phi^2 x\, y^\theta z^\theta = G(G(x, z), y).$$

Accordingly, the representation

$$\phi x y^\theta = f^{-1}\left(f(x) + g(y)\right) \tag{20}$$

---

[2]Note that the standard solutions for Pexider equations are valid when the domain of the equation is an open connected subset of $\mathbb{R}^2$ rather than $\mathbb{R}^2$ itself. Indeed, Aczél (1987, see also Aczél, 2005, Chudziak and Tabor, 2008, and Radó and Baker, 1987) has shown that, in such cases, this equation can be extended to the real plane.

must hold for some functions f and g. One of the two solutions of this functional equation is: for some positive constant a and some constant b

$$f(x) = a \ln x + b,$$
$$g(x) = a \ln \left(\phi y^\theta \right).$$

The other solution is similar.

(d) THE PYTHAGOREAN THEOREM. The function

$$P(x,y) = \sqrt{x^2 + y^2} \qquad (x, y \in \mathbb{R}_{++}), \tag{21}$$

representing the length of the hypotenuse of a right triangle in terms of the lengths of its sides, is a permutable code. We have indeed

$$P(P(x,y),z) = \sqrt{P(x,y)^2 + z^2} = \sqrt{x^2 + y^2 + z^2} = P(P(x,z),y).$$

The other conditions are clearly satisfied, and so is Condition [S1]. Condition [S2] would be achieved by taking an appropriate restriction of the function P as in the case of Examples 1(a) and (b). Notice that the code P is a symmetric function: we have $P(x,y) = P(y,x)$ for all $x, y \in \mathbb{R}_{++}$.
We must solve the equation

$$\sqrt{x^2 + y^2} = f^{-1}\left(f(x) + f(y)\right)$$

or, equivalently,

$$f\left(\sqrt{x^2 + y^2}\right) = f(x) + f(y). \tag{22}$$

With $z = x^2$, $w = y^2$, and defining the function $h(z) = f\left(z^{\frac{1}{2}}\right)$, Equation (22) becomes

$$h(z+w) = h(z) + h(w),$$

a Cauchy equation on the positive reals, with h strictly increasing. It has the unique solution $h(z) = \xi z$, for some positive real number ξ (cf. Aczél 1966, p.31). So, we get $f(x) = \xi x^2$ and

$$f^{-1}(f(x) + f(y)) = \left(\frac{1}{\xi}\left(\xi x^2 + \xi y^2\right)\right)^{\frac{1}{2}} = \sqrt{x^2 + y^2}.$$

(e) THE COUNTEREXAMPLE: VAN DER WAALS EQUATION. One form of this equation is

$$T(p,v) = K\left(p + \frac{a}{v^2}\right)(v - b), \tag{23}$$

in which p is the pressure of a fluid, v is the volume of the container, T is the temperature, and a, b and K are constants; K is the reciprocal of the Boltzmann constant. It can be shown that the function T in (23) is not permutable. (The assumption that T is a permutable function leads to a contradiction.)

Meaningful Collection of Codes

We want to axiomatize a particular type of invariance that must hold for all (quantitative) scientific or geometric laws. The consequence of this axiomatization should be that the form of an expression representing a law should not be altered by changing the units of the variables. As a first step, we consider here the special case of laws which are functions in two real, ratio scalable variables, and moreover, the unit of the output variable—the value of the function—is specified by the units of the two input variables.

Our four examples (a), (b), (c) and (d) satisfy that requirement[3]. Dealing with this particular case may serve as a building block for the general situation. The definition given below is in the spirit of that used by Falmagne (2004) (see also Falmagne and Narens 1983; Narens 2002).

We begin by considering the case of our Example 1(a) involving the Lorentz-Fitz-Gerald Contraction, which we expressed by the equation

$$L(\ell, v) = \ell \sqrt{1 - \left(\frac{v}{c}\right)^2}. \tag{24}$$

The trouble with this notation is its ambiguity: the units of ℓ, which denotes the length of the rod, and of v, for the speed of the observer, are not specified. Writing $L(70, 3)$ has no empirical meaning if one does not specify, for example, that the pair $(70, 3)$ refers to 70 meters and 3 kilometers per second, respectively. Such a parenthetical reference is standard in a scientific context, but is not serviceable for our purpose, which is to express, formally, an invariance with respect to any change in the units[4].

To rectify the ambiguity, we propose to interpret $L(\ell, v)$ as a shorthand notation for $L_{1,1}(\ell, v)$, in which ℓ and L on the one hand, and v on the other hand, are measured in terms of two particular initial or 'anchor' units fixed arbitrarily. Such units could be m (meter) and km/sec, if one wishes. The $1, 1$ indices of $L_{1,1}$ signify these initial units. Describing the phenomenon in terms of other units amounts to multiply ℓ and v in any pair (ℓ, v) by some positive constants α and β, respectively. At the same time, L also gets to be multiplied by α, and the speed of light c by β. Doing so defines a new function $L_{\alpha,\beta}$, which is different from $L = L_{1,1}$ if either $\alpha \neq 1$ or $\beta \neq 1$ (or both), but carries the same information from an empirical standpoint. For example, if our new units are km and m/sec, then the two expressions

$$L_{10^{-3}, 10^3}(.07, 3000) \quad \text{and} \quad L(70, 3) = L_{1,1}(70, 3),$$

while numerically not equal, should describe the same empirical situation. This points to the appropriate definition of $L_{\alpha,\beta}$. We should write:

$$L_{\alpha,\beta}(\ell, v) - \ell \sqrt{1 - \left(\frac{v}{\beta c}\right)^2}. \tag{25}$$

The connection between L and $L_{\alpha,\beta}$ is thus

$$\frac{1}{\alpha} L_{\alpha,\beta}(\alpha\ell, \beta v) = \left(\frac{1}{\alpha}\right) \alpha\ell \sqrt{1 - \left(\frac{\beta v}{\beta c}\right)^2} \tag{26}$$

$$= \ell \sqrt{1 - \left(\frac{v}{c}\right)^2} = L(\ell, v).$$

This implies, for any α, β, ν and μ in $\mathbb{R}_{++}$,

$$\frac{1}{\alpha} L_{\alpha,\beta}(\alpha\ell, \beta v) = \frac{1}{\nu} L_{\nu,\mu}(\nu\ell, \mu v), \tag{27}$$

which is the invariance equation we were looking for, in this case, and which is generalized as Equation (31) in the next definition. Note that the second variable of the function $L_{\alpha,\beta}$ now ranges in the interval $[0, \beta c[$ instead of $]0, c[$. The range of the first variable of L is the non negative reals and so did not change.

---

[3]While the Counterexample 1(e) does not.

[4]A relevant point is made by Suppes (2002, see "Why the Fundamental Equations of Physical Theories Are not Invariant", p. 120).

Consider now the case of the Pythagorean Theorem, represented by the equation

$$P(x,y) = \left(x^2 + y^2\right)^{\frac{1}{2}}. \tag{28}$$

In the manner of our previous example, we regard Equation (28) as a representation of the Pythagorean Theorem in initial units; so P is an abbreviation for $P_{1,1}$. Since the units of x and y are the same, adopting another unit results in transforming P into $P_\alpha = P_{\alpha,\alpha}$, which is defined by

$$P_\alpha(x,y) = \alpha \left(\left(\frac{x}{\alpha}\right)^2 + \left(\frac{y}{\alpha}\right)^2\right)^{\frac{1}{2}} = \alpha P\left(\frac{x}{\alpha}, \frac{y}{\alpha}\right).$$

We obtain

$$\frac{1}{\alpha}P_\alpha(\alpha x, \alpha y) = \left(\left(\frac{\alpha x}{\alpha}\right)^2 + \left(\frac{\alpha y}{\alpha}\right)^2\right)^{\frac{1}{2}} = \left(x^2 + y^2\right)^{\frac{1}{2}} = P(x,y),$$

leading to the meaningfulness equation

$$\frac{1}{\alpha}P_\alpha\left(\alpha x, \alpha y\right) = \frac{1}{\beta}P_\beta\left(\beta x, \beta y\right), \tag{29}$$

which is similar to Equation (27).

It is clear from our discussion of these examples and from Equations (27) and (29) that the definition of 'meaningfulness' must apply to a collection of codes, each of which corresponds to a different choice of units, that is, the choice of (α, β) and (ν, μ) in the case of Equation (27), and of α and β in Equation (29).

Definition 3. *Let $[a_1, a_2[$ and $[b_1, b_2[$ be two half open nonnegative intervals, with a_2 or b_2 possibly equal to ∞, and let*

$$\mathcal{M} = \{M_{\alpha,\beta} \mid (\alpha, \beta) \in \mathbb{R}^2_{++}\} \tag{30}$$

be a collection of codes, with

$$M_{\alpha,\beta} : [\alpha a_1, \alpha a_2[\times [\beta b_1, \beta b_2[\longrightarrow [\alpha a_1, \alpha a_2[.$$

Each of the pairs (α, β) in (30) represents a change of the unit of one or both of the measurement scale, with the unit of $M_{\alpha,\beta}$ equal to $\alpha^{\delta_1}\beta^{\delta_2}$ for some rational numbers δ_1 and δ_2.

The collection of codes $\mathcal{M}$ is (δ_1, δ_2)-meaningful, or meaningful for short, if for any ordered pair $(x,y) \in [a_1, a_2[\times[b_1, b_2[$ and any two ordered pairs $(\alpha, \beta), (\mu, \nu) \in \mathbb{R}^2_{++}$, the following equality holds:

$$\frac{1}{\alpha^{\delta_1}\beta^{\delta_2}}M_{\alpha,\beta}(\alpha x, \beta y) = \frac{1}{\mu^{\delta_1}\nu^{\delta_2}}M_{\mu,\nu}(\mu x, \nu y). \tag{31}$$

So, in particular, we have

$$\frac{1}{\alpha^{\delta_1}\beta^{\delta_2}}M_{\alpha,\beta}(x,y) = M_{1,1}\left(\frac{x}{\alpha}, \frac{y}{\beta}\right) = M\left(\frac{x}{\alpha}, \frac{y}{\beta}\right), \tag{32}$$

with the unit of $M_{\alpha,\beta}$ being the product $\alpha^{\delta_1}\beta^{\delta_2}$.

Let us exercise this definition in the case of some of our four examples. We will see that in one case—the Pythagorean Theorem—the exponents δ_1 and δ_2 in are not integers.

2 Examples

(a) THE LORENTZ-FITZGERALD CONTRACTION. We have a collection $\mathcal{L} = \{L_{\alpha,\beta} \mid (\alpha,\beta) \in \mathbb{R}^2_{++}\}$ of codes. The collection $\mathcal{L}$ is $(1,0)$-meaningful. This implies that

$$L_{\alpha,\beta}(\ell,v) = \alpha\frac{\ell}{\alpha}\sqrt{1 - \left(\frac{v}{\beta c}\right)^2} = \ell\sqrt{1 - \left(\frac{v}{\beta c}\right)^2} = \alpha L\left(\frac{\ell}{\alpha},\frac{v}{\beta}\right). \tag{33}$$

The unit of $L_{\alpha,\beta}$ is $\alpha^1\delta^0 = \alpha$.

(b) BEER'S LAW. The form of this law is similar to the preceding one. We have a collection $\mathcal{I} = \{I_{\alpha,\beta} \mid (\alpha,\beta) \in \mathbb{R}^2_{++}\}$ of codes, which is also $(1,0)$-meaningful. This gives

$$I_{\alpha,\beta}(x,y) = \alpha\frac{x}{\alpha}e^{-\frac{y}{\beta c}} = \alpha I\left(\frac{\ell}{\alpha},\frac{v}{\beta}\right),$$

with the unit of $I_{\alpha,\beta}$ being also $\alpha^1\delta^0 = \alpha$.

(c) THE VOLUME OF A CYLINDER. This example is quite different. We consider the collection of codes $\mathcal{V} = \{V_{\alpha,\alpha} \mid \alpha \in \mathbb{R}^2_{++}\}$, in which the two variables are measured on the same scale. For the initial code V the formula for computing the volume $V(r,h)$ of a cylinder of radius r and height h, that is

$$V(r,h) = \pi h\,r^2.$$

The collection $\mathcal{V}$ must be $(1,2)$-meaningful. We get in general

$$V_\alpha(r,h) = V_{\alpha,\alpha}(r,h) = \alpha\alpha^2 V\left(\frac{r}{\alpha},\frac{h}{\alpha}\right) = \alpha\alpha^2\pi\frac{h}{\alpha}\left(\frac{r}{\alpha}\right)^2 = \pi h\,r^2,$$

with the unit of $V_{\alpha,\beta}$ being $\alpha^1\alpha^2 = \alpha^3$.

(d) THE PYTHAGOREAN THEOREM. Here, we have only one measurement scale, which is the same for the two input variables and for the output variable. We require the collection of codes $\mathcal{P} = \{P_{\alpha,\alpha} \mid \alpha \in \mathbb{R}_{++}\}$ to be $(\frac{1}{2},\frac{1}{2})$-meaningful. We obtain

$$P_\alpha(x,y) = P_{\alpha,\alpha}(x,y) = a^{\frac{1}{2}}a^{\frac{1}{2}}V\left(\frac{x}{\alpha},\frac{y}{\alpha}\right) = a^{\frac{1}{2}}a^{\frac{1}{2}}\sqrt{\left(\frac{x}{\alpha}\right)^2 + \left(\frac{y}{\alpha}\right)^2} = \sqrt{x^2 + y^2}. \tag{34}$$

The unit of $P_{\alpha,\alpha}$ is $\alpha^{\frac{1}{2}}\alpha^{\frac{1}{2}} = \alpha$.

Two Lemmas

We recall a result of Hosszú (1962a,b,c) (cf. also Aczél 1966), which will be instrumental in our proof of Theorem 1.

Lemma 2. (i) *A solvable code $M : J \times J' \to H$ is quasi-permutable if and only if there exists three continuous functions $m : \{f(y) + g(r) \mid x \in J, r \in J'\} \to H$, $f : J \to \mathbb{R}$, and $g : J' \to \mathbb{R}$, with m and f strictly increasing and g strictly monotonic, such that*

$$M(y,r) = m(f(y) + g(r)). \tag{35}$$

(ii) *A solvable code $G : J \times J' \to J$ is a permutable code if and only if, with f and g as above, we have*

$$G(y,r) = f^{-1}(f(y) + g(r)). \tag{36}$$

(iii) *If a solvable code $G : J \times J \to J$ is a symmetric function—that is, $G(x,y) = G(y,x)$ for all $x,y \in J$—then G is permutable if and only if there exists a strictly increasing and continuous function $f : J \to J$ satisfying*

$$G(x,y) = f^{-1}(f(x) + f(y)). \tag{37}$$

The meaningfulness condition introduced in Definition 3 and Equation (31) is a powerful one. In particular, it enables some properties of any of the codes in $\mathcal{M}$ to extend to all the others codes in that collection. The next lemma illustrates this point.

Lemma 3. *Let $\mathcal{F}$ be a (δ_1, δ_2)-meaningful collection of codes; so, all the codes in $\mathcal{F}$ are functions of two variables. Suppose that some code $F_{\alpha,\beta}$ in the collection $\mathcal{F}$ satisfies any of the following five properties:*

(i) $F_{\alpha,\beta}$ is solvable;
(ii) $F_{\alpha,\beta}$ is differentiable in both variables;
(iii) $F_{\alpha,\beta}$ is quasi-permutable;
(iv) $F_{\alpha,\beta}$ is a symmetric function, with $\alpha = \beta$ and $F_{\alpha,\beta} = F_\alpha$;
(v) $F_\alpha = F_{\alpha,\beta}$ is a symmetric, homogeneous function, that is,

$$\gamma F_\alpha(x, r) = F_\alpha(\gamma x, \gamma r)$$

for any $\gamma > 0$.

Then all the codes in $\mathcal{F}$ satisfy the same property. Moreover, Condition (v) implies that $F_\alpha(x, r) = F(x, r)$ for all $\alpha > 0$ and $x, r \geq 0$.

If the initial code F is solvable and permutable, so that by Lemma 2(ii) $F(x, y) = f^{-1}(f(x) + g(y))$ holds for some strictly increasing function f and some strictly monotonic function g, then for any code $F_{\mu,\eta}$ in the collection $\mathcal{F}$, we have

$$F_{\mu,\eta}(x, r) = \mu^{\delta_1} \nu^{\delta_2} f^{-1}\left(f\left(\frac{x}{\mu}\right) + g\left(\frac{r}{\eta}\right) \right). \tag{38}$$

Proof. Without loss of generality, we suppose that $\alpha = \beta = 1$, with $F = F_{1,1}$. As the family $\mathcal{F}$ is (δ_1, δ_2)-meaningful, we have, for all positive real numbers μ and ν and writing $\eta = \mu^{\delta_1} \nu^{\delta_2}$ for simplicity:

$$F_{\mu,\nu}(x, r) = \eta F\left(\frac{x}{\mu}, \frac{r}{\nu}\right) \qquad (x \in [\mu a, \mu a'[\,; r \in [\nu b, \nu b'[\,). \tag{39}$$

(i) Suppose that the code F is solvable. If $F_{\mu,\nu}(x, r) < p$, for some code $F_{\mu,\nu}$ in $\mathcal{F}$, then $F(\frac{x}{\mu}, \frac{r}{\nu}) < \frac{p}{\eta}$ follows from (39). Because the code F satisfies [S1], there must be some $w \in [b, b'[$ such that $F(\frac{x}{\mu}, w) = \frac{p}{\eta}$. Defining $t = \nu w$, we get

$$F_{\mu,\nu}(x, t) = \eta F\left(\frac{x}{\mu}, \frac{t}{\nu}\right) = p.$$

Thus, the code $F_{\mu,\nu}$ also satisfies [S1]. Since F satisfies [S2], there exists some x_0 in $[a, a'[$ such that F is x_0-solvable. Define $y_0 = \mu x_0 \in [\mu a, \mu a'[$ and take any q in the range of the function $F_{\mu,\nu}$. This implies that $\frac{q}{\eta}$ is in the range of F, and by [S2] applied to F, there is some w such that $F(x_0, w) = \frac{q}{\eta}$ or, equivalently with $v = \beta w$,

$$q = \eta F\left(\frac{y_0}{\mu}, \frac{v}{\nu}\right) = F_{\mu,\nu}(y_0, v),$$

by the meaningfulness of the family $\mathcal{F}$. Thus, $F_{\mu,\nu}$ is y_0-solvable.

(ii) The differentiability of $F_{\mu,\nu}$ results from that of F in view of (39).

(iii) Suppose now that F is quasi-permutable. (We do not assume that $\mathcal{F}$ is a self-transforming family.) Thus, there exists a code $G : [a, a'[\times [b, b'[\to [a, a'[$ co-monotonic

MEANINGFUL PERMUTABLE LAWS / 109

with F such that

$$F(G(x,s),t) = F(G(x,t),s) \qquad (x,y \in [a,a'[\,;\, s,t \in [b,b'[\,). \qquad (40)$$

For any pair $\mu, \nu \in \mathbb{R}_{++}$, define the function $G_{\mu,\nu} : [\mu a, \mu a'[\times [\nu b, \nu b'[\to [\mu a, \mu a'[$ by the equation

$$G_{\mu,\nu}(x,r) = \mu G\left(\frac{x}{\mu}, \frac{r}{\nu}\right). \qquad (41)$$

Note that (41) implies that $G_{\mu,\nu}$ is comonotonic with $F_{\mu,\nu}$. Indeed we have:

$$G_{\mu,\nu}(x,r) \leq G_{\mu,\nu}(y,s) \iff \mu G\left(\frac{x}{\mu}, \frac{r}{\nu}\right) \leq \mu G\left(\frac{y}{\mu}, \frac{s}{\nu}\right)$$

$$\iff \eta F\left(\frac{x}{\mu}, \frac{r}{\nu}\right) \leq \eta F\left(\frac{y}{\mu}, \frac{s}{\nu}\right) \quad (F \text{ and } G \text{ comonotonic})$$

$$\iff F_{\mu,\nu}(x,r) \leq F_{\mu,\nu}(y,s) \qquad (\text{by meaningfulness}).$$

Now successively

$$F_{\mu,\nu}(G_{\mu,\nu}(x,r),v) = \eta F\left(\frac{1}{\mu}G_{\mu,\nu}(x,r), \frac{v}{\nu}\right) \quad (\text{by meaningfulness})$$

$$= \eta F\left(G\left(\frac{x}{\mu}, \frac{r}{\nu}\right), \frac{v}{\nu}\right) \quad (\text{by the definition of } G_{\mu,\nu})$$

$$= \eta F\left(G\left(\frac{x}{\mu}, \frac{v}{\nu}\right), \frac{r}{\nu}\right) \quad (\text{by the quasi-permutability of } F)$$

$$= F_{\mu,\nu}(G_{\mu,\nu}(x,v),r) \quad (\text{by symmetry}).$$

Consequently, any code $F_{\mu,\nu}$ is $G_{\mu,\nu}$-permutable.

(iv) This follows from the definition of the (δ_1, δ_2)-meaningfulness of the collection.

(v) This follows from

$$F_\alpha(x,r) = \alpha F\left(\frac{x}{\alpha}, \frac{y}{\alpha}\right) \qquad (\text{by meaningfulness}) \qquad (42)$$

$$= F(x,r) \qquad (\text{by the homogeneity of } F) \qquad (43)$$

yielding

$$F_\alpha(x,r) = F(x,r), \qquad (44)$$

and so F_α is homogeneous.

Equation (38) also results from (δ_1, δ_2)-meaningfulness, together with Lemma 2(ii). $\square$

An Exemplary Result

The main point of this paper is that, in the context of meaningfulness and under reasonable side conditions, permutability, which can be assessed by a thought experiment without any collection of data, may yield the possible forms of scientific or geometric laws. The theorem below is an example.

Quasi-Permutability: Representation Theorem

Theorem 1. Let $\mathcal{F} = \{F_{\alpha,\beta} \mid \alpha, \beta \in \mathbb{R}_{++}\}$ be a meaningful collection of 2-codes, with

$$F_{\alpha,\beta} : \mathbb{R}_{++} \times \mathbb{R}_{++} \xrightarrow{onto} \mathbb{R}_{++} \quad for \ all \quad \alpha, \beta \in \mathbb{R}_{++}.$$

We suppose that the collection is self-transforming, that is, for each of these codes, the measurement unit of the output of the code is the same as the measurement unit of its first variable. Moreover, suppose that one of these codes, say the code $F_{\alpha,\beta}$, is solvable, strictly increasing in both variables, and permutable with respect to the initial code $F = F_{1,1}$. Then the initial code F must have one of the three forms listed as Cases **A**, **B**, *and* **C** *below.*

Either there is a code in $\mathcal{F}$ that is not a symmetric function. Then, for some constant $\theta > 0$, we have:

Case A.
$$F(y,r) = \phi y\, r^\theta\,, \tag{45}$$

$$F_{\alpha,\beta}(y,r) = \phi y \left(\frac{r}{\beta}\right)^\theta\,, \qquad \text{for all } F_{\alpha,\beta} \in \mathcal{F}. \tag{46}$$

Or there is a code in $\mathcal{F}$ that is a symmetric function; so $F_{\alpha,\beta} = F_{\alpha,\alpha} = F_\alpha$ for all codes in $\mathcal{F}$. Then, for some constants $\theta > 0$ and η, we have one of the two possible cases:

Case B.
$$F(y,r) = \theta y r\,; \tag{47}$$

$$F_\alpha(y,r) = \frac{\theta}{\alpha} y r \qquad \text{for all } F_\alpha \in \mathcal{F}. \tag{48}$$

Moreover, if some code in $\mathcal{F}$ is homogeneous, then
$$F_\alpha(y,r) = F(y,r) = \theta y r \qquad \text{for all } F_\alpha \in \mathcal{F}. \tag{49}$$

Case C.
$$F(y,r) = \left(y^\theta + r^\theta + \eta\right)^{\frac{1}{\theta}}\,; \tag{50}$$

$$F_\alpha(y,r) = \left(y^\theta + r^\theta + \alpha^\theta \eta\right)^{\frac{1}{\theta}} \qquad \text{for all } F_\alpha \in \mathcal{F}. \tag{51}$$

Moreover, if some code in $\mathcal{F}$ is homogeneous, then
$$F_\alpha(y,r) = F(y,r) = \left(y^\theta + r^\theta\right)^{\frac{1}{\theta}} \qquad \text{for all } F_\alpha \in \mathcal{F}. \tag{52}$$

One application of Equation (52) is the Pythagorean Theorem. In fact, we show later in this paper that the function $P : (x,y) \mapsto P(x,y)$ measuring the hypotenuse of a right triangle with side length x and y is a quasi-permutable function satisfying the other of Case C of Theorem 1, including homogeneity. So we get $P(x,y) = (x^\theta + y^\theta)^{\frac{1}{\theta}}$, giving us another proof of the Pythagorean Theorem (up to the exponent).

Proof. Case A. By Lemma 3, all the codes in $\mathcal{F}$ are solvable, non symmetric, permutable with respect to the initial code F, and strictly increasing in both variables. Moreover, Lemma 1 implies that F is permutable. Using Lemma 2(ii) and the fact that $\mathcal{F}$ is a meaningful ST-collection, we get for all $F_{\alpha,\beta}$ in $\mathcal{F}$, :

$$F_{\alpha,\beta}(y,r) = \alpha\, F\left(\frac{y}{\alpha}, \frac{r}{\beta}\right) = \alpha f^{-1}\left(f\left(\frac{y}{\alpha}\right) + g\left(\frac{r}{\beta}\right)\right), \tag{53}$$

for some continuous, strictly increasing functions f and g, with in particular

$$F(y,r) = f^{-1}(f(y) + g(r))\,. \tag{54}$$

We get successively

$$F_{\alpha,\beta}(F(y,r),s)$$

$$= F_{\alpha,\beta}(f^{-1}(f(y)+g(r)),s) \qquad \text{(by Lemma 2(ii))} \qquad (55)$$

$$= \alpha F\left(\frac{1}{\alpha}f^{-1}(f(y)+g(r)),\frac{s}{\beta}\right) \qquad \text{(by meaningfulness)} \qquad (56)$$

$$= \alpha f^{-1}\left(f\left(\frac{1}{\alpha}f^{-1}\left(f(y)+g(r)\right)\right)+g\left(\frac{s}{\beta}\right)\right) \qquad \text{(by Lemma 2(ii))} \qquad (57)$$

$$= \alpha f^{-1}\left(f\left(\frac{1}{\alpha}f^{-1}\left(f(y)+g(s)\right)\right)+g\left(\frac{r}{\beta}\right)\right) \qquad \text{(by quasi-permutability).} \qquad (58)$$

Equating the last two right hand sides, canceling the α's, and applying the function f on both sides, we get

$$f\left(\frac{1}{\alpha}f^{-1}\left(f(y)+g(r)\right)\right)+g\left(\frac{s}{\beta}\right) = f\left(\frac{1}{\alpha}f^{-1}\left(f(y)+g(s)\right)\right)+g\left(\frac{r}{\beta}\right). \qquad (59)$$

Setting $s = f(y)$, $t = g(r)$, fixing $s = 1$, and temporarily assuming that $\frac{1}{\alpha} = \frac{1}{\beta} = \nu$, Equation (59) becomes

$$f\left(\nu f^{-1}(s+t)\right) + g(\nu) = f\left(\nu f^{-1}(s+g(1))\right) + g\left(\nu g^{-1}(t)\right). \qquad (60)$$

Defining the functions

$$h_\nu = f \circ \nu f^{-1}, \qquad m_\nu : s \mapsto f\left(\nu f^{-1}(s+g(1))\right), \qquad k_\nu : t \mapsto g\left(\nu g^{-1}(t)\right) - g(\nu),$$

Equation (60) becomes

$$h_\nu(s+t) = m_\nu(s) + k_\nu(t),$$

a Pexider Equation. In view of the conditions on the functions, the solution is

$$h_\nu(s) = p(\nu)s + q(\nu) + w(\nu) \qquad (61)$$

$$m_\nu(s) = p(\nu)s + q(\nu)$$

$$k_\nu(t) = p(\nu)t + w(\nu), \qquad (62)$$

for some constants $p(\nu)$, $q(\nu)$ and $w(\nu)$ possibly varying with ν. Rewriting (61) and (62) in terms of the functions f and g, we obtain, with $v(\nu) = q(\nu) + w(\nu)$,

$$h_\nu(s) = (f \circ \nu f^{-1})(s) = p(\nu)s + v(\nu),$$

$$k_\nu(t) = g\left(\nu g^{-1}(t)\right) - g(\nu) = p(\nu)t + w(\nu) \qquad (63)$$

yielding, with $q(\nu) = w(\nu) + g(\nu)$

$$f(\nu y) = p(\nu)f(y) + v(\nu), \qquad (64)$$

$$g(\nu r) = p(\nu)g(r) + q(\nu). \qquad (65)$$

These are standard functional equations (cf. Aczél 1966, pp. 148–150). In principle, for each of Equations (64) and (65), we have two solutions for the functions f and g depending on whether or not $p(\nu)$ is a constant function. But only the case below is consistent with the hypotheses.

SUPPOSE THAT p IS A CONSTANT FUNCTION. We have then, for some constants $b > 0$, a, $d \neq 0$, and c,

$$f(y) = b \ln y + a, \quad \text{and so} \quad f^{-1}(z) = e^{\frac{z-a}{b}} \qquad (66)$$

$$g(r) = d \ln r + c. \qquad (67)$$

Rewriting F in terms of the solutions (66) and (67) for the functions f and g yields

$$F(y,r) = f^{-1}(f(y) + g(r)) = e^{\frac{f(y)+g(r)-a}{b}}$$

$$= e^{\frac{b\ln y + a + g(r) - a}{b}} = y\, e^{\frac{d\ln r + c}{b}} = e^{\frac{c}{b}} y\, r^{\frac{d}{b}},$$

(68)

and with $\phi = e^{\frac{c}{b}}$ and $\theta = \frac{d}{b}$,

$$F(y,r) = \phi\, y\, r^\theta .$$

(69)

With

$$F_{\alpha,\beta}(y,r) = \alpha F\left(\frac{y}{\alpha}, \frac{r}{\beta}\right),$$

we get

$$F_{\alpha,\beta}(y,r) = \phi y \left(\frac{r}{\beta}\right)^\theta .$$

(70)

It is easily verified that (69) and (70) imply quasi-permutability:

$$F_{\alpha,\beta}(F(y,r),t) = \alpha\phi\left(\frac{F(y,r)}{\alpha}\right)\left(\frac{t}{\beta}\right)^\theta = \phi y r^\theta \left(\frac{t}{\beta}\right)^\theta = \phi\left(\frac{1}{\beta}\right)^\theta r^\theta t^\theta$$

$$= F_{\alpha,\beta}(F(y,t),r).$$

OBSERVATION. If p takes at least two distinct values, then the form obtained for the functions $F_{\alpha,\beta}$ results in a collection $\mathcal{F}$ that is not quasi-permutable. The argument goes as follows. From (64) and (65), we get:

$$f(y) = by^\lambda + a \qquad \text{for some constants } b > 0 \text{ and } a \qquad (71)$$

$$g(r) = dr^\lambda + c \qquad \text{for some constants } d \neq 0 \text{ and } c. \qquad (72)$$

From (71), we obtain

$$f^{-1}(t) = \left(\frac{t-a}{b}\right)^{\frac{1}{\lambda}} .$$

(73)

Computing $F(y,r)$, we obtain from (71), (73) and (72), successively

$$F(y,r) = f^{-1}(f(y) + g(r)) = \left(\frac{f(y) + g(r) - a}{b}\right)^{\frac{1}{\lambda}}$$

$$= \left(\frac{by^\lambda + a + dr^\lambda + c - a}{b}\right)^{\frac{1}{\lambda}} = \left(y^\lambda + \frac{d}{b}r^\lambda + \frac{c}{b}\right)^{\frac{1}{\lambda}}$$

and with $\theta = \lambda$, $\phi = \frac{d}{b}$ and $\mu = \frac{c}{b}$, finally

$$F(y,r) = \left(y^\theta + \phi r^\theta + \mu\right)^{\frac{1}{\theta}} .$$

This gives for any $\alpha > 0$ and $\beta > 0$

$$F_{\alpha,\beta}(y,r) = \alpha\left(\left(\frac{y}{\alpha}\right)^\theta + \phi\left(\frac{r}{\beta}\right)^\theta + \mu\right)^{\frac{1}{\theta}}$$

$$= \left(y^\theta + \phi\left(\frac{\alpha r}{\beta}\right)^\theta + \alpha^\theta \mu\right)^{\frac{1}{\theta}} .$$

Since we have:

$$F_{\alpha,\beta}(F(y,r),t) = \left(y^\theta + \phi r^\theta + \mu + \phi \left(\frac{\alpha t}{\beta} \right)^\theta + \alpha^\theta \mu \right)^{\frac{1}{\theta}},$$

assuming that $F_{\alpha,\beta}(F(y,r),t) = F_{\alpha,\beta}(F(y,t),r)$ leads to $r^\theta = t^\theta$ after simplification.

This completes our proof of Case A.

CASES B AND C. Assume that there is some code $F_\alpha = F_{\alpha,\alpha}$ in $\mathcal{F}$ that is a symmetric function. By Lemma 3 (iii), all the codes in $\mathcal{F}$ are symmetric, and we have, for some strictly increasing and continuous function f.

$$F_\alpha(y,r) = \alpha F \left(\frac{y}{\alpha}, \frac{r}{\alpha} \right) \qquad \text{(by meaningfulness)} \qquad (74)$$

$$= \alpha f^{-1} \left(f \left(\frac{y}{\alpha} \right) + f \left(\frac{r}{\alpha} \right) \right), \qquad \text{(by Lemma 2(iii))} \qquad (75)$$

with in particular

$$F(y,r) = f^{-1}(f(y) + f(r)). \qquad (76)$$

Applying the same derivation as in asymmetric case, namely Equations (55)-(58), we end up with the equation

$$f \left(\frac{1}{\alpha} f^{-1} \left(f(y) + f(r) \right) \right) + f \left(\frac{z}{\alpha} \right) = f \left(\frac{1}{\alpha} f^{-1} \left(f(y) + f(z) \right) \right) + f \left(\frac{r}{\alpha} \right), \qquad (77)$$

replacing Equation (59). We then proceed as in our proof of Case A. Setting $s = f(y)$, $t = f(r)$, $\nu = \frac{1}{\alpha}$, Equation (77) becomes

$$f \left(\nu f^{-1} (s+t) \right) + f(\nu) = f \left(\nu f^{-1} (s + f(1)) \right) + f \left(\nu f^{-1}(t) \right). \qquad (78)$$

Defining the functions

$$h_\nu = f \circ \nu f^{-1},$$
$$k_\nu : s \mapsto h_\nu(s + f(1)) - f(\nu),$$

Equation (78) becomes

$$h_\nu(s+t) = k_\nu(s) + h_\nu(t),$$

a Pexider equation. Because the functions h_ν and κ_ν are defined on a real interval of positive length and are strictly monotonic, its solution is

$$h_\nu(s) = w(\nu)s + v(\nu) \qquad (79)$$

$$k_\nu(s) = w(\nu)s, \qquad (80)$$

for some constants $w(\nu)$ and $v(\nu)$ which may, however, depends on ν. Rewriting now (79) in terms of the function f, we get

$$f(\nu y) = w(\nu)f(y) + v(\nu), \qquad (81)$$

the functional equation encountered in Case A. Here, however, we have two possible solutions for the function f, given in Case B and C below.

CASE B. With w a constant function in (81), we get the solution for f as:

$$f(y) = \phi \ln y + \psi \qquad (\phi > 0). \qquad (82)$$

Replacing f in the representation equation (76) by its form in (82) leads for the code F to the equation

$$F(y, r) = \theta y r, \tag{83}$$

with $\theta = e^{\frac{\psi}{\phi}}$. With $F_\alpha(y, x) = \alpha F\left(\frac{y}{\alpha}, \frac{r}{\alpha}\right)$, we get

$$F_\alpha(y, x) = \frac{\theta}{\alpha} y r. \tag{84}$$

If we assume that some code in $\mathcal{F}$ is homogeneous, then by Lemma 2(v), all the codes are homogeneous. This implies

$$
\begin{aligned}
F_\alpha(y, r) &= \alpha F\left(\frac{y}{\alpha}, \frac{r}{\alpha}\right) && \text{(by meaningfulness)}\\
&= F(y, r) && \text{(by homogeneity)}\\
&= \theta y r,
\end{aligned}
$$

by Equation (83).

CASE C. With w not constant in (81), we get the solution for f:

$$f(y) = \phi y^\theta + \psi, \tag{85}$$

with $\phi\theta > 0$. Replacing f in (76) by its form in (85) and setting $\eta = \frac{\psi}{\phi}$ leads to

$$F(y, r) = \left(y^\theta + x^\theta + \eta\right)^{\frac{1}{\theta}}, \tag{86}$$

with $\theta > 0$. This implies

$$F_\alpha(y, r) = \alpha F\left(\frac{y}{\alpha}, \frac{r}{\alpha}\right) = \left(y^\theta + r^\theta + \alpha^\theta \eta\right)^{\frac{1}{\theta}}. \tag{87}$$

Now if some code of $\mathcal{F}$ is homogeneous, all the codes are, and we obtain

$$
\begin{aligned}
F_\alpha(y, r) &= \alpha F\left(\frac{y}{\alpha}, \frac{r}{\alpha}\right) && \text{(by meaningfulness)}\\
&= \alpha\left(\left(\frac{y}{\alpha}\right)^\theta + \left(\frac{r}{\alpha}\right)^\theta + \eta\right)^{\frac{1}{\theta}} && \text{(by Equation (86)}\\
&= \left(y^\theta + r^\theta + \alpha^\theta \eta\right)^{\frac{1}{\theta}} = F(y, r) && \text{(by homogeneity),}
\end{aligned}
$$

which implies that $\eta = 0$, and so

$$F_\alpha(y, r) = F(y, r) = \left(y^\theta + r^\theta\right)^{\frac{1}{\theta}}. \quad \square$$

Remark 1. *Note for further reference that the formula for the functions L and L_α of the Lorentz-FitzGerald Contraction, represented by our Equations (13) and (25), does not satisfy the quasi-permutability condition, as it is easily checked. A similar remark applies to Beer's Law.*

We argued that permutability could be assessed intuitively, without any experimentation. The case of quasi-permutability is not so clear. However, we show below, in the situation of the Pythagorean Theorem, the quasi-permutability can be deduced by a simple geometrical argument.

The Pythagorean Theorem

With $\theta = 2$, Equation (52) of Case C of Theorem 1 is the formula for the Pythagorean Theorem. In fact, this theorem provides us with still another proof of the Pythagorean theorem, to be added to the several hundreds that already exist.

We suppose that the length $P(x, y)$ of the hypotenuse of a right triangle with leg lengths $x > x_0$ and $y > x_0$ (for some $x_0 > 0$) is a symmetric solvable code[5]; thus $P : [x_0, \infty[\times [x_0, \infty[\to [x_0, \infty[$. We take the function P to be the initial code of a family of codes $\{P_\alpha\}$. We establish the permutability and the quasi-permutability of the code P_α with respect to P, for any $\alpha > 0$, by an elementary geometric argument.

The Permutability of P

A right triangle $\triangle ABC$ with leg lengths x and y and hypotenuse of length $P(x, y)$ is represented in Figure 1A. Thus $AB = x$, $BC = y$ and $P(x, y) = AC$. Another right triangle $\triangle ACD$ is defined by the segment $\overline{CD}$ of length z, which is perpendicular to the plane of $\triangle ABC$. The length of the hypotenuse $\overline{AD}$ of $\triangle ACD$ is thus $P(P(x, y), z) = AD$. Still another right triangle $\triangle EAB$ is defined by the perpendicular $\overline{AE}$ to the plane of $\triangle ABC$. We choose E such that $AE = z = CD$; we have thus $EB = P(x, z)$. Since $\overline{AE}$ is perpendicular to the plane of $\triangle ABC$ and $\triangle ABC$ is a right triangle, $\overline{EB}$ is perpendicular to $\overline{BC}$. The lines $\overline{BC}$ and $\overline{BE}$ are perpendicular. (Indeed, the perpendicular L at the point B to the plane of triangle $\triangle ABC$ is coplanar with $\overline{AE}$. So, as $\overline{BC}$ is perpendicular to both $\overline{AE}$ and L, it must be perpendicular to the plane of $\triangle EAB$, and so it must be perpendicular to $\overline{EB}$.) Accordingly, $EC = P(P(x, z), y)$ is the length of the hypotenuse of the right triangle $\triangle EBC$. It is clear that, by construction, the four points A, C, D and E are coplanar. They define a rectangle whose diagonals $\overline{AD}$ and $\overline{EC}$ must be equal. So, we must have $P(P(x, y), z) = P(P(x, z), y)$, establishing the permutability of the code P.

The Quasi Permutability of P_α

For any positive real number α, the triangle $\triangle A'B'C'$ pictured in Figure 1B, with $C' = c$, A collinear with $A'C'$, B collinear with $B'C'$, and $A'B' = \frac{x}{\alpha}$, $B'C' = \frac{y}{\alpha}$ and $A'C' = \frac{P(x,y)}{\alpha}$, is similar to the triangle $\triangle ABC$ also represented in Figure 1B. So, P is homogeneous and we have

$$P\left(\frac{x}{\alpha}, \frac{y}{\alpha}\right) = \frac{P(x, y)}{\alpha}. \tag{88}$$

The function P is the initial code of the meaningful family of codes $\{P_\alpha\}$. For the code P_α in that family, we get

$$
\begin{aligned}
P_\alpha\left(P(x, y), z\right) &= \alpha P\left(\frac{P(x, y)}{\alpha}, \frac{z}{\alpha}\right) && \text{(by $\left(\tfrac{1}{2}, \tfrac{1}{2}\right)$-meaningfulness)} \\
&= \alpha P\left(P\left(\frac{x}{\alpha}, \frac{y}{\alpha}\right), \frac{z}{\alpha}\right) && \text{(by Equation (88))} \\
&= \alpha P\left(P\left(\frac{x}{\alpha}, \frac{z}{\alpha}\right), \frac{y}{\alpha}\right) && \text{(by the permutability of P)} \\
&= \alpha P\left(\frac{P(x, z)}{\alpha}, \frac{y}{\alpha}\right) && \text{(by Equation (88))} \\
&= P_\alpha\left(P(x, z), y\right) && \text{(by $\left(\tfrac{1}{2}, \tfrac{1}{2}\right)$-meaningfulness).}
\end{aligned}
$$

We conclude that any code P_α in the family $\{P_\alpha\}$ is quasi-permutable with respect to the initial code P.

---

[5]Cf. our discussion of Condition [S2] in the context of Example 1(e).

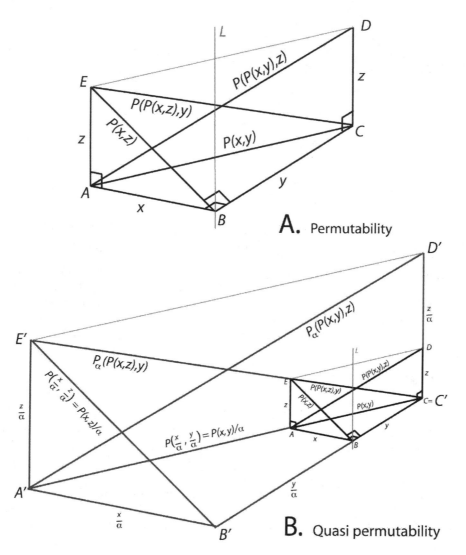

A. Permutability

B. Quasi permutability

FIGURE 1 The upper graph **A** illustrates the permutability condition formalized by the equation $P(P(x, y), z) = P(P(x, z), y)$. The lower graph **B** shows that the quasi-permutability condition formalized by the equation $P_\alpha(P(x, y), z) = P_\alpha(P(x, z), y)$ only involves a rescaling of all the variables pictured in Figure 1A, resulting in a similar figure, with the rectangle $A'B'C'D'$ similar to the rectangle $ABCD$. The measures of the two diagonals of the rectangle $A'B'C'D'$ are $P_\alpha(P(x, y), z)$ and $P_\alpha(P(x, z), y)$

Open problems

Theorem 1 captures only two of the four examples of permutable laws described in Subsection 1. This suggests the following problem.

Is there an interplay between meaningfulness and permutability and/or quasi-permutability, which would imply, possibly up to some parameters, the functional forms of the Lorentz-FitzGerald Contraction and of Beer's Law? This may require a reformulation of the meaningfulness condition taking into account the presence of the constant c in the formulas of both of these laws.

Many scientific laws are not permutable. There are several other equations in the functional equation literature which would be worth investigate in the context of meaningfulness, in the style of this paper.

References

Aczél, J. 1966. *Lectures on Functional Equations and Their Applications.* New York: Academic Press.

—. 1987. *A short course on functional equations based on recent applications to the social and behavioral sciences.* Dordrecht and Boston: Reidel/Kluwer.

—. 2005. Utility of extension of functional equations—when possible. *Journal of Mathematical Psychology* 49(6):445–449. doi:10.1016/j.jmp.2005.05.002.

Chudziak, J. and J. Tabor. 2008. Generalized pexider equation on a restricted domain. *Journal of Mathematical Psychology* 52(6):389–392. doi:10.1016/j.jmp.2008.04.002.

Falmagne, J.-C. 2004. Meaningfulness and order invariance: Two fundamental principles for scientific laws. *Foundations of Physics* 34(9):1341–1384. doi:10.1023/B:FOOP.0000044096.50863.8e.

Falmagne, J.-C. and L. Narens. 1983. Scales and meaningfulness of quantitative laws. *Synthese* 55(3):287–325.

Hosszú, M. 1962a. Algebrai rendszereken értelmezett függvényegyenletek, i. algebrai módszerek a függvényegyenletek elméletében. *Magyar Tud. Akad. Mat. Fiz. Oszt. K* 12:303–315.

—. 1962b. Néhány lineáris függvényegyenletröl. *Mat. Lapok* 13:202.

—. 1962c. Note on commutable mappings. *Publ. Math. Debrecen* 9:105–106.

Narens, L. 2002. *Theories of Meaningfulness.* Mahwah, N.J.: Lawrence Erlbaum Associates.

Radó, F. and J. Baker. 1987. Pexider's equation and aggregation of allocation. *Aequationes Mathematicae* 32(1):227–239. doi:10.1007/BF02311311.

Sedov, L. I. 1959. *Similarity and Dimensional Methods in Mechanics.* New York: Academic Press.

Suppes, P. 2002. *Representation and Invariance of Scientific Structures.* Stanford, CA: CSLI Publications.

Suppes, P. and J. L. Zinnes. 1963. Basic measurement theory. In R. D. Luce, R. R. Bush, and E. Galanter, eds., *Handbook of Mathematical Psychology*, vol. 1. New York and London: John Wiley and Sons.

11

Belief as Qualitative Probability

HANNES LEITGEB

I want to use this opportunity for thanking Patrick Suppes for his generous personal and intellectual support over the years. Happy birthday again, Pat! This paper was triggered by Suppes' work on qualitative probability as well as by a question by one of his colleagues, that is, Johan van Benthem: the question was "What does your joint theory of belief and degrees of belief have to do with qualitative probability?"

There are at least two concepts of belief in epistemology (and beyond): on the one hand, a qualitative or categorical notion of belief, in the sense that either I believe that A is the case, or I believe that $\neg A$ is the case, or I suspend judgement with respect to the whole alternative. And, on the other hand, a quantitative notion of belief, according to which I believe A to a certain numerical degree that measures the strength with which I believe A to be true: assigning a degree of 1 to A represents believing A with maximal possible strength, a degree of 0 means that I believe A with minimal possible strength, and a degree of $\frac{1}{2}$ represents an attitude that is right between the two extremes. These two notions of belief also come associated with two different standards or ideals of normativity: in the qualitative case, the beliefs of a perfectly rational agent are assumed to be consistent and closed under logical consequence; whereas in the quantitative case the degree-of-belief function of a perfectly rational agent is supposed to satisfy the axioms of (subjective) probability.

Although belief in the first qualitative sense has been one of the central notions of traditional epistemology—for it is belief in this sense that is entailed by *knowledge*— in many areas of philosophy in which the concept of belief plays a role these days, those theories that are spelled out in terms of degrees of belief seem to dominate; this is true especially of the more technically advanced corners of epistemology or "formal epistemology". It seems as if the qualitative notion of belief is being regarded as old-fashioned, as something which ought to be replaced by the "enlightened" quantitative notion of belief. In a nutshell: probabilistic theories of reasoning seem to outwit logic-based theories of reasoning. In the eyes of the radical Bayesian, such as e.g. Richard Jeffrey, it would even be justified to simply eliminate the qualitative notion of belief from philosophical and scientific discourse and instead proceed solely in terms of degrees of belief.

Foundations and Methods from Mathematics to Neuroscience.
Colleen E. Crangle, Adolfo García de la Sierra and Helen Longino.

In my view, this would be an unfortunate development. Without arguing for it here, I am convinced that any epistemology in which the concept of belief on the qualitative scale were to be excluded would be seriously incomplete. What we should do instead is to devise joint theories for qualitative and quantitative belief by which the interaction between belief and degrees of belief can be studied and applied, much as in measurement theory the structural relationships between qualitative ("empirical") systems and numerical systems are investigated and applied (cf. Suppes 2002).

The present paper is a précis of such a joint theory of belief and degrees of belief in which pairs

$$\langle P, Bel \rangle$$

of a degree of belief function P and a belief set Bel of one and the same perfectly rational agent (at one fixed point of time) are the central objects of investigation. The theory is *normative* in the sense of describing what an *(inferentially) perfectly rational* agent's beliefs are like on a quantitative and a qualitative scale; the beliefs of actual human agents may only approximate such perfectly rational agents' beliefs, and they ought to do so. Moreover, the theory is a *logical* theory in the sense that its postulates are insensitive to the actual content and subject matter of the propositions to which they are applied (over and above logical aspects of such contents and subject matters). The postulates of the theory are composed of: (i) the axioms of probability for the degree of belief function P, (ii) logical closure conditions for the belief set Bel, and (iii) a bridge principle in which P and Bel figure simultaneously and which will entail ultimately that belief corresponds to high enough probability (the so-called "Lockean thesis" on belief).

We will be able to prove a representation theorem that will give us insight into what such pairs $\langle P, Bel \rangle$ are like if, and only if, they satisfy all of the postulates. If (i)–(iii) are granted to hold for all perfectly rational agents, then the theorem will tell us what P and Bel have to be like when P is the degree of belief function of an arbitrary perfectly rational agent at some point of time, and when Bel is the class of believed propositions of the same agent at the same point of time.

In the first section, we will explain the representation theorem and its consequences. Qualitative belief will emerge as something like a qualitative version of degrees of belief: as *qualitative probability* if one likes to use that term in this context. In the second section we will contrast the theory with more traditional theories of qualitative probability that descend from de Finetti's work. The main differences between the two kinds of qualitative probability will be: belief in the sense of qualitative probability of Section 1 is closed under multi-premise logical consequence and corresponds to an ordering of worlds (or an ordering of the members of the probabilistic sample space), whereas belief in the sense of qualitative probability in the de Finetti (and Suppes) tradition is not closed under multi-premise logical consequence and corresponds to an ordering of sets of worlds that cannot be reduced generally to an ordering of worlds.

In all of this, we will have the pleasure and the honor of following some of Patrick Suppes' footsteps, as we are also going to highlight in the second section of the paper.[1]

---

[1]There is a vast amount of literature on the topic of *belief (or acceptance) vs degrees of belief*. I will not go into any details here, but Hintikka and Suppes (1966) is an excellent starting point as far as the relevant references are concerned.

1 Belief as Qualitative Probability: The Logical Closure of Belief and the Lockean Thesis

Let us understand by a 'proposition' a member of some given σ-algebra $\mathfrak{A}$ of subsets of some given non-empty set W of possible worlds; as usual, a σ-algebra is a Boolean field of sets that is closed also under countably infinitely many unions. In the following, Bel will be a set of propositions in this sense, and P will be a function that assigns real numbers to such propositions. Instead of '$X \in Bel$' we will write: $Bel(X)$. One should think of Bel and P taken together as capturing a perfectly rational agent's belief system at a point of time, but we will suppress any reference to any such agent and point of time. In probabilistic terms, W will be the sample space of P; at the same time W will also represent the trivial tautological proposition (and hence $\varnothing$ its contradictory negation or complement).

Here is what is probably *the* classical bridge principle for belief and degrees of belief (the "Lockean thesis" in the terminology of Foley 1993): if X is a proposition, then

$$X \text{ is believed if and only if } P(X) \geq r.$$

In other words: X is rationally believed if and only if the rational degree of belief assigned to X is "high enough". If the threshold value of 'r' is set to 1, then belief will be postulated to coincide extensionally with probabilistic certainty: however, at least for many cases this would seem much too restrictive, for we seem to believe lots of propositions X on which we would not accept every bet whatsoever and which, therefore, we do not actually believe with probability 1 according to the standard interpretation of subjective probabilities in terms of betting quotients. Instead, the value of 'r' should be thought of as varying with the context, and in many contexts the value will be less than 1.

For the moment, we will concentrate just on the "$\leftarrow$" direction of the Lockean thesis ("if high enough probability, then belief"), and of course we will also presuppose that the corresponding threshold r is greater than $\frac{1}{2}$. That is, if expressed in slightly more formal terms:

$$\text{LT}_{\leftarrow}^{\geq r > \frac{1}{2}} : \quad Bel(X) \text{ if } P(X) \geq r > \frac{1}{2}.$$

Our first goal will be to determine under which conditions $\text{LT}_{\leftarrow}^{\geq r > \frac{1}{2}}$, the *axioms of probability* for P, and the usual *logical closure* conditions for Bel are jointly satisfied.

By the axioms of probability we simply mean the standard Kolmogorov axioms of probability for P as being given relative to the σ-algebra $\mathfrak{A}$, including the axiom of countable or σ-additivity.

The logical closure conditions for Bel are

- $Bel(W)$,
- for all Y, Z: if $Bel(Y)$ and $Y \subseteq Z$, then $Bel(Z)$,
- for all Y, Z: if $Bel(Y)$ and $Bel(Z)$, then $Bel(Y \cap Z)$,
- not $Bel(\varnothing)$,

which are all well-known from doxastic and epistemic logic (cf. Hintikka 1962). Stated briefly: rational belief is closed under logical consequence from finitely many believed premises.

If the algebra $\mathfrak{A}$ of propositions is finite, then any Bel that satisfies these logical closure conditions must be generated from a logically strongest believed proposition: there must be a (uniquely determined) proposition B_W in $\mathfrak{A}$, such that for all $X \in \mathfrak{A}$,

$Bel(X)$ if and only $B_W \subseteq X$; clearly, this proposition B_W is nothing but the conjunction or intersection of all believed propositions, and in the terminology of possible worlds semantics one would regard B_W as the set of doxastically accessible worlds. If $\mathfrak{A}$ is infinite, then postulating in addition to the postulates above also the closure of belief under arbitrary (even arbitrary infinite) conjunctions or intersections will have the same effect of guaranteeing the existence of B_W. The underlying algebra $\mathfrak{A}$ will then also have to be closed under arbitrary conjunctions of believed propositions. Either of these two options being in place, we will assume the existence of such a least believed proposition B_W. And we will be able to move back and forth between Bel and B_W without any loss of information.

Finally, in order to determine under which conditions all of the postulates above are satisfied simultaneously, we require the following probabilistic concept which will turn out to be crucial for our purposes:[2]

Definition 1. *Let P be a probability measure over a σ-algebra $\mathfrak{A}$. For all $X \in \mathfrak{A}$:*
   X is P-stable iff for all $Y \in \mathfrak{A}$ with $Y \cap X \neq \varnothing$ and $P(Y) > 0$: $P(X|Y) > \frac{1}{2}$.

In words: P-stable propositions have stably high probabilities under salient suppositions. That is: a non-empty P-stable proposition X does have an absolute probability above $\frac{1}{2}$—as follows from plugging in W as the value for 'Y' in the definition above—and it continues to have a probability above $\frac{1}{2}$ in all cases in which the agent supposes or learns a proposition Y that is consistent with X (and where conditionalization on Y is defined). For instance, trivially, all X with $P(X) = 1$ are P-stable; but what is maybe more surprising: a probability measure P may also allow for P-stable sets X where $P(X) < 1$ (and where X is non-empty), as we are going to see below.

The importance of this concept of probabilistic stability or resiliency becomes transparent from the following representation theorem (see Leitgeb (2014) for more on this):

Theorem 1. *Let Bel be a class of members of a σ-algebra $\mathfrak{A}$, and let $P : \mathfrak{A} \to [0,1]$. Then the following two statements are equivalent:*

   I. *(i) P is a probability measure, (ii) Bel satisfies logical closure (in the sense of doxastic logic, as explained above), and (iii) $\mathrm{LT}_{\Leftarrow}^{\geq P(B_W) > \frac{1}{2}}$.*

   II. *P is a probability measure, and there is a (uniquely determined) $X \in \mathfrak{A}$, s.t.*
      – X is a non-empty P-stable proposition,
      – if $P(X) = 1$ then X is the least member of $\mathfrak{A}$ with probability 1; and:
      – for all $Y \in \mathfrak{A}$:
$$Bel(Y) \text{ if and only if } Y \supseteq X$$
      (and hence, $B_W = X$, where 'B_W' is defined from Bel as stated before).

From the left to the right, the theorem states that if $\langle P, Bel \rangle$ satisfies all of the desiderata in I., then the logically strongest believed proposition B_W according to Bel must be P-stable according to P. Note that the threshold in the relevant instance of the right-to-left direction of the Lockean thesis is defined by the probability of B_W: in fact it is easy to see that in a context in which one aims to combine the logical closure of Bel with the Lockean thesis, this choice of threshold is really the only one that makes good sense anyway. At the same time, it should be emphasized that for the same reason this choice of threshold cannot any longer be carried out independently of P: different

---

[2]This concept of P-stability is closely related to Brian Skyrms' notion of resiliency as introduced in Skyrms (1977, 1980).

probability measures P may determine different classes of P-stable sets and hence also different possible choices of threshold so that all of our postulates (i)–(iii) in I. are satisfied jointly.

In the other direction, the theorem says that if Bel is generated from a P-stable proposition, then $\langle P, Bel \rangle$ satisfies all of the desiderata in I.

Summing up both directions: P and Bel taken together satisfy all of the desiderata (i)–(iii) just in case belief is determined by some P-stable set. This means that if one is given a probability measure P, then by determining all P-stable sets one has determined all possible ways of generating Bel so that $\langle P, Bel \rangle$ satisfies all of our desiderata.

In either direction of the theorem, if the P-stable proposition in question has probability 1 (as being measured by P), then it must be the least proposition of probability 1 overall, which must exist then in order for the desiderata in I. to be satisfied. The existence of a least set of probability 1 is no restriction at all if $\mathfrak{A}$ is finite, but in the case of infinite $\mathfrak{A}$ it amounts to another restriction on P (which would, e.g., not be satisfied in the case of the Lebesgue measure on the unit interval $[0, 1]$).

Finally, one can also show that either side of the representation theorem above actually implies the *full* Lockean thesis

$$\text{LT}_{\leftrightarrow}^{\geq P(B_W) > \frac{1}{2}} : \quad Bel(X) \text{ iff } P(X) \geq P(B_W) > \frac{1}{2}$$

with an 'if and only if' instead of just the right-to-left direction. Hence, the representation theorem really clarifies also under which conditions one can combine consistently the axioms of probability for P, logical closure for Bel, and *the thesis that belief corresponds to high probability*: they can be combined if, and only if, the logically strongest believed proposition (B_W) has the property of P-stability, and the threshold in the Lockean thesis is given by the probability of that logically strongest believed proposition.

Here are some simple examples: Let $W = \{w_1, w_2, w_3\}$, and let $\mathfrak{A}$ be the power set of W. Now, if, e.g., P' is the probability measure that is determined by $P'(\{w_1\}) = 0.6$, $P'(\{w_2\}) = 0.3$, $P'(\{w_3\}) = 0.1$, then the corresponding P'-stable sets are: $\{w_1\}$, $\{w_1, w_2\}$, $\{w_1, w_2, w_3\}$. If Bel is generated from one of these three sets, then $\langle P', Bel \rangle$ satisfies all of our desiderata, in particular, the logical closure of Bel and an instance of the full Lockean thesis with a threshold of 0.6, 0.9, or 1, respectively. For instance: if B_W is set to $\{w_1, w_2\}$, then precisely all the supersets of $\{w_1, w_2\}$ are believed to be true, and these are precisely the propositions that have a probability of greater-than or equal to $P'(\{w_1, w_2\})$, that is, 0.9. And *only* by generating Bel from one of the three P'-stable sets it is possible to satisfy all of our desiderata jointly.

Now consider instead P'' with $P''(\{w_1\}) = 0.45$, $P''(\{w_2\}) = 0.4$, $P''(\{w_3\}) = 0.15$, in which case the corresponding P''-stable sets are: $\{w_1, w_2\}$, $\{w_1, w_2, w_3\}$. In this case there are only two ways of combining the logical closure of Bel with the Lockean thesis. E.g., if Bel were generated from $\{w_2, w_3\}$, which is *not* P''-stable, then while Bel would still be closed logically, the Lockean thesis would not hold anymore: it would not be the case for all X that $Bel(X)$ iff $P'(X) \geq 0.55$. E.g., $X = \{w_1, w_2\}$ would be a counterexample: it would not be believed since it is not a superset of $\{w_2, w_3\}$, but its probability is greater than 0.55.

Finally, take P''' to be the uniform probability measure on W, so that $P'''(\{w_1\}) = P'''(\{w_2\}) = P'''(\{w_3\}) = \frac{1}{3}$: then the only P'''-stable set is the set W of all worlds, and hence the only way of satisfying all of our desiderata is to choose a threshold of 1

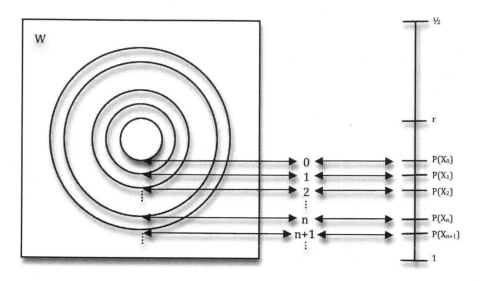

FIGURE 1 *P*-stable sets of probability less than 1

in the Lockean thesis; in this case a proposition is rationally believed if and only if it is probabilistically certain.[3]

We can see that in the case of P' and P'' there are P-stable sets of probability less than 1, while this is no so in the case of the uniform probability measure P'''.

Here is an important general observation on P-stable sets of probability less than 1, if they exist at all (for given P): by the σ-additivity of P, one can prove that this class of P-stable propositions X in $\mathfrak{A}$ with $P(X) < 1$ is always *well-ordered* with respect to the subset relation. In other words: for every two P-stable sets of probability less than 1 one must be a subset of the other, and there cannot be an infinitely descending chain of such P-stable sets. Therefore, if the class of P-stable sets of probability less than 1 is non-empty, it forms a well-founded hierarchy of propositions, or, in the language of David Lewis' possible worlds semantics for counterfactuals, a so-called sphere system that satisfies the limit assumption (cf. Lewis 1973). In terms of a picture, P-stable sets of probability less than 1 look like in Figure 1.

Furthermore, one can show that the least proposition of probability 1—which is P-stable, too, and the existence of which follows from (i)–(iii) according to the representation theorem—is a proper superset of all P-stable propositions of probability less than 1.

Call a P-stable set which has probability less than 1 or which is identical to the least proposition of probability 1 a P-*sphere*: every world w for which $\{w\}$ is in $\mathfrak{A}$ and where $\{w\}$ has positive probability must then be a member of at least one P-sphere. In the following we will concentrate very much on such worlds with positive probabilistic mass.

---

[3]The reason why joining the postulates in condition I. of our representation theorem does not run into the Lottery Paradox (cf. Kyburg 1961) is that the threshold in the Lockean thesis is sensitive to the probability measure P: in the situation of a fair finite lottery it can be shown that the only P-stable set is the set W of all worlds ("tickets") itself, hence the threshold in the Lockean thesis must be 1, and no ticket ends up believed to be a losing ticket.

And, for simplicity, let us presuppose for the rest of this paper that the algebra $\mathfrak{A}$ is the full power set of W; for that reason, for every world $w \in W$ its singleton set $\{w\}$ will always count as a proposition.

For every possible world that has positive probabilistic mass there must be a *first* P-sphere of which it is a member (by the well-ordering property): hence, to every such possible world an ordinal rank of "first appearance in a P-sphere" can be assigned, accordingly. We will not state the formal definition of this ranking function, but it should be obvious how it can be done; let us just have a look at some examples instead.

For instance, as far as P' from above is concerned, the corresponding P'-stable sets were $\{w_1\}$, $\{w_1, w_2\}$, $\{w_1, w_2, w_3\}$, and hence the ordinal rank of w_1 is 0, the rank of w_2 is 1, and the rank of w_2 is 2. In contrast, for P'' the ranks of w_1 and w_2 are both 0, and the rank of w_3 is 1, while in the case of P''', all worlds have the same rank 0.

Equivalently, we may think of every probability measure P as determining "its" corresponding pre-well-ordering $\leq_P$ on the worlds that have positive probabilistic mass, where by a 'pre-well-order' we mean a total pre-order $\leq_P$ of worlds, such that there is no infinitely descending sequence of worlds along $<_P$. The order $\leq_P$ is simply given by the ranks of worlds again, as explained before. Since two distinct worlds may have the same rank, that order is not necessarily a partial order but may well be a pre-order which involves ties between worlds.

For instance: when '$w <_P w'$' is defined by '$w \leq_P w'$ and not $w' \leq_P w$', then the pre-well-orders for P', P'', P''' are given by

$$P': \quad w_1 <_{P'} w_2 <_{P'} w_3$$
$$P'': \quad w_1, w_2 <_{P'} w_3$$
$$P''': \quad w_1, w_2, w_3$$

respectively, where commas separate worlds of the same rank. One can also show that any such $\leq_P$ satisfies the following Sum Condition: if we consider worlds only whose singleton sets have positive probability, then for all such worlds it holds that

$$P(\{w\}) > \sum_{w': \ w <_P w'} P(\{w'\}).$$

Furthermore, $\leq_P$ can be extended naturally to *all* worlds in W whatsoever by: for all $w \in W$ with $P(\{w\}) = 0$, for all $w' \in W$, $w' \leq_P w$.[4]

We may thus read '$w <_P w'$' as: w *is of a probabilistic order of magnitude greater than that of w', as measured by* P. Roughly, the further "down" a world w is in any such ordering $\leq_P$, the more "preferred" it is in the sense that the greater the probability of its singleton set $\{w\}$ as assigned by P, and in fact the *much* greater that probability must be, as $P(\{w\})$ must even be greater than the sum of *all* probabilities of (singletons of) worlds that are higher up or less preferred than w according to $<_P$.

How often is it the case that there P-stable sets of probability less than 1 at all? The answer is: *almost always*.

The triangle in Figure 2 depicts geometrically all probability measures on the power set of $W = \{w_1, w_2, w_3\}$ again; for instance, the vertices correspond to the probability measures that put all of their probabilistic mass on precisely one world (or singleton

---

[4]In theoretical computer science, Benferhat et al. (1999) and Snow (1998) have studied a similar condition such as the Sum Condition, but in application to total partial orders rather than total pre-orders.

126 / Hannes Leitgeb

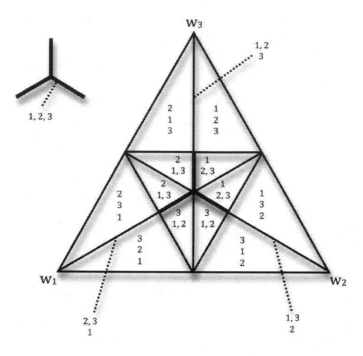

FIGURE 2 Probability measures on three worlds and their P-stable sets

set); P' from above is located within the small sub-triangle in the left bottom area that has the "code"

$$3$$
$$2$$
$$1$$

and P'' occupies a point in the sub-triangle with the "code"

$$3$$
$$1,2$$

in the left half of the overall triangle, whereas P''' is represented by the center of the overall triangle which belongs to the three small boldfaced line segments that have "code"

$$1,2,3$$

The "codes" are to be read from bottom to top and their function is to convey the ordinal ranks of worlds as determined from P, as explained before, and hence to summarize what the P-spheres are like for the probability measures in question: e.g., as mentioned above, for P' the corresponding P'-stable sets are $\{w_1\}$, $\{w_1, w_2\}$, $\{w_1, w_2, w_3\}$, accordingly, the rank of w_1 is 0, the rank of w_2 is 1, and the rank of w_2 is 2, and that is precisely what the "code" of P' expresses, where the numbers in a code represent the indices of worlds and where the position in a code represents the rank of a world; similarly, for all other probability measures and their codes in Figure 2.

It turns out that the only probability measures for which there does not exist a P-stable set with probability less than 1 are those that are represented in Figure 2 by one of the three vertices or by one of the points on the three small boldfaced line segments

Let X be a non-empty set and let events be subsets of X. Then the axioms of what I shall call *weak qualitative probability structures* are the following:

Axiom 1. X is certain.

Axiom 2. If A implies B and A is certain, then B is certain.

Axiom 3. If A implies B and A is more likely than not, then B is more likely than not.

Axiom 4. If A implies B but B does not imply A and A is as likely as not, then B is more likely than not.

Axiom 5. If A is certain, then *not* A is impossible.

Axiom 6. If A is more likely than not, then *not* A is less likely than not.

FIGURE 3 Weak qualitative probability structures

that lead to, and include, the center of the triangle (where the center represents P''' from before) or by one of the center points of the three edges of the large triangle. That is, in terms of the Lebesgue measure on the Euclidean plane: almost everywhere in the triangle one deals with a probability measure P that allows for P-stable sets of probability less than 1. Or in terms of our representation theorem from above: almost every probability measure P on three worlds allows for a belief set *Bel* so that the logical closure of *Bel* can be combined consistently with the Lockean thesis *and the existence of a believed proposition of probability less than 1*. One can show that the same observation applies if W is an arbitrary finite set of possible worlds, and as far as infinite sets W are concerned, one can show that there are probability measures P on infinite σ-algebras on infinite W for which there exist even countably infinitely many P-stable sets of probability less than 1.

Due to the presence of the Lockean thesis, if all of the postulates in condition I. of our representation theorem above are satisfied, belief corresponds to sufficiently high probability. Moreover, all the probability measures that are represented by points in the interior of one and the same small sub-triangle in Figure 2 determine the same class of P-stable sets, hence also the same possible candidates for belief sets, and thus also the same pre-well-orderings of worlds. Combining the axioms of probability with the logical closure of belief and the Lockean thesis has therefore the following consequence: belief ends up to be an abstraction from, or a coarse-grained version of, subjective probability—*rational belief is qualitative subjective probability.*

2 A Comparison with "Traditional" Qualitative Probability

The proposal that belief is something like *subjective probability made qualitative* is obviously not a new one.

First of all, Suppes (1974) on "The Measurement of Belief" suggests to axiomatize categorical notions of *certainty, being more likely than not, being as likely than not, being less likely than not,* and *impossibility* in terms of what he calls 'weak qualitative probability structures': see Figure 3 (which is taken from Suppes (1974, pp. 166f).

Let P be a probability measure on the power set of a non-empty set W, and let all probabilities be determined by P: the intended interpretation of certainty is then of course to have a probability of 1 (although to be identical to W would also be an option), of being more likely than not is to have a probability that is greater than $\frac{1}{2}$, of being as likely as not is to have a probability of exactly $\frac{1}{2}$, of being less likely as not is to have a probability of less than $\frac{1}{2}$, and of impossibility is to have a probability of 0 (although to be identical to $\varnothing$ would be an option, too, if certainty were to represent identity with

W). According to this interpretation, if 'being more likely than not' replaces 'belief' in the Lockean thesis, then obviously the thesis is satisfied with a threshold r of precisely $\frac{1}{2}$. 'Implies' in Suppes' axioms can be understood as expressing the subset relation on propositions (even though Axiom 4 would not be satisfied then on all probability spaces) or as expressing that the conditional probability of a proposition B given a proposition A is 1.

Now, for the sake of comparison, let us take any pair $\langle P, Bel \rangle$ that satisfies all of our desiderata (i)–(iii) in the representation theorem of Section 1 (which we know to entail an instance of the full Lockean thesis), and let us interpret the primitive terms of Suppes' axioms instead in terms of P and Bel, as follows: certainty is to have a probability of 1 again, being more likely than not is to believe a proposition (in the sense of Bel), being as likely as not is to neither believe a proposition nor its negation (its complement with respect to W), being less likely than not is to believe the negation of a proposition, and impossibility is to have a probability of 0. If interpreted in this way, all of Suppes' axioms other than his Axiom 4 are satisfied; Axiom 4, however, does not hold for all of our $\langle P, Bel \rangle$, as can be seen easily by considering examples. At the same time, while we have presupposed the principle of the closure of belief under conjunction (or intersection),

- for all Y, Z: if $Bel(Y)$ and $Bel(Z)$, then $Bel(Y \cap Z)$,

no such principle for Suppes' *being more likely than not* follows from Suppes' axioms, as can be seen by considering examples again. More generally, belief in the sense of the last section is always closed under multi-premise logical consequences, which is not generally the case for belief in sense of Suppes' *more likely than not*.

We conclude that our postulates (i)–(iii) for P and Bel are much like Suppes' on weak qualitative probability, with the exception that ours do not necessarily support his Axiom 4, while in turn his axioms do not necessarily support ours of the closure of belief under conjunction, when we compare Suppes' *more likely than not* with *belief* in the sense of the last section.

Secondly, de Finetti was the first to study orderings $\succeq$ of propositions (or events) the intended interpretation of which is *weakly more probable than*, that is: $A \succeq B$ if and only if $P(A) \geq P(B)$, for all A, B, and for some given probability measure P.

The list of postulates in Figure 4 concerning qualitative probability on the ordinal scale is de Finetti's, but the quote is taken from p. 19 of Suppes' (1994) "Qualitative Theory of Subjective Probability".[5]

Next we want to compare qualitative probability in the sense of de Finetti with qualitative probability in the sense of the last section. In Section 1 we explained how every pair $\langle P, Bel \rangle$ that satisfies all of the desiderata (i)–(iii) in our representation theorem determines a pre-wellordering $\leq_P$ of worlds in W. But such an ordering $\leq_P$ of worlds, or its corresponding strict version $<_P$, can also be "lifted" to an ordering of *propositions* in a manner that is well-known from areas such as belief revision theory or nonmonotonic reasoning or the possible worlds semantics for counterfactuals, which will thus allow us to carry out the intended comparison.

---

[5]For references to de Finetti's work see Suppes' (1994). It is well known that de Finetti's axioms are not sufficient to prove a proper representation theorem for $\succeq$. We will put this to one side here, but see Scott (1964) for a completion of de Finetti's axioms that suffices for the proof of the intended representation theorem.

Definition 1 A structure $\Omega = (\Omega, \mathfrak{F}, \geq)$ is a *qualitative probability structure* if the following axioms are satisfied for all A, B, and C in $\mathfrak{F}$:

S1. $\mathfrak{F}$ is an algebra of sets on Ω;
S2. If $A \geq B$ and $B \geq C$, then $A \geq C$;
S3. $A \geq B$ or $B \geq A$;
S4. If $A \cap C = \varnothing$ and $B \cap C = \varnothing$, then $A \geq B$ if and only if $A \cup C \geq B \cup C$;
S5. $A \geq \varnothing$;
S6. Not $\varnothing \geq \Omega$.

FIGURE 4 Qualitative probability structures

For simplicity, let us again assume that the σ-algebra $\mathfrak{A}$ is the power set of W. If given P, $\leq_P$, and hence also $<_P$, let us define for arbitrary propositions A and B,

- $A \geq_P B$ iff there is a world $w \in A$, such that for all worlds $w' \in B$: $w \leq_P w'$.

It follows from this definition (and the properties of P-stable sets that were mentioned in the last section) that any such $\geq_P$ is now a pre-well-ordering of propositions or sets of worlds, just as $\leq_P$ was a pre-well-ordering of worlds.

$\geq_P$ on propositions is defined by restricting attention only to those members of A which are least according to $\leq_P$—which are most preferred in terms of $\leq_P$ and by determining whether these worlds are more preferred than or equally preferred as the most preferred members of B, as being given by $\leq_P$ again. One can show that if $\geq_P$ on propositions is defined from $\leq_P$ on worlds in this way, where $\leq_P$ is in turn defined as in Section 1 from the rank of appearance of a world in the hierarchy of P-spheres, then by the Sum Condition that applies to $\leq_P$ (as also explained in the last section), it follows: if $A >_P B$ then $P(A \mid A \cup B) > \frac{1}{2}$ and $P(B \mid A \cup B) < \frac{1}{2}$, where '$A >_P B$' is defined from $\geq_P$ by: $A \geq_P B$, but not $B \geq_P A$. Since $>_P$ on propositions derives its properties from $\leq_P$ on worlds, '$A >_P B$' may be read in an analogous way as $<_P$ before: *A is of a probabilistic order of magnitude greater than that of B, as measured by P.* The further "up" a proposition is in any such ordering $\geq_P$, the more "preferred" it is in terms of its probability. On the level of sets of worlds, $\geq_P$ is the natural counterpart of our original relation $<_P$ on worlds.

By turning to $\geq_P$ we are finally able to compare qualitative probability in the sense of de Finetti's $\succeq$ with qualitative probability in the sense of the last section: in contrast with de Finetti's intended interpretation, let '$\succeq$' be interpreted in terms of $\geq_P$ on propositions for a given P. Then all of the axioms in Figure 4 except possibly for S4 are satisfied; S4 does not hold for all P, as can be determined easily by looking at some examples. In compensation, our $\geq_P$ satisfies

- if $A \supseteq B$, then $A \geq_P B$;
- if $A >_P B$ and $A >_P C$, then $A >_P B \cup C$.

The first of these two conditions can be derived also from S1–S6 if '$\geq_P$' is replaced by '$\succeq$' (see Theorem 1 in Suppes (1994)), but the second one cannot be derived, as follows again from considering examples.[6]

We conclude that the theory of qualitative probability as specified in the last section leads naturally to an ordering $\geq_P$ of propositions the properties of which are much like

[6]A complete list of axioms for our $\geq_P$ on propositions, which suffices for the proof of a proper representation theorem, can be found in Halpern (2005), p.46. But we cannot go into any details here.

de Finetti's axioms of qualitative probability, with the exception that our $\geq_P$ do not necessarily satisfy his Axiom S4, while in turn his axioms do not necessarily support the second one of the postulates above which concerns unions in the second argument place of '$>_P$'. Furthermore, each of our orderings $\geq_P$ derives from an ordering of worlds, which is not so in de Finetti's case, and whereas each of de Finetti's orders $\succeq$ represent an ordering of probabilities of propositions, each of our $\geq_P$ represent an ordering of orders of magnitude of probabilities of propositions.

Therefore, the term 'qualitative probability' and the slogan 'Belief is qualitative subjective probability' do not actually uniquely pin down accounts of qualitative probability and joint theories of belief and degrees of belief: while the de Finetti-Suppes understanding of qualitative probability stays closer to probability theory itself, the understanding of qualitative probability as developed in Section 1 takes a step towards doxastic and epistemic logic. Neither of them is "the right" view on qualitative probability, and both of them (and more!) ought to be studied and applied.

References

Benferhat, S., D. Dubois, and H. Prade. 1999. Possibilistic and standard probabilistic semantics of conditional knowledge bases. *Journal of Logic and Computation* 9(6):873–895. doi:10.1093/logcom/9.6.873.

Foley, R. 1993. *Working Without a Net: A Study of Egocentric Epistemology.* Oxford: Oxford University Press.

Halpern, J. 2005. *Reasoning about Uncertainty.* Cambridge, Mass.: MIT Press.

Hintikka, J. 1962. *Knowledge and Belief: An Introduction to the Logic of the Two Notions.* Ithica, N.Y.: Cornell University Press.

Hintikka, J. and P. Suppes. 1966. *Aspects of Inductive Logic.* Amsterdam: North-Holland Pub. Co.

Kyburg Jr., H. E. 1961. *Probability and the logic of rational belief.* Middletown, Conn.: Wesleyan University Press.

Leitgeb, H. 2014. The stability theory of belief. *The Philosophical Review* 123(2):131–171.

Lewis, D. K. 1973. *Counterfactuals.* Oxford: Blackwell.

Scott, D. 1964. Measurement structures and linear inequalities. *Journal of Mathematical Psychology* 1(2):233–247. doi:10.1016/0022-2496(64)90002-1.

Skyrms, B. 1977. Resiliency, propensities, and causal necessity. *The Journal of Philosophy* 74(11):704–713. doi:10.2307/2025774.

—. 1980. *Causal Necessity.* New Haven: Yale University Press.

Snow, P. 1998. Is intelligent belief really beyond logic? In *Proceedings of the Eleventh International Florida Artificial Intelligence Research Society Conference*, pages 430–434. American Association for Artificial Intelligence, AAAI Press.

Suppes, P. 1974. The measurement of belief. *Journal of the Royal Statistical Society. Series B* 36(2):160–191.

—. 1994. Qualitative theory of subjective probability. In G. Wright and P. Ayton, eds., *Subjective Probability*, pages 17–37. Chichester: Wiley.

—. 2002. *Representation and Invariance of Scientific Structures.* Stanford, CA: CSLI Publications.

Part IV

Language, Mind, and Learning Theory

12

Learning to Signal with Two Kinds of Trial and Error

BRIAN SKYRMS

"...and any theory, by which we explain the operations of the understanding, or the origin and connexion of the passions in man, will acquire additional authority, if we find, that the same theory is requisite to explain the same phenomena in all other animals.'"
–David Hume *An Enquiry Concerning Human Understanding*
IX Of the Reason of Animals

1 Low Rationality Game Theory

High rationality game theory is built on idealizations that may be hard to justify such as—at the minimum—common knowledge of strategic interaction and common knowledge of rationality of the interacting agents. Low rationality game theory investigates interactions between more limited agents. At the most modest level, agents may not even be aware that they are in strategic interaction, and may just muddle along with trial and error learning. If low rationality dynamics leads to the same result as high rationality equilibrium analysis, it lends support to high rationality game theory. If they disagree, this raises questions (Suppes and Atkinson 1960; Roth and Erev 1995; Erev and Roth 1998; Young 2004).

It is of special interest to investigate learning to signal with in a low rationality setting for two reasons. The first is that some fairly robust signaling system must be already in place to support (or even approximate) the assumptions of high rationality. Signaling must presumably have its origin in a low rationality setting. The second is that animals incapable of high rationality can learn to signal.

But there is more than one kind of trial and error learning. The term encompasses a whole spectrum of learning dynamics. Here we focus on two very different kinds of trial and error that represent extremes of the spectrum, and compare their success in learning to signal. The first is the simplest form of *reinforcement learning* used by Roth and Erev (1995), Erev and Roth (1998)—which they trace to Herrnstein's (1970) quantification of the law of effect. The second is *probe and adjust* dynamics. A somewhat more complicated form of probe and adjust was introduced by Kimbrough and Murphy (2009) in an analysis of tacit collusion in oligopoly pricing. The simple form used here was introduced by Skyrms (2010) and analyzed by Hutteger and Skyrms (Forthcoming)

Foundations and Methods from Mathematics to Neuroscience.
Colleen E. Crangle, Adolfo García de la Sienra and Helen Longino.
Copyright © 2014, CSLI Publications.

to investigate learning an optimal signaling network. It can be seen as part of the statistical learning theory of Estes (1950).

2 Two Kinds of Trial and Error Learning

2.1 Reinforcement

An individual chooses repeatedly between actions $A_1 \ldots A_n$. At each point in time the probability of choosing A_i is proportional to the accumulated past payoffs for choosing A_i. To get the process started, we assume initial "virtual payoffs" equal to 1 for each act. Thus the learner starts by choosing at random, and then the evolution of choice probabilities is driven by payoffs.

The stochastic process may be realized by an urn scheme. There are balls of colors $C_1 \ldots C_n$ in and urn, initially one ball of each color. A ball is drawn at random from the urn—say of color C_i and replaced, and the corresponding action, A_i, is taken. A payoff (possibly zero, but non-negative) is realized and the number of balls equal to that payoff of color C_i is added to the urn. This is repeated.

2.2 Probe and adjust

An individual *probes* by just trying an act at random, and then *adjusts* by (i) adopting the new act if it has a higher payoff than the old one (ii) going back to the old act if the new one got a lower payoff or (iii) flipping a coin to decide if they are equal. In a repeated choice situation an individual *probes and adjusts* with small probability, e, and just chooses the same as last time with probability $(1 - e)$. We can start the process by just choosing at random.

2.3 Comparison

Although both these processes are kinds of trial and error learning with at least some psychological plausibility, they are quite different in character and perform differently in different learning situations. Their analysis calls for different mathematical methods. The transitions in Roth-Erev reinforcement depend on the number of balls in each urn, and thus require some memory of the whole history of the process. It slows down as the reinforcements pile up, and approximates better and better a mean field deterministic dynamics. The transitions in probe-and-adjust depend only on a comparison of probe and pre-probe payoffs, and so only require a limited memory. Reinforcement learning is analyzed using stochastic approximation theory, Pemantle (2007). Probe and adjust uses Markov chains and analysis is quite straightforward and easy. For an initial comparison, we apply them to a two-armed bandit learning problem.

You have two slot machines, $R; L$, each of which pays off with a different unknown probability. (Trials are independent and identically distributed, and are independent between machines.) Can you learn to play the optimal machine? Roth-Erev reinforcement learning converges to playing the highest paying machine with probability one, Beggs (2005). Here is a sketch of the stochastic approximation approach.

Let L pay 1 with probability p and 0 with probability $(1 - p)$. Let R pay 1 with probability q and 0 with probability $1 - q$. Our learner has an urn with one R ball and one L ball, and proceeds with reinforcement learning. Let N = number of R balls + number of L balls. The probability of choosing R at a given time is then just the number of R balls divided by N.

We start by calculating the expected change in the probability that the learner chooses $R, p(R)$. First calculate the expected value of $p(R)$ after one trial. One of four things can happen: (1) R is chosen and reinforced, (2) R is chosen and not reinforced, (3) L is chosen and reinforced, (4) L is chosen and not reinforced. Accordingly, we calculate the expectation of the next value of $p(R)$ as:

	Chosen?		Reinforced?		New Value of $p(R)$	
1.	$[p(R)$	$*$	q	$*$	$(Np(R)+1)/(N+1)]$	$+$
2.	$[p(R)$	$*$	$(1-q)$	$*$	$p(R)]$	$+$
3.	$[(1-p(R))$	$*$	p	$*$	$(Np(R))/(N+1)]$	$+$
4.	$[(1-p(R)$	$*$	$(1-p)$	$*$	$p(R)]$	

Subtracting the current value, $p(R)$ gives us the expected increment. We get:

$$(1/N+1)p(R)(1-p(R)(q-p)) \quad \text{(expected increment)}$$

The value $1/(N+1)$ is the *step size*. This tells us that the process slows down at a rate such that the stochastic process approximates the mean field dynamics with higher and higher probability as N builds up. The rest of the equation gives us the mean field dynamics:

$$dp(R)/dt = pr(R)(1-pr(R)(q-p)) \quad \text{(mean field dynamics)}$$

Here learning must converge to one of the rest points of the mean field dynamics. If the two machines pay off equally, $q=p$, then every point is a rest point of the mean field dynamics. In this case the urn is a Polya urn and the learner can converge to anything.

If the R pays off more often than L, then there are only two rest points, a stable attracting equilibrium at $p(R)=1$, and an unstable equilibrium at $Pr(R)=0$.

$$L \to\to\to\to\to\to\to\to\to\to\to\to\to\to R$$

Reinforcement learning must converge to one of these points. The instability of the latter equilibrium suggests that learning will never converge to it, and thus will always converge to always playing R. That is correct, but it requires a special argument to show it, Hopkins and Posch (2005).

In probe-and-adjust learning, nothing happens except when there is a probe. We can analyze it by looking only at pre-probe and post-probe states. This embedded sequence is a Markov chain. We can analyze as follows: Suppose bandit L pays off with probability p and bandit R pays off with probability q. The state of playing bandit R transitions to that of playing bandit L with probability:

$p(1-q)$ [R does not pay, probe, L does pay. Switch to L]
$+1/2pq$ [both pay off, flip a coin to decide whether to switch]
$+1/2(1-p)(1-q)$ [neither pays off, flip a coin]

Likewise, L transitions to R with probability:

$$q(1-p) + \frac{1}{2}pq + \frac{1}{2}(1-p)(1-q)$$

This is an ergodic Markov chain with invariant probability distribution:

$$Pr(L) = \frac{1}{2}(1+p-q), Pr(R) = \frac{1}{2}(1-p+q).$$

If, for instance, R pays off 90% of the time and L 50%, probe and adjust plays R 70% of the time. Probe-and-adjust favors the higher paying bandit, but does not learn optimal play.

In the special case where $p + q = 1$, the invariant distribution is:

$$Pr(L) = p, Pr(R) = q.$$

In the long run, the machine is played with the probability of its payoff. This is some version of what psychologists call "*probability matching*". It has been frequently observed in human learning experiments, and discussions of its robustness have generated a large and sometimes contentious literature in psychology and economics, see the survey of Vulkan (2000).

3 Signaling Games

Sender-Receiver signaling games were first introduced by David Lewis in *Convention* (1969). There are two players, a sender and a receiver. The sender observes a situation, which nature chooses at random. She chooses a signal, conditional on the situation observed. The receiver observes the signal, and chooses an act, conditional on the signal observed. Payoffs for sender and receiver are determined by the combination of signal and situation. Signals are cost-free and sender and receiver have pure common interest. In the simplest case, there is an act that is "right" for each situation in that if that act is done in that situation both sender and receiver get a payoff of 1, and otherwise they get a payoff of 0. For simplicity, we can index that situations and acts such that the joint payoff on 1 occurs just in case act A_i is done in situation S_i. (Situations are often called "states", but we reserve this term for the state of a Markov chain.)

Lewis calls an equilibrium in which players always receive a payoff of 1 a *signaling system equilibrium*. There are other equilibria. If the sender sends signals with probabilities independent of the situation and the receiver chooses acts independent of the signals, we have a *total pooling* equilibrium. All the situations are pooled, in that the signal sent carries no information about the situation. It is an equilibrium in that neither sender nor receiver can improve her payoff by changing her behavior. If there are more than two situations, there may be *partial pooling equilibria* in which some but not all states are pooled. For example, suppose there are 3 situations, 3 signals and 3 acts, and that the sender sends:

signal 1 in situations 1 and 2,
signals 2 and 3 in some proportion in situation 3

and the receiver does:

acts 1 and 2 in some proportion for signal 1
act 3 for signals 2 and 3

This is a partial pooling equilibrium in which situations 1 and 2 are pooled.

Many generalizations and variations are of interest, but here we concentrate on the basic signaling game.

4 Learning to Signal with Reinforcement Learning: The Simplest Case

In applying reinforcement learning to signaling games we do not reinforce whole strategies. After all, part of the point of the low rationality approach is that the agents involved may not be thinking strategically at all. Rather we think of just reinforcing responses to stimuli. A sender observes a situation as a stimulus and responds by sending a signal.

For each situation, we equip the sender with an urn, the colored balls corresponding to the signal to send in that situation. And the receiver observes the signal as a stimulus. So for each signal, we equip the receiver with an urn, the colored balls corresponding to the act to take upon receiving that signal.

Consider the simplest Lewis signaling game in which nature flips a fair coin to choose one of two situations, the sender observes the situation and chooses between two signals and the receiver observes the signals and chooses between two acts. Sender and receiver use reinforcement of stimuli. There are now 4 interacting reinforcement processes—two sender's urns and two receiver's urns.

An analysis of these interacting reinforcement processes shows that sender and receiver always learn to signal with probability one, Argiento et al. (2009). Here is a sketch.

There are 8 quantities to keep track of: the numbers of the two types of ball in each of the four urns. But because the receiver is reinforced just when the sender is, and to the same extent, the numbers of balls in the sender's urns contain all the relevant information about the state of the process. Normalizing by dividing by the total number of balls in both sender's urns, we get four quantities that live in a tetrahedron. The mean field dynamics is calculated. A quantity is found that the dynamics always increases, ruling out cycles. The reinforcement learning process must then converge to a rest point of this dynamics. The rest points are shown in Figure 1.

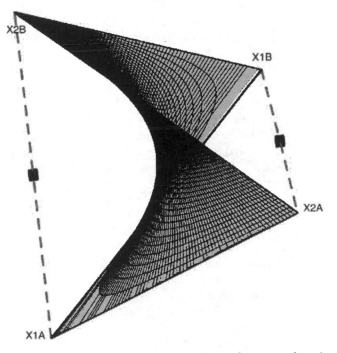

FIGURE 1 Rest points of the mean-field dynamics for reinforcement learning from Argiento et al. (2009)

The two signaling systems are shown as square dots. At a signaling system players are always reinforced. One signaling system ends up with the urn for situation 1 full of signal A balls and the urn for situation 2 full of signal B balls; the other has urn 1 full of signal B balls and urn 2 full of signal A balls. The curved surface consists of pooling equilibria, where the probability of signals is independent of the situation. On one side of the surface, the mean field dynamics leads to one signaling system, on the other to the second signaling system. It is then shown that the probability that reinforcement learning converges to a point on the surface is zero. (The argument requires a separate treatment of the parts of the surface in the interior if the tetrahedron and the parts on the boundary.) The rest points that are left are the signaling systems. With probability one, reinforcement learners learn to signal.

5 Learning to Signal with Probe and Adjust: The Simplest Case

Consider probe and adjust learning applied to the same simple signaling game. As before, we assume that each sender treats each situation as a separate choice stimulus. Each receiver treats each signal received as a separate choice stimulus. In order to apply probe and adjust, individuals must not remember not just what happened last time, but what happened last time for each stimulus. So the sender remembers what he did the last time he was confronted with situation 1 and the payoff he received and likewise for situation 2. The receiver remembers what he did last time and payoff received when observing signal one and likewise for signal 2. Memory requirements are still quite modest.

Most of the time agents just repeat what they did last time for the same stimulus. Every once and a while an agent probes and adjusts. We assume here that only one agent probes and adjusts at a time. The other agent keeps doing the same thing during the probe-and-adjust process. Now we can again find an embedded Markov chain which consists of pre and post probe-and-adjust transitions. Here the *state of the system* of both players consists of a pair: a map from situations to signals for the sender and a map from signals to acts for the receiver. This is just the *memory* of what they did last given the respective stimuli.

(This system state constrains the payoffs that they got last time they did something to an extent sufficient to establish that signaling systems are the only absorbing states, and that there is a positive probability path from any state to a signaling system. When the sender pools, payoffs last time may be underdetermined. We are then dealing with a random, time-inhomogeneous Markov chain. Nevertheless, for each possibility, there is a positive probability path to a signaling system.)

There are 4 possible sender configurations and 4 receiver configurations, so there are 16 possible states of the system, as shown in Figure 2.

The transition probabilities are calculated as follows. Nature chooses the sender or receiver to probe by flipping a fair coin. Nature chooses a situation by flipping a fair coin. If sender is chosen sender probes a new signal for the situation chosen. If receiver is chosen, sender sends the old signal for the situation and receiver probes a new act for the signal. If the probe gave a higher payoff the new configuration is adopted; if it gave the same payoff the new configuration is adopted with probability $\frac{1}{2}$; if it gave a lesser payoff the system remains in the original state.

This Markov chain is no longer ergodic, like that for the bandit problem. States 15 and 16, which correspond to the two signaling systems, are *absorbing states*. Once entered, the Markov chain will not leave them. Any probe will get a smaller payoff, and the agents will adjust back to the original state. Furthermore, as we will see, they are the only absorbing states. If the agents enter one of these states, we will say that they have *learned to signal*. This means that they will continue in the signaling system except

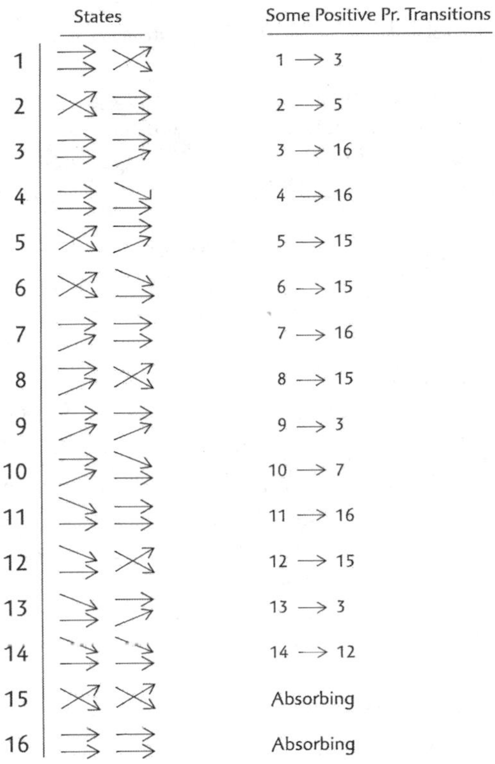

	States	Some Positive Pr. Transitions
1	⇉ ⤬	1 → 3
2	⤬ ⇉	2 → 5
3	⇉ ⇗	3 → 16
4	⇉ ⬎	4 → 16
5	⤬ ⇗	5 → 15
6	⤬ ⇉	6 → 15
7	⇗ ⇉	7 → 16
8	⇗ ⤬	8 → 15
9	⇗ ⇗	9 → 3
10	⇗ ⇉	10 → 7
11	⬎ ⇉	11 → 16
12	⬎ ⤬	12 → 15
13	⬎ ⇗	13 → 3
14	⬎ ⬎	14 → 12
15	⤬ ⤬	Absorbing
16	⇉ ⇉	Absorbing

FIGURE 2 States and transitions: 2 situations, 2 signals, 2 acts

for occasional fruitless probes that will lead them to return to it. Figure 2 shows some (not all) positive probability transitions between states.

States 15 and 16 are signaling systems. States 3, 4, 5, 6, 7, 8, 11 and 12 can move to a signaling system with one probe. Each of the other states can move to these with one probe. From any state there is a (short) positive probability path to a signaling system. Given a positive probability path from each state to an absorbing state, there is a maximum path length, n, and a minimum path probability, e. Starting from any state, the probability of not being absorbed after n probes is at most $(1-e)$. After $m*n$ probes the probability of not being absorbed is $(1-e)^n$. In the limit the probability of not being absorbed is zero. *With probability one, probe and adjust learns to signal.*

We will investigate how this logic generalizes.

6 *Reinforcement*: N equiprobable situations, N signals, N acts

Suppose we keep everything the same except increasing the numbers of situations, signals and acts. Situations are still assumed to be equiprobable and the number of signals and acts matches the number of situations. We know that there is now a new class of equilibria, the partial pooling equilibria. The question is what significance, if any, they have for reinforcement learners. A full analysis of this situation is not quite yet available, but it is known that both convergence to perfect signaling and convergence to partial pooling have positive probability (Hu 2010; Hu et al. 2011). Learning to signal perfectly is no longer guaranteed.

Extensive numerical simulations show that the extent of convergence to partial pooling is not a negligible outcome. Starting from an initial condition of one ball of each color in each urn, Barrett (2007) finds numerical convergence to partial pooling after 10^6 iterations of learning in 10^3 trials as follows:

N	Partial Pooling
3	9.6%
4	21.9%
8	59.4%

7 *Probe and adjust*: N situations, N signals, N acts

As in the simplest case, when we consider transitions between pre-probe-adjust states and post-probe-adjust states, we have an embedded Markov chain. The state of the system for this chain consists of a record of what was done last in each situation by the sender and last for each signal by the receiver. It is thus a pair of sender and receiver functions, $\langle f, g \rangle$. These are functions since each only remembers what she last did in the each decision situation. These need not be one-to-one, as the sender may have sent the same signals in multiple situations and the receiver may have done the same act on receiving multiple signals.

The analysis of the simplest game generalizes. Specifically:

(1) Signaling systems are absorbing states of the Markov chain.
(2) Signaling systems are the unique absorbing states.
(3) From any state, there is a positive probability path to a signaling system.

For details see Skyrms (2012).

Agents always learn to signal perfectly. Here *probe and adjust* learning always achieves optimal signaling in a setting where reinforcement learning sometimes does and sometimes does not. One might ask how this result generalizes. First we look at cases where the number of signals is different from the number of states and acts, keeping states equiprobable. Then we relax the condition of equiprobable states.

8 *Extra Signals*: N states, M signals, N acts (M > N)

If we have excess signals, reinforcement learning can still fall into partial pooling equilibria (Hu 2010; Hu et al. 2011). For probe and adjust, the picture has changed. The embedded Markov chain generated by probe and adjust no longer has any absorbing states. Efficient states—those where $g(f(s))$ is the identity—of the system consist of an N by N by N signaling system together with $M - N$ signals that are not sent. These signals are mapped onto some acts by the receiver's memory. Suppose some unused signal is mapped onto A_i. If the sender probes by sending the unused signal instead of the signal he usually sends in S_i, then the probe does not change his payoff. Then with some probability he adjusts by retaining the new signal in place of the old. (A receiver cannot probe changing an unused signal, since a receiver is not probing a new function, g, but only what to do if confronted with a signal. If a signal is unused, a receiver is not confronted with it.)

For example, in Figure 3, probe and adjust can lead from state A to B and state B to A but not outside the set. Check all other possible probes. They do not lead

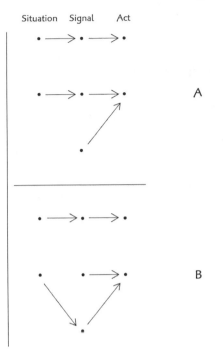

FIGURE 3 Probe and adjust

anywhere. This is an absorbing set (an ergodic set). Within the set, signals 2 and 3 can be thought of as sequential synonyms. One is used for situation 2 for a while and then the other is. After the system has been absorbed into this set, there is a long-term invariant distribution within the set. Signals 2 and 3 are each used half the time for situation 2. There are lots of such ergodic efficient sets.

It is still true that from any state there is a positive probability path to an efficient ergodic set. The same algorithm as before works for the same reasons. Applying the algorithm until one runs out of states give a path to a member of an efficient ergodic set.

Probe and adjust still learns to signal with probability one.

9 *Too Few Signals*: N **states**, M **signals**, N **acts** $(M < N)$

In this case, partial pooling is the best that can be done. There are not enough signals for a signaling system to map each state onto the appropriate act. The best average payoff achievable is (M/N), gotten when for each signal, s, $g(s)$ in $f^{-1}(s)$. The structure of these efficient signaling equilibria is investigated in Donaldson et al. (2007). Consider the case of $M = 2, N = 3$. In the efficient equilibria the sender pools 2 situations. There are 3 choices of situations to pool. For each, there is a choice which signal to assign to the pooled states. So there are 6 sender's strategies that are components of efficient equilibria. On the receiver's side, for such a sender's strategy there are two situations pooled, so it makes no payoff difference if the receiver does the right act for one or for the other. Thus 2 receiver's strategies pair with each of the 6 sender's strategies to make 12 optimal equilibria. This is shown in Figure 4.

signalling strategies

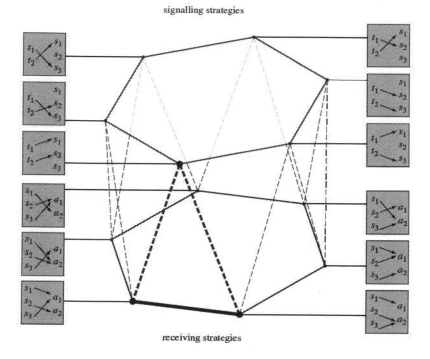

receiving strategies

FIGURE 4 Ergodic set of efficient equilibria from Donaldson et al. (2007)

Probe and adjust can move from one of these equilibria to another. Adopting the notation of the figure (s for situation, t for signal, a for act), suppose that we start at:

	Sender	Receiver
E1:	$s1 => t1$	$t1 => a1$
	$s2 => t2$	$t2 => a2$
	$s3 => t2$	

The sender pools situations 2 and 3. Suppose nature chooses $s3$, sender sends $t2$ and receiver probes $a3$. Receiver gets a payoff of 1 that, at worst, ties previous payoff for doing $a2$ upon seeing $t2$. So with positive probability receiver switches, taking us to:

	Sender	Receiver
E2:	$s1 => t1$	$t1 => a1$
	$s2 => t2$	$t2 => a3$
	$s3 => t2$	

This step is reversible; a probe can take us back to E1.

Now the receiver never does $a2$, so it does not matter if the sender pools $s2$ with $s1$ or with $s3$. Suppose nature chooses $s2$, and sender probes $t1$. This leads to 0 payoff which matches sender's previous payoff in $s2$, so with positive probability sender switches, leading to:

	Sender	Receiver
E3:	$s1 => t1$	$t1 => a1$
	$s2 => t1$	$t2 => a3$
	$s3 => t2$	

(This step is also reversible.) But now sender is pooling $s1$ and $s2$, so by the same logic as before a receiver's probe can lead to:

	Sender	Receiver
E4:	$s1 => t1$	$t1 => a2$
	$s2 => t1$	$t2 => a3$
	$s3 => t2$	

And now receiver never does $a1$, in turn setting up a sender's probe that can lead to:

	Sender	Receiver
E5:	$s1 => t2$	$t1 => a2$
	$s2 => t1$	$t2 => a3$
	$s3 => t2$	

The process continues through a cycle of the 12 efficient equilibria. All these efficient states form one ergodic set. As probe and adjust moves through this set, signals shift their meaning.

There is a positive probability path from any state to a state in the efficient ergodic set (Skyrms 2012).

Probe and adjust leads to efficient signaling.

10   Situations with Unequal Probabilities

So far, states have been supposed to have equal probabilities. There is to date no rigorous analysis of reinforcement learning in situations with unequal probabilities. But preliminary analyses and computer simulations point to the conclusion that in this case reinforcement learning can lead to *total pooling equilibria*. Consider the case of two situations, signals, and acts in which situation 1 is highly probable. Then a total pooling equilibrium in which the receiver just does act 1 and ignores the signals, and the sender ignores the state and always sends the same signal, is not so implausible. Players usually get paid off without bothering much with signaling. Simulations show reinforcement learning sometimes converging to a signaling system, sometimes to total pooling.

How does *probe and adjust* do with unequal probabilities? Notice that none of our foregoing analysis made use of the assumption of equal situation probabilities. We only need that situation probabilities are all positive in order to construct the positive probability paths leading to absorbing sets or states. Probe and adjust dynamics learns to signal perfectly in N situation, M signal, N act signaling games when there are enough signals ($M >= N$).

In the case where there are too few signals, $M < N$, efficiency imposes an extra requirement. Since there are not enough signals, states have to be pooled. In an efficient configuration, the highest probability states must be serviced in a way that maximizes expected payoff.

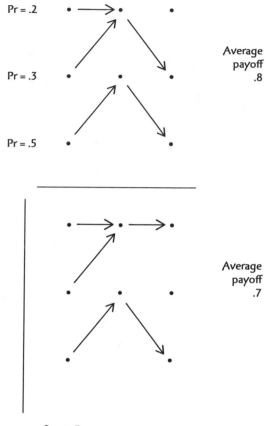

Pr = .2

Pr = .3

Pr = .5

Average
payoff
.8

Average
payoff
.7

figure 5

FIGURE 5 Inefficient state of probe and adjust

For example, suppose that $M = 2$, $N = 3$, and states 1, 2, 3 have probabilities .2, .3, .5 respectively. One efficient configuration will have the sender map states 1 and 2 onto one signal that the receiver maps to state 2, and have the sender map state 3 onto the other signal, which the receiver maps to act 3. Then the average payoff is .8. But probe and adjust may lead from this efficient state to and inefficient one, as shown in Figure 5.

In situation 1 sender sends signal 1 and receiver probes act 1. This gets a payoff of 1, which matches the receiver's memory of a payoff of 1 for doing act 2 on receiving signal 1. The tie is broken by a coin flip, so with some probability, receiver stays with the probe. The receiver has no memory of the *frequencies* of payoffs—only of the magnitude of the last payoff. Probe and adjust can move around the whole ergodic set of the preceding section. This weakness of probe and adjust here is just the other side of the coin from its strength.

11 Rapprochement

Probe and adjust has a constant probability of probing. A natural modification makes the probability of probes payoff-dependent: the worse the payoff, the more likely a probe.

One special case deserves note. Suppose that our trial and error learner has an idea of an optimal payoff. Then she may not probe at all if the last payoff was optimal.

In our signaling games, suppose a player does not probe at all if the last trial was a success, but probes with some probability (greater than zero and less than one) if the last trial was a failure. Then the positive results of the preceding discussion are retained, but residual probing is eliminated. (And it is no longer necessary to include the assumption that there are no simultaneous probes.) These agents always find a signaling system by trial and error, and then stick to it.

On the other hand, reinforcement learners become more likely to stick with successes if their initial propensities are small (or equivalently the rewards for success are large.) Pushing this to the extreme, suppose that we start out with infinitesimal initial propensities. Another way of saying this is that we start out with the urns empty, with the rule that for empty urns you choose at random, but for non-empty urns you use Roth-Erev reinforcement. An initial reinforcement then causes sender and receiver to lock-in the actions that produced the reinforcement. This is like the lock-in of the modified *probe and adjust* described in the previous paragraph. With this dynamics such learners *always learn to signal* (Barrett and Zollman 2009).

This modification weakens exploration. It does very poorly in bandit problems. But reinforcement with very small initial propensities is almost as good at learning to signal. Simulations with very small propensities always learn to signal in all the signaling games discussed here. And small initial propensities still guarantee asymptotic optimality in bandit problems.

12 Conclusion

Is it possible to learn to signal by trial and error? The answer is robustly positive. Both kinds of trial and error that we have considered sometimes learn to signal in all signaling games, and always learn to signal in the simplest signaling game. In addition, probe and adjust always learns to signal in a wide variety of settings.

13 Related Work

The need for low rationality game theory is emphasized in Roth and Erev (1995); Erev and Roth (1998); Fudenberg and Levine (1998); Young (2004, 2009). *Probe and adjust* can be seen as a version of the one-element stimulus sampling model of Estes (1950). Estes learning theory was applied to game theory in the pioneering work of Suppes and Atkinson (1960). There are more complicated versions of probe and adjust type learning, with more memory and payoff-dependent probes, which perform well very generally (Marden et al. 2009; Young 2009).

Equilibrium structure of signaling games with many or few signals is analyzed in Donaldson et al. (2007). Two-population replicator dynamics of a 2 situation, 2 signal, 2 act signaling game where the situations have unequal probabilities is analyzed, with and without mutation, in Hofbauer and Huttegger (2008).

Blume et al. (1998, 2002) present experimental evidence for the emergence of signaling by learning in a 2 situation, 2 signal, 2 act signaling game. Where subjects were only given own payoffs as feedback, experimental results were consistent with reinforcement learning.

References

Argiento, R., R. Pemantle, B. Skyrms, and S. Volkov. 2009. Learning to signal: Analysis of a micro-level reinforcement model. *Stochastic Processes and their Applications* 119(2):373–390. doi:10.1016/j.spa.2008.02.014.

Barrett, J. and K. J. S. Zollman. 2009. The role of foregetting in the evolution and learning of language. *Journal of Experimental and Theoretical Artificial Intelligence* 21(4):293–309. doi:10.1080/09528130902823656.

Barrett, J. A. 2007. Dynamic partitioning and the conventionality of kinds. *Philosophy of Science* 74(4):527–546.

Beggs, A. W. 2005. On the convergence of reinforcement learning. *Journal of Economic Theory* 122(1):1–36. doi:10.1016/j.jet.2004.03.008.

Blume, A., D. V. DeJong, Y. G. Kim, and G. B. Sprinkle. 1998. Experimental evidence on the evolution of the meaning of messages in sender-receiver games. *American Economic Review* 88(5):1323–1340.

Blume, A., D. V. DeJong, G. R. Neumann, and N. E. Savin. 2002. Learning and communication in sender-receiver games: An econometric investigation. *Journal of Applied Econometrics* 17(3):225–247.

Donaldson, M. C., M. Lachmann, and C. T. Bergstrom. 2007. The evolution of functionally referential meaning in a structured world. *Journal of Theoretical Biology* 246(2):225–233. doi:10.1016/j.jtbi.2006.12.031.

Erev, I. and A. Roth. 1998. Predicting how people play games: Reinforcement learning in games with unique mixed-strategy equilibria. *American Economic Review* 88:848–881.

Estes, W. K. 1950. Toward a statistical theory of learning. *Psychological Review* 57:94–107.

Fudenberg, D. and D. K. Levine. 1998. *The Theory of Learning in Games.* Cambridge, Mass.: MIT Press.

Herrnstein, R. J. 1970. On the law of effect. *Journal of the Experimental Analysis of Behavior* 13(2):243–266. doi:10.1901/jeab.1970.13-243.

Hofbauer, J. and S. M. Huttegger. 2008. Feasibility of communication in binary signaling games. *Journal of Theoretical Biology* 254(4):843–849. doi:10.1016/j.jtbi.2008.07.010.

Hopkins, E. and M. Posch. 2005. Attainability of boundary points under reinforcement learning. *Game and Economic Behavior* 53(1):110–125. doi:10.1016/j.geb.2004.08.002.

Hu, Y. 2010. *Essays on random processes with reinforcement.* Ph.D. thesis, St. Anne's College, University of Oxford.

Hu, Y., B. Skyrms, and P. Tarrés. 2011. Reinforcement learning in signaling games. ArXiv:1103.5818v1[math.PR].

Huttegger, S. and B. Skyrms. Forthcoming. Emergence of a signaling network with *Probe and Adjust.* In *Cooperation, Complexity, and Signaling*, Ed. B. Calcott, R. Joyce and K. Sterelney. Cambridge, Mass.: MIT Press.

Kimbrough, S. O. and F. H. Murphy. 2009. Learning to collude tacitly on production levels by oligopolistic agents. *Computational Economics* 33(1):47–78.

Lewis, D. K. 1969. *Convention: A Philosophical Study.* Cambridge, Mass.: Harvard University Press.

Marden, J. P., H. P. Young, G. Arslan, and J. S. Shamma. 2009. Payoff-based dynamics for multiplayer weakly acyclic games. *SIAM Journal on Control and Optimization* 48(1):373–396.

Pemantle, R. 2007. A survey of random processes with reinforcement. *Probability Surveys* 4:1–79.

Roth, A. E. and I. Erev. 1995. Learning in extensive form games: Experimental data and simple dynamic models in the intermediate term. *Games and Economic Behavior* 8(1):164–212. doi:10.1016/S0899-8256(05)80020-X.

Skyrms, B. 2010. *Signals: Evolution, Learning and Information.* Oxford: Oxford University Press.

—. 2012. Learning to signal with 'probe and adjust'. *Episteme* 9(2):139–150. doi:10.1017/epi. 2012.5.

Suppes, P. and R. C. Atkinson. 1960. *Markov Learning Models for Multiperson Interactions.* Stanford, CA: Stanford University Press.

Vulkan, N. 2000. An economist's perspective on probability matching. *Journal of Economic Surveys* 14(1):101–118.

Young, H. P. 2004. *Strategic learning and its limits.* Oxford: Oxford University Press.

—. 2009. Learning by trial and error. *Games and Economic Behavior* 65(2):626–643. doi: 10.1016/j.geb.2008.02.011.

13

From Rousseau to Suppes. On Diaries and Probabilistic Grammars

Willem J. M. Levelt

Jean-Jacques Rousseau was born in 1712, three centuries ago. This year (2012) we also celebrate the 250th anniversary of his book *Émile ou de l'éducation* of 1762. Although Rousseau was an almost exact contemporary of David Hume, the *Émile* was not an Enlightenment treatise, but in fact an early Romantic one. Not rationality, but nature should be our lead in creating a society in which the Enlightenment virtues of human goodness, freedom and equality will prosper. With the *Émile*, Rousseau intended to revolutionize educational practice. It is nature, not human drill, by which the child should be educated. A first requirement to make this work is for teachers to carefully observe their children: "Hence, begin by better studying your pupils, because surely you don't know them at all" (Rousseau (1762, p. 3)). It is a condition for a natural education of the child. And that was certainly a tough type of education: "Observe nature, follow the route that it traces for you. Nature exercises children continually, it hardens their temperament by all kinds of difficulties, it teaches them early the meaning of pain and sorrow." (p. 65); "Accustom them therefore to the hardships they will have to face; train them to endure extremes of temperature, climate, and condition, hunger, thirst, and weariness." (p. 66). (All children Rousseau fathered himself were dropped in children's homes, completely deserted by Jean-Jacques).

Although Rousseau's own observations on children's development were rather limited, many pedagogues took his advice seriously and began collecting observations on children's development, including language development. Rousseau's theories became especially influential in German pedagogical circles. Educational practice should turn back to nature became the new principle. Joachim Heinrich Campe, enlightened theologian and pedagogue in Hamburg, organized a "Society of Practical Pedagogues" who jointly published, in the period from 1785 to 1793, a "General Revision" of the educational system in 16 volumes.

In his preface, Campe pleaded for well-off philanthropists to make available a thousand thaler for a competition on diary writing. Such a diary on "the bodily and mental changes of a child" from birth would "indescribably" enrich our knowledge of the growing-

Foundations and Methods from Mathematics to Neuroscience.
Colleen E. Crangle, Adolfo García de la Sienra and Helen Longino.

up child. (Campe (1785-1792, pp. xxiv–xxv)). There is no evidence that Campe's competition materialized, but a few attempts in diary writing followed his plea.

Dietrich Tiedemann, a philosopher, had kept his diary from 1781 to 1784 and published it in 1787. It contains, well-counted, 15 observation's on son Friedrich's speech during the 30 month observation period since birth. Around 7 months, for instance, Friedrich began to imitate spoken sounds, such as *ma*. At 8 months he would point to X when asked "where is X?" At 19 months he would produce a variety of words, but the productions were monosyllabic, usually the word's last or stressed syllable. At 21 months the first sentences appeared, mostly combining an infinitive verb and a nominal noun; there were no inflections or articles.

Only one year later, mathematician Moritz Adolph von Winterfeld, heavily inspired by Rousseau, published another diary. It announced to relate the "the gradual formation of the quite peculiar language, the very simple children's grammar." (von Winterfeld (1789, 1791, p. 405)). However, it mostly concerns the bodily development of daughter Amalie Louise (born January 13, 1785), far less her mental development. As for linguistic observations, there is mention of a few first words, there is mention of the impossibility to pronounce *h*, "although this is one of the easiest letters" , and there is mention of a first negation, *nicht all* at 32 months. That is all. But Winterfeld was certainly enlightened. He inoculated his child with his own hands with the puss of cow's smallpox. The child got very ill, but then survived the following smallpox epidemic.

A second small wave of diary keeping emerged around the midst of the 19th century. Four diaries survive from that period (by Goltz, Löbisch, Eschericht and Sigismund). Most extensive and most cited became the 1856 monograph by Berthold Sigismund, a family doctor and teacher in Rudolfstadt. Sigismund dedicated some 50 quite perceptive pages to language acquisition in the second year of life. Linguistic observations on the first year are few, and for us remarkable for their underestimation of the child's capacities. I will not go over much interesting detail, but just mention two important theoretical claims, that would play a long-lasting role in language acquisition research. The first one is Sigismund's claim about the function of all first words:

> "That the little speaker uses the first uttered words at once, mainly or maybe exclusively, as expressions of will", "The protolanguage is nothing but a will made audible."
> (pp. 112–113)

This idea would be picked up, almost half a century later by Wundt's student Meumann and then became canonical in the literature. The second one concerns the child's early phonology. It is the notion of 'least effort'. Easy speech sounds, such as *b* and *m*, *n*, *d*, and *s*, come before the harder ones, such as *g*, *w*, followed by *f*, *ch*, *k*, with *l*, and *sch*, with *r* closing the ranks. This 'least effort' notion was going to play a major role in evolutionary explanations.

All in all, however, in mid-19th century, diary making had lost its intellectual appeal. All diary keepers were isolated teachers without any link to science or academic circles. That changed drastically 30 years later, in 1886/87. Then, what had been no more than marginal, scattered business, was suddenly drawn into an explosive scientific development. The French man of letters Hippolyte Taine provided the fuse and Charles Darwin set fire to it. In 1876 Taine had published a report of the diary notes he had made on his infant daughter's language development, a report making ample reference to evolution theory:

"Speaking generally, the child presents in a passing state the mental characteristics that are found in a fixed state in primitive civilizations, very much as the human embryo presents in a passing state the physical characteristics that are found in a fixed state in the classes of inferior animals." (p. 259)

The next year the new journal *Mind* published an English translation thereof. This triggered Charles Darwin to publish, in the same year 1877 and the same journal, a 10-page *Biographical sketch* of his own son William's development as an infant. The sketch was based on copious notes Darwin had made between 1839 (upon William's birth) and 1841. Clearly, after reading Taine's paper, Darwin didn't want to repeat the Wallace affair. He had been the first to keep a diary, over 30 years before Taine, and the world should know. Celebrity Darwin's paper appeared the same year also in French, German and Russian, not failing to promote the diary business on a grand scale. From now on, keeping diaries on child development was real science. A tsunami of diary keeping emerged, which reverberates till the present day.

Darwin's sketch includes some observations on the development of William's language skills, hardly more than the 15 observations Tiedemann had provided almost a century earlier and substantially less than Sigismund's extensive records. Darwin stressed in particular the invention of first words, such as *mum* to express the wish for food. He also noticed the "instinctive" use of intonation patterns, "voice modulation" , to express various modes, such as interrogation and exclamation. Here he concluded, repeating what he expressed in *The Descent of Man*, that "before man used articulate language, he uttered notes in a true musical scale" (p. 293), the singing origins of language, which never stopped echoing in the literature. The importance of Darwin's paper was not so much in its content. But in one swoop it made the study of child development a respectable branch of human biology. Diaries now appeared at an accelerated rate, and in various languages, see Table 1. My book on the history of psycholinguistics (Levelt (2013)) provides much detail on the history of child language diaries.

TABLE 1 Nineteenth century diaries

Goltz (1847), German	Sayce (1889), Arabic
Löbisch (1851), German	Chamberlain (1890), Algonkin
Eschericht (1852), German	Gabriel Deville (1890/91), French
Sigismund (1853), German	Garbini (1892), Italian
Baudouin de Courtenay (1869), Polish	Compayré (1893), French
Taine (1876, 1877), French	Balassa (1893), Hungarian
Darwin (1877), English	Frederic Tracy (1894), English
Perez (1878, 1886), French	Paola Lombroso (1894), Italian
Strümpell (1880), German	Preyer (1896), German
Sikorsky (1883), Russian	Kathreen Moore (1896), English
Blagovescenskij (1886), Russian	Milicent Washburn (1898), English
Machado y Álvarez (1885–1887), Spanish	Ament (1899), German

It is from these early diaries that the first child language statistics was derived. Doran (1907) was the first to publish an overall statistics on vocabulary size (based on over 100 children), see Fig. 1.

This diary industry continued all over the 20th century. Table 2 presents an overview of 20th century diaries before 1960. Here Clara and William Stern's 1907 extensive study

Vocabularies of Children.

Table I.

Vocabularies of Children.

Age. Month.	Number of Words, and Reference. No.	Ref.	No.	Ref.	No.	Ref.	No.	Ref.	No.	Ref.	Av.
8	1	51									1
9	1	51	9	107							5
10	3	51	10	107	3	40b					5.3
11	12	51									12
12	4	107	8	107	10	40	17	30	200	107	13.8
	24	51									
13	22	40	38	51							25
14	50	40	58	51							54
15	4	40	60	44	75	40	106	51			61.2
16	9	40	89	40	199	51					99
17	18	40	35	107	151	40	232	51			109
18	33	40	74	110	80	44	113	40	144	30	112.8
	233	40									
19	83	40	115	30	144	107	281	40			155.7
20	204	40	374	40							289
21	177	107	369	40	579	40					375
22	28	107	69	107	377	40	465	40	665	40	320.8
23	136	107	489	40	612	40	710	40			486.7
24	10	101	36	107	50	81	115	41	139	107	
	173	56	285c	107	397	82	399	56	400c	41	455.5
	483	56	578	42	614	42	729	40	741	40	
	783	40	1050	41	1121	58	d	78			
25	250	107	832	40							341
26	923	40									923
27	171	107	1032	40	1079	40					760.2
28	451	107	667	107							559
29	1187	40									1187
30	327	107	480	110	629	41	751	41	769	41	869.8
	1050c	44	1432	40	1519	40					
32	642	88									642
36	1176	42									1176
42	839	44									837
45	1009	44									1009
60	830	e	1600	42	1600	107					1399.2
66	1528	88	1529	53							1528.5
72	960	e	2000	78f							1480

Notes.

a. All references (51) indicate the same child at different ages.

b. All references (40) refer to one of three children in the same family at different ages, shown separately in Tables II and III.

c. Exact age uncertain.

d. For the record of Viola Olerich, "the baby scholar," see p. 426.

e. Pres. E. D. Murdaugh.

f. Estimated.

FIGURE 1 Doran's (1907) child vocabulary statistics

set the new standards for the decades to come. They reported in much detail on the language development of their three children Hilde, Günther and Eva. Vocabulary development is only one aspect of this study. A major part of the book is dedicated to syntactic development. In the decades to follow, a rich statistics was collected on syntactic complexity, from mere utterance length to the variety of syntactic types, coordination and subordination.

When famous, but Jewish William Stern was dismissed from his Hamburg professorship in 1933, Clara and William moved to Duke University, where William died in 1938. Clara died in 1945 in New York. Their former student Gorden Allport took care that the diaries, the largest ever created, were deposited in the Widener library. However, nobody showed any interest in them. Youngest daughter Eva then moved them to Hebrew University. With Eva's help we transcribed the full diaries at my Max Planck Institute and made them digitally available to the world, then the largest corpus of German language acquisition data.[1]

[1] `http://www.mpi.nl/resources/data/stern-diaries`

TABLE 2 Twentieth century diary studies before 1960

Clara & William Stern (1907), German	Grégoire (1937, 1947), French
O'Shea (1907), English	Wawroska (1938), Polish
Gheorgov (1908, 1910), Bulgarian	Velten (1943), English
Ronjat (1913), French, German	Frontali (1943, 1944), Italian
Pavlovitsch (1920), Serbian	Gvozdev (1948, 1949), Russian
Bolin & Bolin, Swedish	Skorupka (1949), Polish
Jesperson (1916), Danish	Leopold (1939, 1949), English, German
Van Ginneken (1917), Dutch	Chao (1951), Cantonese
Kenyeres (1926), Hungarian	Cohen (1952), French
David & Rosa Katz (1928), German	Kaczmarek (1953), Polish
Ohwaki (1933), Japanese	Burling (1959), Garo
Lewis (1936), English	Bar-Adon (1959), Hebrew

The intellectual break with the rich German tradition was complete after the war. Roger Brown doesn't even mention the Sterns' monumental work in his famous 1973 book *A first language*, to which I will return below.

But first I should commemorate another occasion, dear to me. Four decades ago, in 1972, a few months after his 50th birthday, Patrick Suppes lectured in a NUFFIC summer course, which my former supervisor John van de Geer had organized in The Hague. I was on the organizing committee and Professor Suppes lectured on formal grammars and automata. I had just returned from the Institute for Advanced Study in Princeton (I will be eternally grateful to Duncan Luce who had invited me there). During the year I had written my treatise on formal grammars (Levelt (1974)), so I was all tuned in for Professor Suppes' course. One thing he discussed was his work on probabilistic grammars. In my book I had included a chapter on probabilistic grammars and further chapters on their (potential) application in linguistics and psycholinguistics. In the dominant Chomskyan linguistic community of the time, this was absolutely not done. This is what Chomsky himself had to say about it:

> "It must be recognized that the notion 'probability of a sentence' is an entirely useless one, under any known interpretation of this term. On empirical grounds, the probability of my producing some given sentence of English -- say, this sentence, or the sentence "birds fly" or "Tuesday follows Monday" , or whatever -- is indistinguishable from the probability of my producing a given sentence of Japanese."   (Chomsky 1969, p. 57)

Patrick Suppes not only pertinently and repeatedly argued against that curious position, but also set the example. In the early 1970s he and his research team were the first to do serious work on probabilistic grammars for early children's speech. It was the only empirical work on probabilistic grammars available when I wrote my book. Patrick Suppes' first applications were to the Adam corpus of utterances, collected by Roger Brown and co-workers, on which *A first language* is partly based. Brown had, unknowingly, continued the work by the Sterns, in particular their work on syntactic and semantic development.

The classic contribution Patrick Suppes (Suppes 1970b,a, 1971a,b; Suppes and Feldman 1971; Suppes 1974; Léveillé and Suppes 1976; Suppes 1976b,a; Suppes and Macken 1978) made to the study of language acquisition was two-pronged. He was the first to construct probabilistic grammars for a range of child language corpora. Not only Adam's corpus, but also corpora collected by his own team (especially Madeleine Léveillé and

Production Rule	Probability
1. NP → N	a_1
2. NP → AdjP	a_2
3. NP → AdjP + N	a_3
4. NP → Pro	a_4
5. NP → NP + NP	a_5
6. AdjP → AdjP + Adj	b_1
7. AdjP → Adj	b_2

Estimated Parameter Values

$a_1 = .6391$ $b_1 = .0581$

$a_2 = .0529$ $b_2 = .9419$

$a_3 = .0497$

$a_4 = .1439$

$a_5 = .1144$

TABLE I

Probabilistic Noun-Phrase Grammar for Adam I

Noun phrase	Observed frequency	Theoretical frequency
N	1445	1555.6
P	388	350.1
NN	231	113 7
AN	135	114 0
A	114	121.3
PN	31	25.6
NA	19	8.9
NNN	12	8.3
AA	10	7.1
NAN	8	8.3
AP	6	2.0
PPN	6	.4
ANN	5	8.3
AAN	4	6.6
PA	4	2.0
ANA	3	.7
APN	3	.1
AAA	2	.4
APA	2	.0
NPP	2	.4
PAA	2	.1
PAN	2	1.9

FIGURE 2 The very first probabilistic grammar for a child language corpus. The noun phrase grammar for the Adam I corpus, (From Suppes (1973))

Robert Smith (Suppes et al. 1974) such as Nina's corpus (23–39 months), 102.230 tokens, Philippe's French corpus (25–39 months), 56.982 tokens, Erica's corpus and a small Chinese corpus.

Figure 2 depicts the very first probabilistic grammar for Adam's corpus. It was only the beginning. When you read all subsequent papers, you get impressed not only by the sheer amount of detailed work, but by the enormous constraints imposed by the probabilistic paradigm on feasible syntactic rules. Rules that seem obvious to the ordinary linguist just do not work, whereas others that are considered trivial explain large degrees of variance.

The second important innovation Suppes introduced was to supply these grammars with a compositional, model-theoretic semantics (Fig. 3). That was the other thing not done in the Chomskyan linguistics of the day. Syntax was the thing, semantics was eschewed. Suppes supplied each syntactic production rule with a semantic function (such as identity, intersection, intensification, etc). These functions then combined, following the syntax, to compose the meaning of the noun phrase as a whole. And again, the experience was that these semantic functions put further constraints on what could be a possible grammar for the corpus.

To conclude, the aim of this paper was to acknowledge the important innovative twist Patrick Suppes gave to the now 250 year old tradition (since Rousseau's Émile) of collecting data on children's spontaneous speech. It was the introduction of probabilistic grammars and semantics to full corpora of children's speech. This innovation went much against the current in the linguistics of the late 1960s and early 1970s. Indeed, it took decades before the application of probabilistic grammars to large scale corpora really took off. It is an established, booming field now, both for developmental and adult corpora, and in many languages. Its combination with model-theoretic semantics, however,

Production Rule	Probability	Semantic function
1. NP → N	a_1	identity
2. NP → AdjP	a_2	identity
3. NP → AdjP + N	a_3	intersection
4. NP → Pro	a_4	identity
5. NP → NP + NP	a_5	choice function
6. AdjP → AdjP + Adj	b_1	intersection
7. AdjP → Adj	b_2	identity

FIGURE 3 Model-theoretic semantics for Adam I probabilistic noun phrase grammar (From Suppes 1973).

is still a rare commodity. For completeness' sake, the *References* below also lists the original papers of the Suppes team during the 1970s.

References

Brown, R. 1973. *A First Language: The Early Stages.* Cambridge, MA. Harvard University Press.

Campe, J. H. 1785-1792. *Allgemeine Revision des gesammten Schul- und Erziehungswesens von einer Gesellschaft praktischer Erzieher*, vol. 16. Hamburg: Carl Ernst Bohn.

Chomsky, N. 1969. Quine's empirical assumptions. In D. D., W. V. Quine, and J. Hintikka, eds., *Words and objections: Essays on the work of W. V. Quine*, pages 53–68. Dordrecht: D. Reidel.

Darwin, C. 1877. A biographical sketch of an infant. *Mind* 2(7):285–294. doi:10.1093/mind/os-2.7.285.

Doran, E. W. 1907. A study of vocabularies. *Pedagogical Seminary* 14(4):401–438.

Léveillé, M. and P. Suppes. 1976. La compréhension des marques d'appartenance par les enfants. *Enfance* 29(3):309 318.

Levelt, W. J. M. 1974. *Formal grammars in linguistics and psycholinguistics*, vol. 3. The Hague: Mouton, new edition (2008). Amsterdam: John Benjamins edn.

—. 2013. *A History of Psycholinguistics. The Pre-Chomskyan Era.* Oxford: Oxford University Press.

Rousseau, J. J. 1762. *Émile ou de l'éducation.* Amsterdam: Chez Jean Néaulme.

Sigismund, B. 1856. *Kind und Welt.* Braunschweig: F. Vieweg und Sohn, new, undated edition by F. Forster. Leipzig: Jaeger edn.

Stern, C. and W. Stern. 1907. *Die Kindersprache. Eine psychologische und sprachtheoretische Untersuchung.* Leipzig: Johan Ambrosius Barth.

Suppes, P. 1970a. Probabilistic grammars for natural languages. *Synthese* 22(1/2):95–116.

—. 1970b. Semantics of context free fragments of natural languages. In J. Hintikka, J. Maravesik, and P. Suppes, eds., *Approaches to natural language*, pages 370–394. Dordrecht: Reidel.

—. 1971a. Introduction to automata and formal languages. Manuscript, Stanford University.

—. 1971b. Semantics of context-free fragments of natural languages. Psychology Series 171, Institute for Mathematical Studies in the Social Sciences, Stanford University, Stanford, CA.

—. 1973. Theory of automata and its applications to psychology. In G. J. Dalenoort, ed., *Process models for psychology: Lecture notes of the NUFFIC international summer course, 1972.* Rotterdam: Rotterdam University Press.

—. 1974. The semantics of children's language. *American Psychologist* 29(2):103–114. doi: 10.1037/h0036026.

—. 1976a. Elimination of quantifiers in the semantics of natural language by use of extended algebras. *Revue Internationale de Philosophie* 30:243–259.

—. 1976b. Syntax and semantics of children's language. In S. R. Harnad, H. Steklis, and J. Lancaster, eds., *Origins and evolution of language and speech*, vol. 280, pages 227–237. New York: New York Academic of Sciences.

Suppes, P. and S. Feldman. 1971. Young children's comprehension of logical connectives. *Journal of Experimental Child Psychology* 12(3):304–317. doi:10.1016/0022-0965(71)90027-0.

Suppes, P. and E. Macken. 1978. Steps toward a variable-free semantics of attributive adjectives, possessives, and intensifying adverbs. In K. Nelson, ed., *Children's language*, vol. 1, pages 81–115. New York: Gardner Press.

Suppes, P., R. Smith, and M. Léveillé. 1974. The French syntax of a child's noun phrases. *Archives de Psychologie* 42(166):207–269.

Taine, H. 1877. Acquisition of language by children. *Mind* 2(6):252–259. doi:10.1093/mind/os-2.6.252.

Tiedemann, D. 1787. Beobachtungen über die Entwicklung der Seelenfähigkeiten bei Kindern. In *Hessische Beitrage zur Gelehrsamkeit und Kunst*, vol. (Bd. II. S. 313 ff. S. 486 ff.). Oskar Bonde, new edition by C. Ufer (1897) edn.

von Winterfeld, M. A. 1789. Tagebuch eines Vaters über sein neugebohrenes Kind. *Braunschweigisches Journal philosophischen, philologischen und padagogischen Inhalts* 4:404–441.

—. 1791. Tagebuch eines Vaters über sein neugebohrenes Kind. *Braunschweigisches Journal philosophischen, philologischen und padagogischen Inhalts* 6:476–484.

14

Suppes and Husserl

DAGFINN FØLLESDAL

I have known Patrick Suppes for almost fifty years, from 1964, when I met him at the World Congress for Logic, Methodology and Philosophy of Science in Jerusalem and he invited me to come to Stanford. Since then I have greatly appreciated him as a friend and colleague. Over the years we have taught numerous seminars together at Stanford on a large variety of topics. I have participated in the celebration of several of his life's events, including most of his many anniversaries and the numerous celebrations connected with his receiving prizes and other recognitions. On such occasions I have often given speeches about him and his work. Two of these went into volumes in his honor, Føllesdal (1994, 2008), see alsoFrauchiger (2013, pp. 283ff.), and I will avoid repeating what I have said there. However, there is always more to tell, Suppes continues to be productive in ever new fields.

In 1990, when Suppes received the U.S. National Medal of Science, the brief statement attached to the award summarized his work as follows:

"For his broad efforts to deepen the theoretical and empirical understanding of four major areas:

the measurement of subjective probability and utility in uncertain situations;
the development and testing of general learning theory;
the semantics and syntax of natural language; and
the use of interactive computer programs for instruction."

Every attempt to group Suppes' work into categories is bound to be arbitrary. A far more detailed grouping was given by Suppes himself in his autobiography in 1978. However, in the bibliography on his present home page Suppes arranges his writings in a way that I think cuts his huge production better at its joints:

Methodology, Probability and Measurement
Psychology
Foundations of Physics
Language and Logic
Computers and Education
Mind and Brain

Foundations and Methods from Mathematics to Neuroscience.
Colleen E. Crangle, Adolfo García de la Sienra and Helen Longino.
Copyright © 2014, CSLI Publications.

In my earlier tributes I have praised the combination of enthusiastic openness and critical questions about evidence that Suppes brings to all these fields, and I have focused particularly on his contributions to the philosophy of language.

The last field on the list, "Mind and Brain" is a late addition. Suppes' first publications in this field came in 1997 and it is now his most active field of research, with several papers published during the past three years. I will discuss two of Suppes' latest papers in this field: "Neuropsychological Foundations of Philosophy" (2009a) and "Neurophysiological Foundations of Phenomenology: Is It Possible?" (2013). In these two papers, and particularly in the latter, he focuses on themes that we have discussed over many years and that have been the topic of several of our joint seminars. The key philosophers involved are Aristotle, Thomas Aquinas, Hume, James and Husserl.

One important theme in all these philosophers is the central role they give to association. This is also the topic that ties their views to neuroscience. Suppes' work in neuroscience focuses on associative networks. These play an important role in all human activity and their complexity is a great challenge in neuroscience. Suppes points out that associations are an important part of human brain activity, and are the mechanism of learning in species very far down the scale of evolutionary development. He has in particular studied these networks in connection with learning and use of language. In a recent article he even has brought these networks into the study of poetry: "Rhythm and Meaning in Poetry" (2009b). He here points out that Hume in Book II of the *Treatise*, "Of the Passions," gives a broader and detailed discussion of his theory of association, where also our passions, or emotions, come in. This broader theory is then further developed by Suppes and is used by him to throw light on the importance of rhythm in poetry.

What is remarkable, is that Husserl, who from 1997 on called himself a transcendental philosopher and in many ways was close to Kant,[1] belongs in this group of philosophers. Suppes has explored this in his "Neurophysiological Foundations of Phenomenology: Is It Possible?" In this article I will follow this up further and explore some of the similarities and differences between Husserl and the other philosophers I listed. Very many of the connections are due to Husserl's teacher Franz Brentano, who was a main reason why Husserl in 1884 changed to philosophy, one year after he got his doctorate in mathematics. Husserl writes that Brentano's lectures "created in him the conviction that ... philosophy, too, is a field of serious work, which can be treated in the manner of a strict science and therefore must be treated that way."[2] (Something similar also happened to Freud, who followed Brentano's lectures and under his influence almost decided to study for a doctorate in philosophy.) Brentano was an Aristotle and Aquinas scholar, and his theory of intentionality, which Husserl took over and developed further, is inspired by these philosophers. Particularly important were Aristotle's *De Anima* and Thomas Aquinas' commentary on this work, which Suppes and I used as texts for one of our seminars.

Brentano taught his students to emphasize argument and empirical evidence. He advised them to take courses in the natural sciences, since he regarded both the methods and the findings of science important to philosophy. He also encouraged his students to study the British empiricists instead of steering them toward German idealism, which was common in Germany and Austria at that time. Freud translated some of John

[1] Kern (1964) is an excellent book on this subject.
[2] Husserl, *Erinnerungen an Franz Brentano*, p. 154.

Stuart Mill's work into German, and in Freud's *On Aphasia* (1891) there are references to both *An Examination of Sir William Hamilton's Philosophy* and to *A System of Logic*. Freud makes use of the idea that words acquire meaning by being linked to an object-presentation through a chain of associations. This idea recurs in "The Unconscious" (1915), where it plays a crucial role in differentiating the conscious from the unconscious.

Brentano regarded Hume very highly, and so did Husserl. Husserl preserved this high esteem for Hume throughout his life. He wrote in *Erste Philosophie* (1923–24):

"The one who completes Berkeley, but far outranks him in his immanent naturalism, is David Hume. His unique importance in the history of philosophy lies in this: he saw in Berkeley's theories and criticism the breakthrough of a new psychology in which he recognizes the foundation for all possible sciences ... and tries to work it out systematically, in the style of an immanent naturalism with the sharpest stringency."

Erste Philosophie, p. 155

Husserl finds, however, that Hume has been misunderstood, by Kant and others, who have not grasped the full subtlety of his position. He writes in *Formal and Transcendental Logic* (1929):

"Hume's greatness (a greatness still unrecognized in this, its most important aspect) lies in the fact that ... he was the first to grasp the universal concrete problem of transcendental philosophy."

Formal and Transcendental Logic, Cairns'translation, p. 256

Husserl continues at length his praise for Hume and explains what he found missing in Kant. However, now to the next philosopher on our list: William James. Both in our seminars and in his publications Suppes has made it clear that James is one of his heroes. Remarkably, James is also the one on our list who comes closest to Husserl. Husserl learned about James from Carl Stumpf, one of Brentano's earlier students, for whom Husserl wrote his habilitation dissertation, on the recommendation of Brentano. James refers to Stumpf several times in *The Principles of Psychology*, and comparing Stumpf to a number of other authors from whom he has "derived most aid and comfort" James writes that "Stumpf was the most philosophical and profound of all these writers; and I owe him much." James (1890, Harvard edition, p. 911)

Husserl read parts of *The Principles of Psychology* in 1891–92, and in 1894 he studied it carefully. His copy of the work, which is preserved in the Husserl Archive in Leuven, is intensively marked in Chapters IX (Stream of Thought), XI (Attention), and XII (Conception). Husserl was so impressed by James that he told Ralph Barton Perry that he had abandoned his plan of writing a psychology of his own.

Several central ideas and themes from James were taken up by Husserl in his later work: the notion of fringes, elements in his philosophy of time, the origins of our conception of reality, and many, many others—above all the central role of association.

I will now take up some of these themes that are suitable for bringing out the many connections between Husserl, James and the two recent papers of Suppes on neurophysiology that I mentioned in my introduction.

Like Suppes I find that perception is a good starting point. Aristotle held that when we perceive an object, the form of the object is transferred from the object through the medium and via our sensory organs to our passive intellect. He expressed it briefly in his famous statement (*De Anima* 424a18–21) that "a sense is what has the power of receiving into itself the sensible forms of things without the matter." Suppes and I both find this congenial, especially when interpreted in the way Thomas Aquinas does it in his commentary on *De Anima*. However, the view has its problems. I will mention two, which both can be illustrated by the duck/rabbit picture. First, in its normal use, this picture illustrates that what reaches the retina is not sufficient to determine what we see – we can vacillate between seeing a duck and a rabbit. Secondly, a person who has grown up seeing lots of rabbits but never seeing a duck, will probably not experience the picture as ambiguous, but see it as a picture of a rabbit. Conversely with people growing up seeing ducks but no rabbits.

Explaining perception by help of entrenched associations takes care of this. Husserl develops this view in great detail, starting when he was still working with Stumpf. He never uses the duck/rabbit example, which became popular with the rise of Gestalt psychology twenty years later. However, the duck/rabbit picture originated at the time Husserl worked on these problems, it was first published in *Fliegende Blätter*, a German humor magazine (Oct. 23, 1892, p. 147)[3] and was made famous by Joseph Jastrow, the Gestalt psychologists and Wittgenstein in his *Philosophical Investigations*.

Husserl kept on developing his view during the rest of his life. His presentation in *Ideas* (1950) is particularly important. He there introduces the three notions of *noesis, hyle and noema*, which are the key elements not only in his theory of perception, but in his phenomenology as a whole.

Briefly, the *noesis* is the complex set of anticipations that is involved in all acts, not just acts of perception. These are largely established through past experience, Husserl often calls them *sedimentations* [Niederschläge] from past experience. The notion of habit, or habitualities, is also central in his work, and it is important that these are not just cognitive anticipations, but involve all aspects of human life, practical, social, normative, etc. I have not seen any development of the associationist view that is as wide-ranging as Husserl's. In his discussion of sedimentation Husserl often recurs to how sedimentations are interrelated and make up a net where past sedimentations are modified by new ones. Here is a typical passage from his discussion of intersubjectivity in 1920:

> "Each act that takes place will immediately awaken the old sediments and these will determine their incorporation in a different way."
>
> Husserl (1973) Text Nr. 16 (1920). §8, 454.22–24

An important feature of Husserl's theory of perception is that he explores the central role of self-movement in perception, how our associations and anticipations relate not only to our senses, but also our bodily movements, our changing location in space and time, etc. Husserl discusses in detail how:

> "Kinesthetic experiences play a role not only in connection with the movements of our bodies, but also in connection with the ever-changing perspectives of an object when we move relative to it, for example when we move around a cube and see the cube from ever new sides."

And further:

> "The 'sensations of movement' ... play an essential role in the appearance of every external thing, but they are not themselves apprehended in such a way that they make representable either a proper or an improper matter; they do not belong to the 'projection' of the thing. Nothing qualitative corresponds to them in the thing, nor do they

---

[3] Weisstein, Eric W. "Rabbit-Duck Illusion." From MathWorld—A Wolfram Web Resource. http://mathworld.wolfram.com/Rabbit-DuckIllusion.html

adumbrate bodies or present them by way of projection. And yet without their cooperation there is no body there, no thing."

Husserl (1997), p. 160, p. 136 of the Collected Works translation

The first of these observations is a point of similarity between Husserl and Helmholtz. Husserl heard Brentano lecture on Helmholtz's "Die Tatsachen der Wahrnehmung" in 1885–86, and from 1889 on Husserl lectured on Helmholtz himself. However, in these lectures he concentrated mainly on the Riemann-Helmholtz theory of space.

This emphasis on practical activity and habits is a point of contact between Husserl and Suppes, likewise the stress on the importance of apprenticeship both for our intellectual and our practical life. A further parallel between Suppes and Husserl is that they both hold that almost all of this structuring activity takes place unnoticed by us. We are aware only of the tip of an iceberg. Given the richness and complexity of this vast associative network such unawareness is required for our survival.

Husserl held that the sensuous experience is the source of our conception of reality. What we perceive makes a claim to reality, and this is due to the fact that in perception the noesis has to adjust to what is happening at our sense organs. The sensory experience constrains what we can perceive. The duck-rabbit case is an example. While I can see a duck or a rabbit, various other configurations are also possible, I cannot see, for example a locomotive. The anticipations involved in seeing a locomotive do not fit in with what meets the eye. The "thetic character" of perception is a claim to reality, and Husserl finds that this is connected with the constraints that our sensory experiences put on the anticipations in the act of perception.

There is a similarity here between Husserl and William James, who writes: "Sensible vividness or pungency is then the vital factor in reality...", James (1890, Ch. XXI, Vol. 2. p. 930 of the Harvard edition, 1983). However, while James has only short remarks on this topic, Husserl explored and developed it rather fully. Especially in his later works he emphasized the role that the body and our bodily activity play for our conception of reality:

"To answer these questions I shall look for the ultimate source which ... makes it possible that I consciously find a factually existing world of physical things confronting me and that I ascribe to myself a body in that world and now am able to assign myself a place there. Obviously, this ultimate source is *sensuous experience*."

Husserl (1950), §39, Husserliana III, 88.12–23, Husserliana III.1, 80.33–39[4]

Now to the *hyle*. These are the experiences we typically undergo when our sense organs are affected, but we can also have them when we are affected by illness, drugs, etc. The kinesthetic experiences are examples of hyle, and they are good examples, since they block the interpretation of hyle as "sense data" of the kind that are appealed to in some theories of perception, that hold that we first encounter sense data and then through inference or some other process proceed to physical objects.

When we are now familiar with the main features of the notions of noesis and hyle it is easy to define the notion of the *noema*. The noema is a set of constraints on the object of the act, corresponding to the anticipations in the noesis. Acts can have a variety of objects: concrete, abstract, living, lifeless, and so on. However, sometimes we discover that an act has no object, we may be seriously misperceiving, hallucinating without being aware of it, and so on. However, even in such case, the act is as *if* it has an

---

[4]In the first Husserliana version some lines are included that have been deleted in the second version and are skipped here.

object. Phenomenology is to a large extent a study of this 'as if'. What is the pattern of anticipations involved in my present act, for example an act where I see a person?

There is hence a parallelism between the noema and the noesis: to the various anticipations in the noesis there will be corresponding features in the noema and conversely. However, there are also important differences. The noeses are temporal, they are *experiences* that begin at a certain time, and end at a later time. We can therefore not have the same noesis twice. The noema is not temporal, it is not tied to any particular time or agent. If two acts had totally similar noeses they would have the same noema. However, there will never be two acts with totally similar noeses. One obvious problem would be that even when the same person experiences the same object in the same setting, from the same point of view etc. the two experiences of the object will not be totally alike, new sedimentations take place whenever we experience something and they will modify the later experiences.

Husserl returned again and again to the noetic-noematic parallelism. He found that these two notions were what was missed by Brentano when he went too quickly from intentionality to object. Here is a passage from *Formal and Transcendental Logic* (1929):

> "There was no consistently correlative observing of noeses and noema, of *cogito* and *cogitatum qua cogitatum*. There was no unravelling of the intentionalities involved, no uncovering of the „multiplicities" in which the ‚unity' becomes constituted."

> Husserl (1929), pp. §100. 268.37–269.4 of the Husserliana edition

What I have sketched here is one interpretation of Husserl. However, alternative interpretations have been proposed. In particular, many have contested my key thesis:

> The noema of an act is *not* the object of the act (i.e. the object toward which the act is directed).

In earlier work I have given textual evidence for this thesis. This is not the place to go into all the passages where Husserl discusses the noema. Let me, however, mention some of the *systematic* reasons for not identifying the noema with the object. Here are four such reasons, which also may serve to make us more familiar with the notion of the noema:

First we have the problem that I mentioned earlier concerning acts that have no object. For example an act of hallucination. Like all acts, it has a noema. However, does the act of hallucinating a sword have an object? If not, what does it then mean in such a case that the act's noema is identical to this object? And what about cases where one searches for something and finds out that there is no such thing? Both ordinary life and the sciences are full of examples. In early mathematics many attempts were made to find the greatest prime number. Here the noesis has two main components, the anticipation of being a prime number and that of being greater than any other prime number. These two anticipations are reflected in the noema. However, which is the number that is identical to this noema?

A second problem arises when we reflect on how the objects of acts may belong to any number of different categories, not just physical objects, but persons, animals, numbers, etc. Are there equally many different categories of noemata? If my friend, who is the object of my act, is hungry, is the noema of my act hungry? If he is sad, is the noema sad? And so on.

Here is a third problem: As we have noted, the noesis of an act contains a large number of anticipations of which we are not aware. And correspondingly the noema is tremendously rich. However, the object of the act has features which go far beyond those that are explicitly or tacitly anticipated in the noesis and reflected in the noema. What we anticipate is, as we have noted, largely based on past experience, and the objects in the world around us have lots of features that we have never had any experience of and never even thought about. How can the noema be identical to the object and still fail to have many of the properties of the object? According to Leibniz's law, if a is identical to b, then all properties of a are properties of b and conversely. So what is meant by the noema being identical to the object?

A fourth problem is the opposite of this, but again an application of Leibniz's law. In many cases some of our anticipations concerning the act's object are wrong, but not so radically wrong that we think that the act has some other object or no object at all. I may, for example, believe that my friend is hungry, but learn that he is not. The noesis of the act in which I perceive or think of my friend, contains the anticipation that my friend is hungry. A corresponding constraint is contained in the noema. However, in spite of my small mistake, my friend is the object of my act. In fact, the notion of *error* often requires just this: that we attribute a feature to an object that it does not have. So here the noema seems to have a property that the object lacks. This is a problem Husserl addresses. Every noema has an important component, that he calls "the determinable X." This serves several functions. One of them is to indicate that our act is of some specific object, although we are aware that we may have wrong expectations about this object. The object of the act is therefore not whatever uniquely satisfies our anticipations, if there should happen to be such a thing. The identity of the object depends on more complicated features of the noema, and the determinable X here plays a crucial role. Husserl had very interesting thoughts about these issues, relating to indexicals, demonstratives and reference. Of particular interest is his discussion, in a manuscript from 1911, of what he calls the twin world problem, and where he anticipates by sixty years Putnam's discussion of the same problem:

> "However, if on two celestial bodies two people whose surroundings seem to be totally similar, conceive of "the same" objects and adjust "the same" utterances accordingly? Does not the 'this' in both cases have a different meaning?"
>
> Husserl (1987), supplement XIX, pp. 211–212.[56]

Husserl uses the notion of the determinable X in order to clear up this puzzle. This notion and the closely related notions of object and reference play a crucial role in his approach. There is no room to go into his solution here, but the interplay between these notions is of crucial importance for how associations get entrenched. They are hence highly relevant for Husserl's and Pat's study of the vast associative networks that underlie both perception and thought.[7]

---

[5]A recent discussion of these issues may be found in Beyer (2013).

[6]Wie aber, wenn auf zwei Himmelskörpern zwei Menschen in völlig gleicher Umgebungserscheinung „dieselben" Gegenstände vorstellen und danach „dieselben" Aussagen orientieren? Hat das „dies" in beiden Fällen nicht eine verschiedene Bedeutung? (Husserliana XXVI, Beilage XIX, 211.44–212.2)

[7]I am grateful to Willem Levelt for several helpful comments. Levelt points out that in the last sentence in the last citation from Husserl above, "Does not the 'this' in both cases have a different meaning?" there is no 'this' earlier in the citation – what does it refer to? Levelt is right, there is no earlier 'this' in the citation, nor is there an earlier 'dies' in the German original. Husserl's intention is clear, he considers a person who utters 'this' in the twin situation which he describes. But as it stands, the sentence is ungrammatical.

References

Beyer, C. 2013. Noema and reference. In M. Frauchinger, ed., *Reference, Rationality and Phenomenology: Themes from Føllesdal*, vol. 2 of *Lauener Library of Analytical Philosophy*, pages 73–88. Frankfurt: Ontos Verlag.

Føllesdal, D. 1994. "Patrick Suppes' contribution to the philosophy of language." symposium in honor of Patrick Suppes, Venice, June 16-21, 1992. In P. Humphreys, ed., *Patrick Suppes, Mathematical Philosopher, Vol. 3: Philosophy of Language and Logic, Learning and Action Theory*, pages 3–15. Dordrecht: Kluwer. Comments by Patrick Suppes: pp. 15-18.

—. 2008. "Laudatio." The Lauener Symposium in honor of Patrick Suppes, Bern, September 9-10, 2004. In M. Frauchiger and W. K. Essler, eds., *Representation, Evidence, and Justification: Themes from Suppes*, Lauener Library of Analytical Philosophy, Vol. 1, pages 9–18. Frankfurt: Ontos Verlag.

Frauchiger, M., ed. 2013. *Reference, Rationality and Phenomenology: Themes from Føllesdal*, vol. 2 of *Lauener Library of Analytical Philosophy*. Frankfurt: Ontos Verlag.

Husserl, E. 1929. Formale and transzendentale logik. versuch einer kritik der logischen vernunft. In P. Janssen, ed., *Husserliana 14: Jahrbuch für Philosophie und phänomenologische Forschung, 10. Halle a.d.S.*, vol. 14. Haag: Martinus Nijhoff. Translation: Formal and Transcendental Logic. Translated by Dorion Cairns. The Hague, Netherlands: Martinus Nijhoff, 1969.

—. 1950. Ideen zu einer reinen phänomenologie und phänomenlogischen philosophie. In K. Schuhmann, ed., *Husserliana 3: Erstes Buch: Allgemeine Einführung in die reine Phänomenologie*. The Hague: Martinus Nijhoff. New edition Husserliana III-1, The Hague: Nijhoff, 1977.

—. 1973. Zur phänomenologie der intersubjektivität. In I. Kern, ed., *Husserliana 13: Texte aus dem Nachlass. Erster Teil. Part 1: 1905-1920*, vol. 13. Haag: Martinus Nijhoff. Translation: Formal and transcendental logic. Attempt at a critique of logical reason.

—. 1987. Vorlesungen zur Bedeutungslehre. Sommersemester 1908 [Lectures on the doctrine of meaning: summer semester 1908]. In U. Panzer, ed., *Husserliana XXVI*. The Hague, Netherlands: Martinus Nijhoff.

—. 1997. Thing and space (lectures of 1907). In *Edmund Husserl Collected Works, Vol. VII*. Dordrecht, Holland: Kluwer. Translated by Rojeewicz, Richard.

James, W. 1890. *The Principles of Psychology*, vol. 2. New York: Holt. Here quoted from the Harvard edition, 1981.

Kern, I. 1964. *Husserl und Kant. Eine Untersuchung über Husserls Verhältnis zu Kant und zum Neukantianismus*. Phaenomenologica; 16. Den Haag: Martinus Nijhoff.

Suppes, P. 2009a. Neurophysiological foundations of philosophy. In A. Hieke and H. Leitgeb, eds., *Reduction. Between the Mind and the Brain*, pages 137–176. Frankfurt: Ontos Verlag.

—. 2009b. Rhythm and meaning in poetry. In P. A. French and H. K. Wettstein, eds., *Midwest Studies in Philosophy Volume No. XXXIII, Philosophy and Poetry*, pages 159–166. Boston, MA and Oxford UK: Blackwell Publishing.

—. 2013. Neurophysiological foundations of phenomenology: Is it possible? In *Reference, Rationality and Phenomenology: Themes from Føllesdal*, vol. 2 of *Lauener Library of Analytical Philosophy*, pages 33–48. Frankfurt, Paris, Lancaster, New Brunswick: Ontos Verlag.

15

Mirroring and Moral Psychology

RUSSELL HARDIN

For Patrick Suppes on his 90th birthday

"...The human countenance, says HORACE, borrows smiles or tears from the human countenance.'" David Hume (EPM5.2.3)

Sympathy and moral sentiments are two important terms of art in Scottish Enlightenment moral and political theory, as in the work of David Hume and Adam Smith. Sympathy is about communication of personal feelings of various kinds (Árdal (1966, pp. 152–156)). Moral sentiments are judgments or evaluations. Unfortunately, these writers worked under the disadvantage of having little formal account of psychology and especially, of course, neuropsychology. Perhaps therefore they are ambiguous in the ways they use these and many other important terms, but especially the term sentiments. They sometimes seem clearly to mean emotion or passion, and sometimes judgment or opinion (Jones (1982, p. 98)). The first usage seems to run the term into sympathy. Let us focus on Hume's accounts, which sometimes make sympathy more inclusive. For example, he says that sympathy is the source of moral approbation (T3.2.2.24), and again that it "is the chief source of moral distinctions" (T3.3.6.1). Sentiments should have this role. Most of the time in Hume's usage sympathy is simply communication.

I will keep sympathy and moral sentiments separate because the ideas at the core of the Scottish discussion are communication and judgment, these are clearly different and, when Hume uses these latter two terms, he is not ambiguous. He generally associates sympathy with communication and judgment with moral sentiments. Indeed, he often includes sympathy and sentiment in the same passage, and when he does he allows that one of the things we can communicate is our sentiment about the rightness or wrongness of some action. It does not much matter how we use the terms, except that it pays to keep them clearly focused and not to run them together.

1 Sympathy and Moral Sentiments

Hume first introduces the notion of sympathy early in his discussion of the passions (T2.1.11). He needs it there to explain the otherwise odd fact that our virtues, beauty, and riches "have little influence when not seconded by the opinions and sentiments of others." This requirement of approbation by others is only natural for such things as our

Foundations and Methods from Mathematics to Neuroscience.
Colleen E. Crangle, Adolfo García de la Sienra and Helen Longino.

good reputation, character, and name, but it is explicable for virtues, beauty, and riches only through the mediating effects of sympathy. Why do we have sympathy? Hume must stop somewhere in explaining motivations, and merely accept human nature as it is in order to investigate our behavior and beliefs. Here he wants a psychological account of judgments from sympathy and of their social role. *We*, however, might be interested in explaining the fact of sympathy as a product of evolution through dyadic interactions that benefit us individually and pairwise.

The third Earl of Shaftesbury, an early advocate of sympathy as the font of most human pleasure, says:

> How many the Pleasures are, *of sharing contentment and delight with others*; of receiving it in Fellowship and Company; and gathering it, in a manner, from the pleas'd and happy States of those around us, from accounts and relations of such Happinesses, from the very Countenances, Gestures, Voices and Sounds, even of Creatures foreign to our Kind, whose Signs of Joy and Contentment we can anyway discern. So insinuating are these Pleasures of Sympathy, and so widely diffus'd thro' our whole Lives, that there is hardly such a thing as Satisfaction or Contentment, of which they make not an essential part.
>
> (Shaftesbury 2001, II: p. 62)

Hume shares Shaftesbury's evident pleasure in the company of others. Indeed, Henry Aiken supposes that in his doctrine of sympathy, Hume is emphasizing that humans are pre-eminently social beings in the sense that "whatever others do, their joys and sorrows, loves and hates, have an immediate and continuous impact upon our own sentiments. It is this capacity for reciprocity of feeling which renders possible a common moral life."[1] For Hume, that capacity, which he calls sympathy, is definitive.

. Note that for Shaftesbury, sympathy does not define or explain morality or our moral views. Rather, we have a moral sense that handles all of this, and Shaftesbury's moral sense just knows the good and the right. Hume rejects any such view that moral judgments are somehow planted in our minds, as they would be if they were grounded in a moral sense such as Shaftesbury's. Hume focuses on explaining an important part of our knowledge about others as the result of sympathy. *Sympathy gives me knowledge of others because they are like me in fundamental ways*, so that I can read from their reactions what they feel to a large extent, as Shaftesbury does.

We might ask, how does sympathy work to let me feel what another feels? This is not a question Hume answers. He merely describes its working and puts the fact of it to use in explaining our moral judgments. He develops the idea in Book 2, where it is useful in understanding the passions and how they are communicated from one person to another. There he refers to it as "the principle of sympathy or communication" (T2.3.6.8). In first introducing the idea, he says, "No quality of human nature is more remarkable, both in itself and in its consequences, than that propensity we have to sympathize with others, and to receive by communication their inclinations and sentiments, however different from, or even contrary to our own" (T2.1.11.2). He reintroduces the idea in Book 3 in discussing the origin of the natural virtues and vices, well after presenting his main political views (T3.3.1.4-18).

When we have finished working through the idea and we contemplate how fully and dramatically we can sense what another is thinking or feeling, we too are apt to find it a remarkable human capacity. The fundamental fact that makes all of this possible is our similarity to others and, indeed, it is a stronger phenomenon in dealings with those

---

[1] Aiken (1948), "Introduction," p. xxiii.

who are more similar to us. "All human creatures are related to us by resemblance," especially in their capacity to feel affliction and sorrow (T2.2.7.2; also see T2.2.7.4).[2] Maybe sympathy is partially hard-wired in us (it may also partially be socially inculcated, at least in its objects, which must differ from one culture to another). Hume cites Horace, in the epigraph for this paper, as recognizing that we tend to reflect the moods of others.[3] For Hume, this fact of human behavior is a starting point; he does not attempt to explain it but only to characterize it and to put it to use in explaining other things.

Consider briefly Hume's summary of some of the effects of sympathy. Sympathy "is the chief source of moral distinctions" (T3.3.6.1).[4] This implies, incidentally, that *it is actions toward others that are the concern of morality*, although we may indirectly judge someone by their character and its propensity for the right kind of action. This is, as always, a psychological claim about the way we think about and react to things. The public good is indifferent to us except insofar as sympathy interests us in it (ibid.). Moreover, Hume goes on to argue that the same must be true finally also of the natural virtues and our approbation of them when we see them in others — so that these are also indirect or functional.[5] Earlier he says that a means to an end can only be agreeable where the end is agreeable; and as the good of society where our own interest or that of our friends is not concerned pleases us only by sympathy, it follows that sympathy is the source of the esteem which we pay to all the artificial virtues (T3.3.1.9).

Rawls thinks that sympathy is the wrong term in our time for what Hume wants here. Rawls refers to "imparted feeling," which is passive, so that we are not moved to action by it.[6] This is a feature that we want for Hume's idea here because my sympathy for your pleasures or pains can be purely contemplative and it need not provoke me to any action at all. Rawls's phrase is unlikely to catch on, not least because it is unclear. If I have sympathy *for you*, some feeling you have is imparted *to me*. One almost has to know that this phrase is a substitute for sympathy to get its meaning. We might also wonder just what feeling is imparted to us; surely it is not the literal kind of feeling the other has.

Oddly, however, Rawls criticizes Hume's term for seeming to imply that I feel your fever when I sympathize with your illness; Stroud makes a similar criticism.[7] This reading is no part of Hume's meaning nor should anyone read him this way. Bill Clinton on the campaign trail said, "I feel your pain." He did not say he felt any particular pain,

---

[2] But contrary to Hume's claim here, when I hear of someone reacting violently to a smear on his supposed honor, I cannot feel their emotions at all. For example, the father of Fadime Sahindal tells a Swedish court that they must sympathize with the depth of the insult he felt when his daughter refused to marry the man her family had chosen for her, an insult so painful that it justified his killing her. I cannot have any Humean sympathy for his action or his feelings. They are outside my ken so thoroughly that they might as well be expressed by a strange creature from another planet. My sense is of horror at him, not of empathy with him. A psychiatric expert said he is "cognitively underdeveloped and lacking in empathy" (Wikan (2004), "Deadly Distrust: Honor Killings and Swedish Multiculturalism," p. 200). That is probably wrong, at least in Hume's vocabulary. The elder Sahindal has moral sentiments that differ grossly from those of the cosmopolitan Swedish court. The court ruled that he had committed murder. He was inerrantly convinced that he had salvaged his family's honor, that he deserved high praise.

[3] See also Árdal (1966), *Passion and Value* in Hume's Treatise, pp. 57–8.

[4] The two terms, sympathy and sentiments, are somewhat confused here.

[5] As noted earlier, for Hume virtue is a means to an end (T3.3.6.2).

[6] Rawls (2000), *Lectures on the History of Moral Philosophy*, pp. 86–7.

[7] Rawls (2000), *Lectures on the History of Moral Philosophy*, pp. 86–7; Stroud (1977), *Hume*, pp. 197–8.

such as shortage of funds to pay medical bills, but only a very general pain. I do not literally have your experience, but I do share the pleasurableness or the painfulness of your experience. For example, suppose I see a child ecstatically enjoying something that I do not even like — say, peanut butter or hard candy. I do not share the tastes of the child but only the pleasure, and I therefore experience pleasure. Often I could not even know what you taste or experience although I could know that you are experiencing pleasure or pain.

Having sympathy allows us to have knowledge of others beyond ourselves. Sympathy is about knowledge and its communication from one person to another. There may be no method of inquiry to discover knowledge of another's mind, but through sympathy we simply gain it, apparently almost directly as though we could read it from facial expressions and ways of expressing things. Rawls is right to suppose that sympathy is not an ideal term here, but it has become a term of philosophical art and we are likely stuck with it. Unfortunately, it is hard to come up with a single term that is both clear and apt, perhaps because the idea behind the term is not part of vernacular understanding or language. The term we would want must convey several things. Hume's sympathy is partial in that it is stronger for those like ourselves. It is emotional rather than reasoned and it can be both very lively and very calm. Finally, however, we might well prefer to continue to use Hume's sympathy as a philosophical term of art. One who reads much of Hume must soon enough make the term include these senses.

It is not clear that we always have direct access to our own emotions. Sometimes we discover what they are from their effect on our behavior or our reactions, or even from the apparent understanding of others who read us better than we do. For example, we notoriously sometimes discover how deeply we love or care about someone only when we might be about to lose the person. Jane Austen is a master of showing examples of such lacks of self-awareness. In *Pride and Prejudice*, she says that Elizabeth "rather *knew* that she was happy, than *felt* herself to be."[8] Witold Gombrowicz similarly says that Isabel, "knowing that when one is in love, one is happy, was happy."[9] As Hume notes, "Our predominant motive or intention is, indeed, frequently concealed from ourselves" (EPM Appendix 2.7).

In *Emma* Austen says of Knightley that he "had been in love with Emma, and jealous of Frank Churchill, from about the same period, one sentiment having probably enlightened him as to the other."[10] Earlier, Emma herself has to attempt to infer whether she is falling in love with Churchill from her sudden sense of listlessness when he departs for London.[11] The reader knows well enough that she is not falling for him in any serious way. He is merely lively in sufficiently different ways as to be interesting to her in her somewhat sequestered life. He is virtually a foreigner to her extremely close and limited community and the appeal of his novelty will pass soon enough. If the reader knows such things, that is because Austen has communicated them, but she has done so only subtly and tangentially so that it is through our sympathy that we receive a communication that is not articulate. Austen's extraordinary gift for such quiet communication lets us flatter ourselves with our own sensibility. We know that Emma will, as did Austen herself, stay in her community and that she likely will, as Austen did not succeed in doing, marry within that community.

---

[8] Austen (1952), *Pride and Prejudice*, Chapter 59, p. 337.
[9] Gombrowicz, *Ferdydurke* (1961), pp. 267–8.
[10] Austen (1985), *Emma*, Chapter 49, p. 419.
[11] Austen (1985), *Emma*, Chapter 31, p. 266.

Hume's moral sentiments are a matter of moral judgment or opinion; they are among the things we can communicate via sympathy. From sympathy I identify somewhat with your pleasure or pain. From sentiment, I approve or disapprove of some action or state or character. I know the effect of an action on another person from sympathy, but I judge from moral sentiment. Our moral sentiments invest what we see with approbation or disapprobation. These sentiments are passions and they are the psychology of moral judgment. In the writings of many philosophers of Hume's time, the moral sentiments are somehow in our heads or emotions, as in the moral sense school of Shaftesbury and others. For Hume they get there through our reaction to things as though they caused us pleasure or pain. When they are coupled with our capacity for sympathy, they lead us to react on behalf of others. It is not hard to see why I might react to your doing something beneficial or harmful to me. Your action affects my interests, and I either like it or dislike it, provoking my sentiments so that I then think well or ill of you, or at least of your action. It is an interesting fact that Hume (1751/1998), who thinks his *Treatise* a failure in reaching the public, relegates much of the discussion of the sentiments to an appendix in the later *Enquiry* (EPM Appx.1). Perhaps he thought that deleting that analysis from the body of the text makes that text more approachable.

2 Mirroring

What might stand behind the phenomenon that Hume recognizes and uses to ground his claims for sympathy, but that he does not explain? There was not strictly any need for him to explain; he could observe the phenomenon and could start from the observed fact of it (EPM5.17n19). There are now fMRI (functional magnetic resonance imaging) studies of the brain's reaction to others' sensations that corroborate Horace's and Hume's observed facts and that, in a sense, seem to show the phenomenon of *seeing* another's emotions at work. The fMRI studies do not do much more than Hume already did — they suggest that mirroring happens, although they are more definitive than Hume's singular testimony. The part of the brain that perceives a smile is evidently the part that engineers a smile of our own, so that Horace's observation may be a biologically hard-wired fact of our brains. Smiles evoke smiles. The evolution of this feature of our brains might be explained by the benefits of smiling in gaining the good graces of others, especially when we are too young to survive on our own.[12] Smiling may enable humans to enjoy very long periods of infancy, childhood, and adolescence so that we can develop extraordinary abilities that set us apart from other animals.

Hume (1739-40/2000) further notes "we may remark, that the minds of men are mirrors to one another, not only because they reflect each other's emotions, but also because those rays of passions, sentiments and opinions may be often reverberated, and may decay away by insensible degrees" (T2.2.5.21; see also T3.3.1,7 and EPM5.18).[13] What in twentieth century philosophy was the problem of other minds (how can we know another's mind?) is assumed away in limited part by Hume. For this too there may now be neurophysiological evidence from fMRI studies. Sympathy, these studies suggest, is a form of direct, non-verbal communication and the evocation of relevant feelings.

---

[12] There are reputedly recent studies that suggest other connections. Those who yawn when another yawns seem to score higher on empathy tests than those who do not mirror the yawns of others. (Henry Fountain, "Tarzan, Cheetah and the Contagious Yawn," *New York Times*, 24 August 2004, F1).

[13] See further, Penelhum (1993), "Hume's Moral Psychology," p. 143.

It is psychological mirroring that leads me to like or dislike something that is done to you, by letting me sense what you enjoy or suffer. Contemporary neurophysiological findings seem to strengthen Hume's claims for the moral psychology of mirroring, although the mechanism is not yet clear. Those readers who have had difficulty accepting this part of Hume's argument might soon find it easy to accept. Hume appears to be right on the psychology here. The only question that might remain for some is that of his general claims for morality psychologized. Do we have moral reactions (approbation or disapprobation) to the feelings we get from mirroring? Those would be moral reactions on behalf of another. That is to say, the important and very difficult trick Hume needs to complete his explanatory theory is to evoke *my* sentiments — that is, a moral judgment — in response to actions that affect *your* interests.

From the fMRI data it appears possible that these two phenomena — sympathy and moral sentiments — are at least partially run together in our brains.[14] Hence, Hume's theory is complete but in a way that he apparently did not see. The knowledge and the feeling, the sympathy and the sentiments, may come in a single package. There is no mediating interpretation that our brains have to make. A nearly brand new baby smiles back at our smile. It is implausible to suppose that the baby is interpreting our kindness or good will in its first days of life; it is reacting from an apparently hard-wired capacity. Empathy seems to "mirror" another person's emotional responses in one's own brain.[15] Happily, "mirror" is Hume's (T2.2.5.21) word and also the terminology of contemporary neurophysiological science. Mimicry has usually been explained as a two-step process. Our perceptions of, say, a smile stimulate thoughts, which guide our behavioral response: smiling back. Studies of brain activity, as measured by fMRI) brain scans, suggest that the whole reaction is immediate in a single step, not mediated by thought. The part of my brain that recognizes a smile also forces or stimulates my own smile and my own feeling of pleasure.[16] Seeing your smile triggers mine. We are to a degree hard-wired to each other.

Hume seems to have grasped the nature of this phenomenon to a sufficient degree as to make it the foundation of his moral psychology. He does not attempt an explanation of the phenomenon but merely starts from it to explain morality as a matter of fellow feeling. In fact, of course, he had no way to prove his assertion of the nature of this psychological trick other than to elicit our agreement that we too have the experience he describes. The technology of fMRI now seems to give us some direct entrée to the phenomenon.

Incidentally, mirroring seems to be very weak in those with autism.[17] Hence, David Owen is apparently right in saying that the renowned autistic woman, Temple Grandin, is lacking in Humean sympathy.[18] Her lack is organic. Her reason functions very well so that she is able to see herself as like, in her words, an anthropologist on Mars, where she would have little in common with and no sympathy for others — this is the condition she is in on Earth.

---

[14] See also, Árdal (1966), *Passion and Value* in Hume's Treatise, 47n.

[15] The German psychologist, Theodore Lipps, coined the German term for empathy in 1903, and he described the phenomenon of mirroring (Bower (2003, p. 330), "Repeat after Me: Imitation is the Sincerest Form of Perception".

[16] See various contributions to Meltzhoff and Prinz (2002), editors, *The Imitative Mind*. Also see Miller (2005), "Reflecting on Another's Mind."

[17] Miller (2005), "Reflecting on Another's Mind," p. 947.

[18] Owen (1994), *"Reason, Reflection, and Reductios,"* p. 195. On Temple Grandin, see Sacks (1993), "An Anthropologist on Mars."

Sympathy in the fMRI studies appears to be a form of direct, non-verbal communication and the evocation of relevant feelings. In another context, Hume dismissively says of the possibility of an innate sense of rules of property, "We may as well expect to discover, in the body, new senses, which had before escaped the observation of all mankind" (EPM3.40). In actual fact, he may well have discovered, along with some others, including Horace and recent psychologists, what we might come to call a sense: the sense of sympathy. It is a sense that is apparently much more acute in Hume than in most people but that is clearly evident in large numbers of people, including new born babies, and apparently other species as well.[19] It may be as hard-wired as the sense of taste or smell.

In his dismissal of an innate sense of rules of property, Hume has implicitly rejected the possibility of rule utilitarianism long before its articulation less than a century ago. Such a theory or any other rule morality is stillborn.

Hume says our sympathy for those on a ship sinking off shore will be greatly heightened if they are close enough for us to see their faces and their frightened responses. He does not explain this fully but only says that contiguity makes their suffering clearer to us (T3.3.2.5; also see T2.1.11.6 and 8). The fMRI studies suggest that the issue is not that we have to see their expressions in order to understand their emotions; our reason is adequate for such understanding. The issue is that *we have to see their expressions in order to trigger the mirroring of our own similar emotions.* Again, this is a phenomenon that is not mediated by thought or reason, and perhaps it cannot be replaced by thought or reason when the actual visions are not available.

Suppose we accept this entire account of our moral sentiments and of their apparent mirroring. If they are merely a fact of our psychology, should they determine our morality? Yes, in Hume's functional way. That is, our sentiments about others evoke responses from us that are responses to the utility, pleasure, emotions or pain of those others. What typically brings pleasure to others is their own benefit, which is *good for them.* We cannot go further to say it is good per se unless we go so far as to say that something like utilitarianism is the right moral theory. Hume does not make this claim, but in his analysis of the motivating force of mirrored reactions he does imply that he and we are psychologically utilitarian. One of the things we can tell about another through mirroring is how something affects their welfare, pleasure, or pain. This fact is important if we are psychologically utilitarian — and mirroring virtually makes us be, as though evolution has produced utilitarianism as our moral response. Psychological utilitarianism connects observation to judgment. These facts do not make utilitarianism the true moral theory; they merely characterize our psychology as moralized through mirroring. This psychology gives us a science of moral beliefs and approbations; it cannot additionally justify those approbations or make them right. This is the sense of Hume's assertions that his *Treatise* gives us a science of morals.

Mirroring is a major discovery for Hume despite the fact that seemingly all people experience it, so that it might well have been a matter of widespread common knowledge. It apparently remains unconscious and inarticulate to most people even while it often regulates their emotions and behavior. Hume is sufficiently perceptive that, once he has noticed the phenomenon, he finds mirroring to be a fundamental part of the psychology of sympathy and therefore a fundamental part of distinctively moral psy-

---

[19] Chimpanzees and Macaque monkeys, even in infancy at three days, apparently mirror emotions of others (Bower (2006), "Copycat Monkeys: Macaque Babies Ape Adults' Facial Feats").

chology. Mirroring makes Hume's theory psychologically richer than any of the then contemporary moral sense and sentiments theories, which are inherently psychological in their foundations, but not neurological. Their proponents are generally content to stop their inquiries at the point of asserting that we just do know right from wrong, that reason can determine these, or that god has given us such knowledge. Hume has empirically observed — and supposes we can all observe — the phenomenon of mirroring and sympathy.

There is a further trick that might still be difficult: evoking an emotional response of a similar kind in response to the interests *of society*. Identification with the interests of society must be very weak psychologically and it is grounded in reason more than in sympathy. The neurophysiological studies probably cannot address such an abstract phenomenon as responding psychologically to the interests of society. Seeing the interests of society forwarded or abused is not comparable to the visual cue of a smile or frown. Society does not smile back at us.

Implications

Hume's account of mirroring yields several insights or conclusions:

(1) It underlies Hume's morality psychologized;
(2) It can readily be fitted to an evolutionary account of how it works;
(3) It gives some entrée to the black box of another's mind;
(4) This implication may be strengthened by fitting it to Patrick Suppes' work on reading another's mind for content (not moral theory);
(5) It suggests that sympathy and sentiments are partly run together; hence there is no need of a connecting move because they are not mediated by thought;
(6) Sympathy is a form of direct (non-verbal?) communication;
(7) Basing morality in mirroring yields a science of morals, not a substantive moral theory;
(8) Mirroring weakly entails utilitarianism.

References

Aiken, H. 1948. Introduction. In H. Aiken, ed., *Hume's Moral and Political Philosophy*, pages ix–li. New York: Hafner Press.

Árdal, P. 1966. *Passion and Value in Hume's Treatise*. Edinburgh: Edinburgh University Press.

Austen, J. 1952. *Pride and Prejudice*. London: Collins.

—. 1985. *Emma*. London: Penguin.

Bower, B. 2003. Repeat after me: Imitation is the sincerest form of perception. *Science News* 163(21):330–332. doi:10.2307/4014561.

—. 2006. Copycat monkeys: Macaque babies ape adults' facial feats. *Science News* 170(11):163. doi:10.2307/4017167.

Gombrowicz, W. 1961. *Ferdydurke*. New York: HHarcourt, Brace& World.

Hume, D. 1739-40/2000. *A Treatise of Human Nature*. Oxford: Oxford University Press. D. F. Norton and M. J. Norton (eds). Cited in the text as T followed by book, part, section, and paragraph numbers.

—. 1751/1998. *An Enquiry Concerning the Principles of Morals*. Oxford: Oxford University Press. T. L. Beauchamp (ed.). Cited in the text as EPM followed by section and paragraph numbers.

Jones, P. 1982. *Hume's Sentiments: Their Ciceronian and French Context*. Edinburgh: Edinburgh University Press.

Meltzoff, A. N. and W. Prinz. 2002. *The Imitative Mind: Development, Evolution, and Brain Bases.* Cambridge: Cambridge University Press.

Miller, G. 2005. Reflecting on another's mind. *Science* 5724(308):945–947.

Owen, D. 1994. Reason, reflection, and reductios. *Hume Studies* 20(2):195–210. doi:10.1353/hms.2011.0065.

Penelhum, T. 1993. Hume's moral psychology. In D. F. Norton, ed., *The Cambridge Companion to Hume*, pages 117–147. Cambridge: Cambridge University Press.

Rawls, J. 2000. *Lectures on the History of Moral Philosophy.* Cambridge, Mass: Harvard University Press. B. Herman (ed.).

Sacks, O. 1993. An anthropologist on mars. *New Yorker* 69(44):106–125.

Shaftesbury, A. A. C. 2001. *Characteristics of Men, Manners, Opinions, Times.* Indianapolis: Liberty Fund. D. D. Uyl (ed.).

Stroud, B. 1977. *Hume.* London: Routledge and Kegan Paul. T. Honderich (ed.).

Wikan, U. 2004. Deadly distrust: Honor killings and Swedish multiculturalism. In R. Hardin, ed., *Distrust*, pages 192–204. New York: Russell Sage Foundation.

Part V

Neuroscicnce

16

Response Selection Using Neural Phase Oscillators

José Acacio de Barros, and Gary Oas

It is an honor for Acacio de Barros and Gary Oas to participate in a volume for Patrick Suppes. It is especially rewarding to do so with a paper where we discuss our most recent work with him, a model of brain processes using neural oscillators. We are, as he would say, "true blue physicists." So, for us, collaborating with Suppes on this truly interdisciplinary subject is an example not only of his intellectual influence, but also of his friendship and mentorship. We are happy to dedicate this paper to Pat. Happy Birthday Pat!

Abstract

In a recent paper, Suppes, de Barros, & Oas (2012) used neural oscillators to create a model, based on reasonable neurophysiological assumptions, of the behavioral stimulus-response (SR) theory. In this paper, we describe the main characteristics of the model, emphasizing its physical and intuitive aspects.

1 Introduction

The work we present here started more than ten years ago, when Suppes and one of the authors (JAB) begun thinking about how to model in a physically plausible way collections of neurons in terms of oscillators. In one of his known intuitions, Suppes kept insisting that the brain "has got to use oscillators." Of course, as is often the case, his "intuition" was based on hard work and detailed empirical data that he collected working on the EEG of words and sentences. Nevertheless, as we kept trying to make our model work (and we had many failures, and some successes; see Vassilieva et al. (2011) for an example), He kept insisting: we should understand the brain computations with oscillators. We are pleased to say that, despite our initial skepticism, we now have a reasonable model that not only do we feel is grounded on neurophysiologically sound evidence, but that also reproduces quite well some empirical behavioral data (Suppes et al., 2012).

In this paper we attempt to describe the main features of this model by focusing on the physical processes underlying neural computations. We chose to do so for the

Foundations and Methods from Mathematics to Neuroscience.
Colleen E. Crangle, Adolfo García de la Sienra and Helen Longino.
Copyright © 2014, CSLI Publications.

following reasons. First, because of its interdisciplinarity, our model requires concepts from many different areas (neurophysiology, physics, psychology, etc). Such concepts are not necessarily complex, but are often unfamiliar to most researchers. Second, we are confident that our model is relevant to cognitive psychologists, as it may explain some unconventional mathematical models showing good empirical fit (de Barros 2012a,b). So, we believe that this paper can provide a clearer and intuitive view of the main physical features of our model for those thinking about applying it, supplementing the discussions found in Suppes, de Barros, & Oas (2012).

Let us start our discussion with the broad problem of understanding how the brain processes information. This is perhaps the most challenging current scientific endeavors, mainly due to the fact that our brain is tremendously complicated, as it is constituted of many different components that are, by themselves, complex, but that also seem to sometimes interact holistically with each other. Among the approaches to try and understand the brain, the most prominent ones are the top-down and bottom-up. In the top-down approach, we start with the higher-level functions and go to their underlying mechanisms. An example of such approach would be the field of cognitive neuroscience, where often one starts with experiments in cognitive psychology and tries to understand them from principles in neuroscience (Adolphs, 2003). In the bottom-up approach, by studying how each elementary component works, such as neurons, one tries to start with neurophysiological processes to see how higher functions arise from such components and their interaction (Jessell et al. 2000).

Each of those approaches have their shortcomings. For example, one of the main issues is what we may call a problem of scale. When trying to understand a complex system, the first question that arises is how detailed we need to be. In the case of the brain, some researchers say that we need to go all the way down to the chemical reactions in the synapses. Others argue that individual neurons hold the key to understanding brain computation. Yet another view is that collections of neurons are important. So, when trying to understand how the brain works, our first problem is where to begin. Regardless of what scale is chosen and where we start, ultimately we would need to understand the whole process if we were to claim to have understood the brain.

The main problem with connecting a higher scale with a lower one is due to its complexity. For example, evidence exists that higher cognitive processes involve tens to hundreds of thousands of neurons, interacting with each other in very complex ways. Modeling such processes require the use of powerful computers. But, even when a model is shown to work from the underlying neuronal dynamics, the use of massive computer simulations helps little in understanding, in an intuitive or conceptual way, what is actually happening. The system is simply too complex.

To deal with the issue of complexity, different approaches can be taken. One possible route is to find physically plausible arguments that impose constraints on the system's dynamics, therefore reducing it to fewer degrees of freedom. This is the approach taken by Suppes et al. (2012). In this paper, a large number of independent neurons was modeled by a single dynamical parameter determined by the phase of a neural oscillator. It was then showed that under certain reasonable assumptions, the main characteristics of behavioral stimulus-response (SR) theory could be described by neural oscillators. The use of neural oscillators thus provided a significant reduction on the number of degrees of freedom, allowing for the physical interpretation of many different parameters in the model.

Here we present the work of Suppes, de Barros, & Oas (2012), with emphasis on the physics and intuition behind the model. Our goal is to make this model more understandable, as many of the concepts used in our previous paper are not well-known to certain audiences. For example, while all physicists have an excellent knowledge of oscillations and interference and could easily follow the arguments leading from neurons to oscillators, only a few would feel comfortable with the mathematical learning theories used. Neuroscientists or psychologists, on the other hand, would probably feel at home with neurons and learning theories, but not so much with oscillators and interference. Here we focus on the intuitions behind the physics, with the hopes that, in conjunction with the oscillator model, psychologists and neuroscientists could benefit more from the insights gained.

2 A Brief Review of SR Theory

Stimulus-response theory (or SR theory; see Suppes and Atkinson (1960)) is one of the most successful behavioral learning theories in psychology. Though it has decreased in importance in current psychology, we chose to model SR theory for the following reasons. First, it is based on a rigid trial structure, which permits its concepts to be formally axiomatized, resulting in many important non-trivial but illuminating representation theorems (Suppes 2002). In fact, the theory is rich enough to represent language in it. Second, despite its few parameters (the learning probability c and the number of stimuli), it has been shown to fit well to empirical data in a variety of experiments. Finally, as we showed in Suppes, de Barros, & Oas (2012), SR theory seems to have natural counterparts at a neuronal level, and is, in some sense still used by neuroscientists (though, sadly, not in its mathematical form).

Here we present the mathematical version of SR theory for a continuum of responses, formalized in terms of a stochastic process (we follow Suppes et al. (2012)). Let $(\Omega, \mathcal{F}, P)$ be a probability space, and let $\mathbf{Z}$, $\mathbf{S}$, $\mathbf{R}$, and $\mathbf{E}$ be random variables, with $\mathbf{Z} : \Omega \to E^{|S|}$, $\mathbf{S} : \Omega \to S$, $\mathbf{R} : \Omega \to R$, and $\mathbf{E} : \Omega \to E$, where S is the set of stimuli, R the set of responses, and E the set of reinforcements. Then a trial in SR theory has the following structure:

$$\mathbf{Z}_n \to \mathbf{S}_n \to \mathbf{R}_n \to \mathbf{E}_n \to \mathbf{Z}_{n+1}. \tag{1}$$

The trial structure works the following way. Trial n starts with a certain state of conditioning and a sampled stimulus. Once a stimulus is sampled, a response is computed according to the state of conditioning. Then, reinforcement follows, which can lead (with probability c) to a new state of conditioning for trial $n + 1$. In more detail, at the beginning of a trial, the state of conditioning is represented by the random variable

$$\mathbf{Z}_n = \left(z_1^{(n)}, \ldots, z_m^{(n)} \right).$$

The vector

$$\left(z_1^{(n)}, \ldots, z_m^{(n)} \right)$$

associates to each stimuli $s_i \in S$, $i = 1, \ldots, m$, where $m = |S|$ is the cardinality of S, a value $z_i^{(n)}$ on trial n. Once a stimulus $\mathbf{S}_n = s_i$ is sampled with probability

$$P\left(\mathbf{S}_n = s_i | s_i \epsilon S \right) = \frac{1}{m},$$

its corresponding $z_i^{(n)}$ determines the probability of responses in R by the probability distribution $K(r|z_i^{(n)})$, i.e.

$$P\left(a_1 \leq \mathbf{R}_n \leq a_2 | \mathbf{S}_n = s_i, \mathbf{Z}_{n,i} = z_i^{(n)}\right) = \int_{a_1}^{a_2} k\left(x|z_i^{(n)}\right) dx,$$

where $k(x|z_i^{(n)})$ is the probability density associated to the distribution, and where $\mathbf{Z}_{n,i}$ is the i-th component of the vector $(z_1^{(n)}, \ldots, z_m^{(n)})$. The probability distribution $K(r|z_i^{(n)})$ is the smearing distribution, and it is determined by its variance and mode $z_i^{(n)}$. The next step is the reinforcement $\mathbf{E}_n$, which is effective with probability c, i.e.

$$P\left(\mathbf{Z}_{n+1,i} = y | \mathbf{S}_n = s_i, \mathbf{E}_n = y, \mathbf{Z}_{n,i} = z_i^{(n)}\right) = c$$

and

$$P\left(\mathbf{Z}_{n+1,i} = z_i^{(n)} | \mathbf{S}_n = s_i, \mathbf{E}_n = y, \mathbf{Z}_{n,i} = z_i^{(n)}\right) = 1 - c.$$

The trial ends with a new (with probability c) state of conditioning $\mathbf{Z}_{n+1}$.

3 Oscillator Model

In this section we will describe intuitively the oscillator model. We start by arguing for the use of neural oscillators as a way to model the brain at a system level. We then discuss how we can represent in a mathematically sensible way these oscillators. Finally, we show how response computations and learning can be modeled using this theoretical apparatus. Readers interested in more detail are referred to Suppes, de Barros, & Oas (2012).

There are many different ways in which researchers try to figure out how the brain works. For example, in cognitive neuroscience, among the most popular research techniques are fMRI (functional magnetic resonance imaging), MEG (magnetoencephalogram), and EEG (electroencephalogram). MEG and EEG measure the electrical activities in the brain, whereas fMRI measures changes in blood flow associated with higher metabolic rates. While fMRI's popularity is due to its better spatial resolution, MEG and EEG present significantly better time resolution. However, what these techniques have in common is that, in order to measure a signal from the brain, they require a large numbers of neurons to fire synchronously. To make our point, let us focus on EEG (though MEG would be adequate too). There are many experiments (see Carvalhaes et al. (2014) and references) showing that the EEG data allow a good representation of language or visual imagery. Thus, neurophysiological evidence points toward language being an activity involving large collections of synchronizing neurons, and we will center our model exactly on this.

Before we show how to describe such collections of synchronizing neurons mathematically, it is useful to think about the physical mechanisms of synchronization. Let us look first at individual neurons, and then think about ensembles of neurons. Figure 1 shows the qualitative behavior of two neurons n_A and n_B firing periodically, with $T_B < T_A$. What happens if we now couple n_A to an excitatory synapse coming from neuron n_B? Because $t_0 + T_B < t_0 + T_A$, the excitatory coupling will increase the membrane potential of neuron n_A before $t_0 + T_A$, causing n_A to fire a little earlier than it would if it were not connected to n_B. So, excitatory synaptic couplings between neurons can change the timing of coupling, and this timing is changed such that the firings of both neurons approach (in this case, the firing of n_A approaches that of n_B). In other words, excitatory

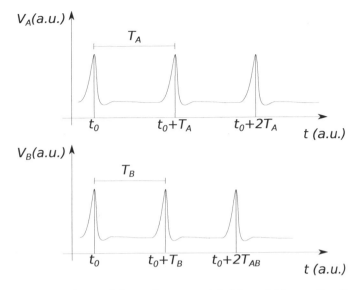

FIGURE 1 Approximate shape of the action potentials V_A and V_B as a function of time t for two uncoupled neurons n_A and n_B firing periodically, with periods T_A and T_B. For simplicity, we chose a t_0 when both neurons fire simultaneously

couplings push n_A and n_B toward synchronization. In fact, it is possible to prove mathematically that if the number of neurons is large enough, the sum of the many weak synaptic interactions can cause a strong effect, making all neurons fire closer together (Izhikevich 2007); even when weakly coupled, ensembles of periodically firing neurons synchronize. It is interesting to note that the argument shown above can be scaled up to distinct collections of neurons. Imagine we have two ensembles of neurons, N_A and N_B, such that neurons in them synchronize. If neurons in N_A and N_B become coupled, then the same mechanism as discussed above will be at play, and the ensembles will synchronize among themselves. We will come back to this point later, when we talk about response mechanisms.

We are now in good shape to introduce the intuition behind the mathematical description for the dynamics of synchronization. One of the main simplifying assumptions we make is that the relevant information coded in the brain is represented by the synchronization of an ensemble of neurons. This ensemble may include tens of thousands of neurons, but because they are synchronized, we can represent them, at least in first approximation, by a single dynamical variable. To understand this, let us think about the simplest case, where an oscillator $O(t)$ can be represented by a sine function[1]:

$$O(t) = A \sin \omega t, \tag{2}$$

where $\omega = \omega(t)$ is its time-dependent frequency. Since ω may be a function of time, the value of $O(t)$ is completely determined by the argument of the sine, i.e. by $\varphi = \omega t$. The quantity φ is the *phase* of the oscillator $O(t) = A \sin \varphi(t)$. Since collections of firing neurons have very little variability in its intensity (except, as we see below, when they interfere), we can describe a neural oscillator by its phase. The interaction of a neural oscillator with other neural oscillators may change the evolution of its phase.

---

[1]We use a sine function for simplicity, but the following argument is valid for periodic functions.

We emphasize that there is a certain invariance of scale in the above argument: it somehow does not matter how many neurons we have; all that matters is that their amplitude does not vary, that their couplings are strong enough to produce synchronization, and that their dynamics is encoded in the phase. Furthermore, in the same way that individual oscillating neurons synchronize to each other, a collection of coherent neurons can also synchronize to another collection of coherent neurons. Since neurons firing coherently may be described approximately by their phase, we can focus on the phase dynamics, instead of being concerned about the full description of the very complex dynamical system.

Now, let us look a little more into the details of the mathematics of two synchronizing oscillators. Let us start with two oscillators, $O_1(t)$ and $O_2(t)$, described by their phases φ_1 and φ_2. If the two oscillators are uncoupled and their frequency ω is constant, then it is clear from equation (2) that they should satisfy the following set of differential equations,

$$\frac{d\varphi_1}{dt} = \omega_1, \tag{3}$$

$$\frac{d\varphi_2}{dt} = \omega_2, \tag{4}$$

where ω_i, $i = 1, 2$, are their natural frequencies. However, if they are weakly coupled, such that their interaction does not affect the overall form of the oscillations given by $O_1(t)$ and $O_2(t)$ but affects their phase, then equations (3) and (4) need to be modified to include changes to the phase. Furthermore, if the underlying interaction is such that it will make the phases approach each other, such as in the case of synaptically coupled neurons, then it is possible to show that, in first approximation, the modified dynamical equations become

$$\frac{d\varphi_1}{dt} = \omega_1 - k_{12}\sin(\varphi_1 - \varphi_2), \tag{5}$$

$$\frac{d\varphi_2}{dt} = \omega_2 - k_{21}\sin(\varphi_2 - \varphi_1), \tag{6}$$

where k_{ij} are the phase coupling strengths. If we extend this to allow for N oscillators, equations (5) and (6) then become

$$\frac{d\varphi_i}{dt} = \omega_i - \sum_{j \neq i} k_{ij}\sin(\varphi_i - \varphi_j). \tag{7}$$

Equation (7) is known as the Kuramoto equation (Kuramoto 2003), and it is widely used to describe complex systems with emergent synchronization. The strength and usefulness of Kuramoto's equation comes from two main points. First, it can be solved under certain symmetric conditions and in the limit of large N, yielding significant insight into the nature of emerging synchronization. Second, a set of weakly-coupled oscillating dynamical systems close to a Andronov-Hopf bifurcation can be described, in first approximation, by Kuramoto-like equations (see Izhikevich (2007)). For our purpose, Kuramoto's equations are a good approximation for the dynamics of coupled neural oscillators.

So, we now turn to the discussion of how we can think of stimulus and response as modeled by oscillators, and in particular by Kuramoto's equations. The basic idea is simple. Once a distal stimulus is presented, the perceptual system activates an ensemble of brain neurons, N_s, associated with it. This system itself is described by Kuramoto's equations, and, because it synchronizes, we use its average phase to describe its mean dynamics. If this stimulus elicits a response, the activation of the response neurons via synaptic couplings follows. Responses, as stimuli, are also represented by synchronously firing ensemble of neurons. The selection of a particular response happens when the stimulus oscillator synchronizes in phase with it, and such phase is determined by the relative couplings between stimulus and response oscillators. Let us now look more into the details.

The simplest stimulus-response neural oscillator model requires three oscillators, O_s, O_{r_1}, and O_{r_2}. O_s is the oscillator representing firing neurons corresponding to the sampling of a stimulus, and O_{r_1} and O_{r_2} are the response oscillators. Their phases are φ_s, φ_{r_1}, and φ_{r_2}. Before we describe their dynamics, let us go through the process of a response computation. Whenever O_s is activated, and subsequently O_{r_1} and O_{r_2}, then the intensity of firings (i.e., the rate of firing, as the individual neuron amplitudes are reasonably stable) in each response oscillator is not only due to its firing, but also to the firings of O_s. As we mentioned earlier, a collection of firing neurons may interfere, and in this case, interference means higher coherence when in phase, and lower coherence when off of phase. Let us analyze this with a mathematically simple example of equal intensity harmonic oscillators, given by

$$O_s(t) = A\cos(\omega_0 t) = A\cos(\varphi_s(t)), \tag{8}$$
$$O_{r_1}(t) = A\cos(\omega_0 t + \delta\phi_1) = A\cos(\varphi_{r_1}(t)), \tag{9}$$
$$O_{r_2}(t) = A\cos(\omega_0 t + \delta\phi_2) = A\cos(\varphi_{r_2}(t)). \tag{10}$$

Equations (8)–(10) represent the case where the oscillators are already synchronized with the same frequency ω_0 but with relative and constant phase differences $\delta\phi_1$ and $\delta\phi_2$. The mean intensity give us a measure of the excitation carried by the oscillations, and for the superposition of $O_s(t)$ and $O_{r_1}(t)$ it is given by

$$I_1 = \left\langle (O_s(t) + O_{r_1}(t))^2 \right\rangle_t$$
$$= \left\langle O_s(t)^2 \right\rangle_t + \left\langle O_{r_1}(t)^2 \right\rangle_t + \left\langle 2O_s(t)O_{r_1}(t) \right\rangle_t,$$

where $\langle f(t) \rangle_{t_0} = \frac{1}{\Delta T} \int_{t_0}^{t_0+\Delta T} f(t)\, dt$ ($\Delta T \gg 1/\omega_0$) is the time average. A quick computation yields

$$I_1 = A^2(1 + \cos(\delta\phi_1)),$$

and, similarly for I_2,

$$I_2 = A^2(1 + \cos(\delta\phi_2)).$$

Therefore, the intensity depends on the phase difference between the response-computation oscillators and the stimulus oscillator.

Now, the maximum intensity of I_1 and I_2 is $2A^2$, whereas their minimum intensity is zero. If we think of I_1 and I_2 as competing possible responses, the maximum difference between them happens when one of their relative phases (with respect to the stimulus oscillator) is zero while the other is π. It is standard to use the contrast, defined by

$$b = \frac{I_1 - I_2}{I_1 + I_2}, \tag{11}$$

as a measure of how different the intensities are. From its definition, b takes values between -1 and 1. When I_1 and I_2 are as different as possible, $|b| = 1$; if, on the other hand, I_1 and I_2 are the same, $b = 0$.

The contrast provides us with a useful way to think about responses that are between r_1 and r_2. To see this, let us impose

$$\delta\phi_1 = \delta\phi_2 + \pi \equiv \delta\phi, \tag{12}$$

which results in

$$I_1 = A^2 \left(1 + \cos\left(\delta\phi\right)\right), \tag{13}$$

and

$$I_2 = A^2 \left(1 - \cos\left(\delta\phi\right)\right). \tag{14}$$

In this case, the single parameter $\delta\phi$ is sufficient to determine the contrast, as

$$b = \cos\left(\delta\phi\right), \tag{15}$$

$0 \leq \delta\varphi \leq \pi$. So, the phase difference $\delta\phi$ between stimulus and response oscillators codes a continuum of responses between -1 and 1 (more precisely, because $\delta\varphi$ is a phase, the interval is on the unit circle $\mathbb{T}$, and not in a compact interval in $\mathbb{R}$). For arbitrary intervals (ζ_1, ζ_2), all that is required is a re-scaling of b.

To summarize the above arguments, when a stimulus and a response oscillator activate, they fire periodically, leading to a synchronization with constant phase relation. This phase relation causes interference, which in turn determines the relative strength of the intensities for each response. Thus, responses are determined by the interference of oscillators, which is itself affected by the neural oscillators' couplings.

We now examine in more detail the mathematics of the stimulus and response model. Let us look at each step of (1).

Sampling

When a stimulus s_n is sampled, a collection of neurons start firing synchronously, corresponding to the activation of a neural oscillator, O_{s_n}. Such activation leads to a spreading of activation to oscillators coupled to the stimulus oscillator, including the response O_{r_1} and O_{r_2}. Since the selection and activation of O_{s_n} involves the perceptual system, we do not attempt to model with neural oscillators this step, but simply assume their activation in a way that is consistent with the stochastic process represented in SR theory by the random variable $\mathbf{S}_n$. Furthermore, though it would be important to develop a detailed theory of spreading activation, we do not, as for our current purposes it suffices to simply assume the activation of O_{r_1} and O_{r_2}.

Response

After the stimulus s_n is sampled, the active oscillators evolve for the time interval Δt_r, the time it takes to compute a response, according to the following set of Kuramoto differential equations.

$$\frac{d\varphi_i}{dt} = \omega_i - \sum_{i \neq j} k_{ij} \sin\left(\varphi_i - \varphi_j + \delta_{ij}\right), \tag{16}$$

where k_{ij} is the coupling constant between oscillators i and j, and δ_{ij} is an antisymmetric matrix representing phase differences, and i and j can be either O_{s_n}, O_{r_1}, or O_{r_2}. Here we use the notation where O_i corresponds to a neural oscillator and φ_i to its phase. Equation (16) can be rewritten as

$$\frac{d\varphi_i}{dt} = \omega_i - \sum_{j} \left[k_{ij}^E \sin\left(\varphi_i - \varphi_j\right) + k_{ij}^I \cos\left(\varphi_i - \varphi_j\right)\right], \tag{17}$$

where $k_{ij}^E = k_{ij} \cos{(\delta_{ij})}$ and $k_{ij}^I = k_{ij} \sin{(\delta_{ij})}$, which has an immediate physical interpretation: k_{ij}^E corresponds to excitatory couplings, whereas k_{ij}^I corresponds to inhibitory ones. These are the $4N$ excitatory and inhibitory coupling strengths between oscillators.

$$\frac{d\varphi_i}{dt} = \omega_i - \sum_{i \neq j} \left[k_{i,j}^E \sin{(\varphi_i - \varphi_j)} - k_{i,j}^I \cos{(\varphi_i - \varphi_j)} \right], \qquad (18)$$

where ω_i is their natural frequency. The solutions to (18) and the initial conditions randomly distributed at activation give us the phases at time $t_{r,n} = t_{s,n} + \Delta t_r$. The coupling strengths between oscillators determine their relative phase locking, which in turn corresponds to the computation of a given response, according to equation (11).

Reinforcement and Conditioning

As we saw above, the computation of a response depends on the inhibitory and excitatory couplings between neural oscillators. Therefore, when an effective reinforcement $\mathbf{Y}_n$ corresponding to changes in the conditioning $\mathbf{Z}_{n+1}$ occurs, the coupling strengths change. As with stimulus and responses, we represent a reinforcement by a neural oscillator. Such oscillator, with frequency ω_e, is activated during reinforcement, and we assume that it forces the reinforced response-computation and stimulus oscillators to synchronize with the same phase difference of $\delta\varphi$, while the two response-computation oscillators are kept synchronized with a phase difference of π. Let the reinforcement oscillator be activated on trial n at time $t_{e,n}$, $t_{r,n+1} > t_{e,n} > t_{r,n}$, for an interval of time Δt_e. Let K_0 be the coupling strength between the reinforcement oscillator and the stimulus and response-computation oscillators. In order to match the probabilistic SR axiom governing the effectiveness of reinforcement, we also assume that there is a normal probability distribution governing the coupling strength K_0 between the reinforcement and the other active oscillators with probability density

$$f(K_0) = \frac{1}{\sigma_{K_0}\sqrt{2\pi}} \exp\left\{ -\frac{1}{2\sigma_{K_0}^2} \left(K_0 - \overline{K}_0 \right)^2 \right\}. \qquad (19)$$

When a reinforcement is effective, all active oscillators phase-reset at $t_{e,n}$, and during reinforcement the phases of the active oscillators evolve according to the following set of differential equations.

$$\frac{d\varphi_i}{dt} = \omega_i - \sum_{i \neq j} \left[k_{i,j}^E \sin{(\varphi_i - \varphi_j)} - k_{i,j}^I \cos{(\varphi_i - \varphi_j)} \right]$$
$$- K_0 \sin{(\varphi_i - \omega_e t + \Phi_i)}, \qquad (20)$$

where $\Phi_{s_n} - \Phi_{r_1} = \delta\varphi$ and $\Phi_{r_1} - \Phi_{r_2} = \pi$. The excitatory couplings are reinforced if the oscillators are in phase with each other, according to the following equations.

$$\frac{dk_{i,j}^E}{dt} = \epsilon(K_0) \left[\alpha \cos{(\varphi_i - \varphi_j)} - k_{i,j}^E \right]. \qquad (21)$$

Similarly, for inhibitory connections, if two oscillators are perfectly off sync, then we have a reinforcement of the inhibitory connections.

$$\frac{dk_{i,j}^I}{dt} = \epsilon(K_0) \left[\alpha \sin{(\varphi_i - \varphi_j)} - k_{i,j}^I \right], \qquad (22)$$

In the above equations,

$$\epsilon(K_0) = \begin{cases} 0 & \text{if } K_0 < K' \\ \epsilon_0 & \text{otherwise,} \end{cases} \qquad (23)$$

where $\epsilon_0 \ll \omega_i$, α and K_0 are constant during Δt_e, and K' is a threshold constant throughout all trials. We can think of K' as a threshold below which the reinforcement oscillator has no (or very little) effect on the stimulus and response-computation oscillators. For large enough values of Δt_e, the behavioral probability parameter c of effective reinforcement mentioned above is, from (19) and (23), reflected in the equation:

$$c = \int_{K'}^{\infty} f(K_0) \, dK_0. \tag{24}$$

This relationship comes from the fact that, if $K_0 < K'$, there is no effective learning from reinforcement, since there are no changes to the couplings due to (21)–(22), and (18) describing the oscillators' behavior. Intuitively K' is the effectiveness parameter: the larger it is, the smaller the probability of effective reinforcement.

4   Final Remarks

In this paper we described the neural oscillator model presented in Suppes, de Barros, & Oas (2012), with particular emphasis to the physics and intuition behind many of the processes represented by equations (18). To summarize, the coded phase differences were used to model a continuum of responses within SR theory in the following way. At the beginning of a trial a stimulus oscillator is activated, and with it the response oscillators. Then, the coupled oscillator system evolves according to (18) if no reinforcement is present, and according to (20)–(22) if reinforcement is present. The coupling constants and the conditioning of stimuli are not reset at the beginning of each trial, and changes to couplings correspond to effective reinforcement. Because of the finite amount of time for a response, the probabilistic characteristics of the initial conditions lead to the smearing of the phase differences after a certain time, with an effect similar to that of the smearing distribution in the SR model for a continuum of responses (Suppes 1959).

We emphasize that in this paper we focused mainly on the physical basis of our model, and did not go much into mathematical detail. Furthermore, in Suppes, de Barros, & Oas (2012) we applied the neural oscillator model to many different experimental situations illustrated in the literature, whereas here we did not address in detail any empirical data. Interested readers are referred to our original paper.

SR theory has enjoyed tremendous success in the past, and, in a certain sense, its main features are still present in modern day neuroscience. We believe that by showing how neurons may result in theoretical structures that are somewhat similar to SR ones, as done in Suppes, de Barros, & Oas (2012), we can provide the basis for an extension of SR theory that could be considered more realistic. For example, in our model, many parameters, such as time of response, frequency of oscillations, coupling strengths, etc., were fixed based on reasonable assumptions. However, a more detailed and systematic study should be able to relate such parameters to either underlying physiological constraints or to behavioral variations, thus opening up the possibilities for new empirical studies that go beyond SR theory. Also, in our model we postulated many features without showing or proving their dynamics from underlying neuronal dynamics. This was the case for the activation of a stimulus and the spreading of activation of a stimulus and responses. A more detailed theory based on neural oscillators of such dynamics would certainly provide interesting empirical tests.

Finally, the use of neural oscillators and interference may also help explain certain aspects of cognition that are considered "non-classical." The distinction between classical and quantum behavior is a subtle one, and still not yet understood. For example, a well studied quantum-like decision making process is the violation of Savage's sure-thing principle, shown in a series of experiments by Tversky and Shafir (Shafir and Tversky 1992; Tversky and Shafir 1992). Similar violations do not need any quantum-like representation in the form of a Hilbert space, as proposed in the literature, but instead can be obtained by interference of neural oscillators (de Barros 2012b). Furthermore, the use of neural oscillator interference even leads to predictions that are not compatible with a Hilbert space structure (de Barros 2012a), suggesting that the use of quantum-like processes is not as quantum as many would wish.

References

Adolphs, R. 2003. Cognitive neuroscience of human social behaviour. *Nature Reviews Neuroscience* 4(3):165–178. ISSN 1471-003X.

Carvalhaes, C. G., J. A. de Barros, M. Perreau-Guimaraes, and P. Suppes. 2014. The joint use of tangential electric field and surface Laplacian in EEG classifcation. *Brain Topography* 27(1):84–94.

de Barros, J. A. 2012a. Joint probabilities and quantum cognition. *AIP Conference Proceedings* 1508(98):98–107. ISSN 0094243X.

—. 2012b. Quantum-like model of behavioral response computation using neural oscillators. *Biosystems* 110(3):171–182. ISSN 0303-2647.

Izhikevich, E. M. 2007. *Dynamical Systems in Neuroscience: The Geometry of Excitability and Bursting*. Cambridge, Mass.: The MIT Press.

Jessell, T. M., E. R. Kandel, and J. H. Schwartz. 2000. *Principles of Neural Science*. New York, NY: McGraw-Hill Health Professions Division, 4th edn.

Kuramoto, Y. 2003. *Chemical Oscillations, Waves, and Turbulence*. Mineola, N.Y.: Dov.

Shafir, E. and A. Tversky. 1992. Thinking through uncertainty: Nonconsequential reasoning and choice. *Cognitive Psychology* 24(4):449–474.

Suppes, P. 1959. Stimulus sampling theory for a continuum of responses. In K. J. Arrow, S. Karlin, and P. Suppes, eds., *Mathematical Methods in the Social Sciences*, pages 348–365. Stanford, California: Stanford University Press.

—. 2002. *Representation and Invariance of Scientific Structures*. Stanford, CA: CSLI Publications.

Suppes, P. and R. C. Atkinson. 1960. *Markov Learning Models for Multiperson Interactions*. Stanford, CA: Stanford University Press.

Suppes, P., J. A. de Barros, and G. Oas. 2012. Phase-oscillator computations as neural models of stimulus–response conditioning and response selection. *Journal of Mathematical Psychology* 56(2):95–117. doi:10.1016/j.jmp.2012.01.001.

Tversky, A. and E. Shafir. 1992. The disjunction effect in choice under uncertainty. *Psychological Science* 3(5):305–309.

Vassilieva, E., G. Pinto, J. A. de Barros, and P. Suppes. 2011. Learning pattern recognition through quasi-synchronization of phase oscillators. *Nueral Networks, IEEE Transactions on* 22(1):84–95.

17

Using the Scalp Electric Field to Recognize Brainwaves

CLAUDIO G. CARVALHAES

Abstract

This paper discusses a new approach to the challenging problem of EEG classification using the scalp electric field. The field is estimated from the given data by applying a numerical technique, but getting the field normal component requires an approximate treatment. The performance of this approach is illustrated with EEG data from two experiments, the first involving visual perception and the second mental imagery.

1 Introduction

Decoding brain activity using non-invasive EEG is a challenging goal in neuroscience. The brain-machine interface enabling people to control computers by thought is just one of many applications envisioned for this research, see Lotte et al. (2007). However, EEG signals are prone to interference, and establishing meaningful relationships between them and underlying neuronal activity is not trivial. The brain activity generates a complex pattern of electric currents that propagate through multiple layers in the extracellular volume of the head before reaching the scalp surface. Along this path the electrical activity is greatly modified by a number of factors that complicate interpretation. Notably important is the skull resistivity which significantly attenuates flow lines, creating biases that can hardly be accounted for by signal process techniques.

Often overlooked, but also important is the nonlocal relation between EEG potentials and underlying current flows. According to Ohm's law for linear conductors, because brain processes are slow time-varying with oscillations ranging typically below $1\,\mathrm{kHz}$, the *current density* is proportional to the gradient of the electric potential at each time instant, with the proportionality constant depending on the medium. By inverting this relation one obtains that the electrical potential integrates the current density over all space, thereby providing a *nonlocal* picture of spatial frequency content of brain electrical activity.

In the Ohmic regime the negative gradient of the electric potential corresponds to the *electric field*, which is a vector quantity that is locally related to the current density. This

Foundations and Methods from Mathematics to Neuroscience.
Colleen E. Crangle, Adolfo García de la Sienra and Helen Longino.
Copyright © 2014, CSLI Publications.

property of locality makes the electric field promising enough to be worth considering more deeply. In addition, the electric field is reference-free, while the electric potential depends on the choice of a reference. But unfortunately, it is the electric potential, not the electric field, which is recorded in conventional EEG. In order to study the electric field we have to estimate it from the available potential distribution. This requires the application of a numerical technique to spatially differentiate the EEG signal. A major problem with that is the determination of the field's normal component. This component expresses the rate of change of the potential in the direction normal to the scalp surface, but scalp recordings are performed along the scalp surface only.

This paper discusses an approach to estimate the scalp electric field which involves an approximation for the field's normal component using the surface Laplacian of the electric potential. The Laplacian differentiation is a well-known technique for enhancing the spatial resolution of EEG data. This technique is now supported by a substantial body of work with numerous applications in clinical and research environments, see Hjorth (1975); Perrin et al. (1989); Babiloni et al. (1995); Nunez and Srinivasan (2006). Numerical issues in the spatial differentiation of the data will be reviewed, focusing mainly on computational aspects that are not well discussed in the literature. By way of illustration, I will summarize some results of a recent work carried out in collaboration with my colleagues from the Suppes Brain Lab in Carvalhaes et al. (2014).

2 Reconstructing the Scalp Electric Field

2.1 Preliminary Considerations

The propagation of electrical activity in the head can be accounted for by the *quasistatic* approximation of the macroscopic Maxwell's equations in continuous media, see Haus and Melcher (1989). This approximation neglects the reciprocal action of electric and magnetic fields, which holds for the brain for two major reasons. First, the characteristic length scale of magnetic induction in biological tissues is too large, typically above 50 m, thereby not affecting scalp recordings. Second, the displacement currents that ordinarily interact with the magnetic field are obliterated by Ohmic currents in the head, and ignoring them has a negligible effect in practice. Typical figures showing the validity of the quasistatic approximation in bioelectric phenomena are given in Plonsey (1969, chap. 5), Hämäläinen et al. (1993) and Nunez and Srinivasan (2006, app. B), where the quasistatic Maxwell's equations are shown explicitly.

Electroquasistatic fields have the important property of being *curl-free* at each instant, meaning that they can be expressed as the (negative) gradient of an electric potential. Denoting the vector electric field by $\mathbf{E}$ and the scalar potential by Φ, we can write

$$\mathbf{E} = -\operatorname{grad}\Phi. \tag{1}$$

(Both $\mathbf{E}$ and Φ depend on space and time, but this dependence will not be written explicitly unless necessary.) Ordinarily, EEG experiments provide us with samplings of instantaneous potentials recorded from many different scalp locations and from several participants. To obtain the scalp electric field all we have to do is apply a numerical technique to partially differentiate these data. This task encompasses two major problems. First, the geometry of the head must be known in order to construct the gradient operator, but this information is generally not available. An easy way to address this problem consists of modeling the scalp as a perfect sphere. This approximation is widely used for practical and theoretical purposes, and it is not overly simplistic to the point of leading to erroneous conclusions. Most importantly, no conceptual obstacle is apparent which precludes an extension to more realistic models of the head such as the ellipsoidal model, see Carvalhaes and Suppes (2011). For convenience, I will map the spherical scalp using the spherical coordinate system (r, θ, φ) following the convention of Griffiths (1999, p. 38), with the polar angle (θ) increasing down from the vertex and the azimuth (φ) varying counterclockwise around the vertex. For future reference, I recall the formulas for the divergence, gradient, and Laplacian operators in spherical coordinates:

$$\operatorname{div} \mathbf{v} = \frac{1}{r^2} \frac{\partial}{\partial r} \left(r^2 v_r \right) + \frac{1}{r \sin \theta} \frac{\partial}{\partial \theta} \left(\sin \theta \, v_\theta \right) + \frac{1}{r \sin \theta} \frac{\partial v_\varphi}{\partial \varphi}, \tag{2a}$$

$$\operatorname{grad} f = \frac{\partial f}{\partial r} \hat{\mathbf{r}} + \frac{1}{r} \frac{\partial f}{\partial \theta} \hat{\boldsymbol{\theta}} + \frac{1}{r \sin \theta} \frac{\partial f}{\partial \varphi} \hat{\boldsymbol{\varphi}}, \tag{2b}$$

$$\operatorname{Lap} f = \frac{1}{r^2} \frac{\partial}{\partial r} \left(r^2 \frac{\partial f}{\partial r} \right) + \frac{1}{r^2 \sin \theta} \frac{\partial}{\partial \theta} \left(\sin \theta \frac{\partial f}{\partial \theta} \right) + \frac{1}{r^2 \sin^2 \theta} \frac{\partial^2 f}{\partial \varphi^2}, \tag{2c}$$

where $\hat{\mathbf{r}}$, $\hat{\boldsymbol{\theta}}$, and $\hat{\boldsymbol{\varphi}}$ are unit vectors in the coordinate directions.

The second problem in the estimation of the scalp electric field is to obtain the field's normal component. This component gives the rate of change of the electric potential along the normal direction, information that is not available in conventional EEG. The solution of this problem requires an in-depth examination of current flow in the head and is addressed in the following sections.

2.2 The Volume Current Density

Neuronal activation is a complex process that involves the flow of capacitive and diffusion currents in electrically active cells and their immediate vicinity in the brain. Conceptually, this process is accounted for by an *impressed current density,* $\mathbf{J}^i$, which is the source for the electric field in the extracellular space. The action of the electric field on charge carriers will characterize an Ohmic current, called *volume current,* governed by Ohm's law:

$$\mathbf{J}^v = \boldsymbol{\sigma} \mathbf{E}. \tag{3}$$

Here, $\mathbf{J}^v$ is the volume current density and $\boldsymbol{\sigma}$ denotes the electrical conductivity tensor, which is a rank-2 symmetric tensor. The tensor representation of $\boldsymbol{\sigma}$ is convenient to account for directional dependency (anisotropy) and location dependency (inhomogeneity) in the volume conductor, but let us assume for a moment such effects can be neglected so that σ is a single-value constant (the bold-face notation can be dropped in this particular case). Substituting $\mathbf{E} = -\operatorname{grad} \Phi$ in (3) and integrating the result from an arbitrary endpoint $\mathbf{r}$ to a reference point $\mathbf{r}_{\mathrm{ref}}$ leads to $\Phi(\mathbf{r}) = \Phi(\mathbf{r}_{\mathrm{ref}}) + \int_{\mathbf{r}}^{\mathbf{r}_{\mathrm{ref}}} \mathbf{J}^v \cdot ds/\sigma$, where ds is a line element tangent to an arbitrary path from $\mathbf{r}$ and $\mathbf{r}_{\mathrm{ref}}$. This means the value of Φ at $\mathbf{r}$ depends on a reference $\Phi(\mathbf{r}_{\mathrm{ref}})$ and on all values assumed by $\mathbf{J}^v$ along a path. In performing this integration we may get a map of spatial frequencies that obscure interpretation. In contrast the relation between $\mathbf{E}$ and $\mathbf{J}^v$ is local, which is a compelling reason for considering the scalp electric field.

Another remark about Eq. (3) is that it predicts that the lower the conductivity of a region, the higher the electric field strength. Hence, driving currents in the poorly-conductive skull requires much more strength to the electric field than through gray and white matter, cerebrospinal fluid (CSF), or the scalp. In principle, this means that the most efficient way to measure the electric field invasively is by placing microelectrodes in the skull, rather on the cortical surface, as is done in conventional electrocorticogram (ECoG).

In its more general form, $\boldsymbol{\sigma} = \boldsymbol{\sigma}(\mathbf{r})$ is a 2nd-rank, symmetric tensor of complex-valued elements. The tensor representation accounts for directional dependency in the medium's structure, see Tuch et al. (2001). For instance, the tangential:normal conductivity anisotropy of the scalp is regard to be $\sim$1.5:1.0, see Abascal et al. (2008); Petrov (2012), which can be easily accounted for by using the tensor representation. Moreover, the location dependency of $\boldsymbol{\sigma}$ is definitely not negligible in the human head, varying up to a ratio of 80:1 between adjacent layers, see Huiskamp et al. (1999).

The real part of the conductivity describes resistive effect, while the imaginary part is associated with capacitive effect. Capacitive effects are generally regarded as negligible in large-scale phenomena such as those underlying EEG generation, but they can be crucial at smaller scales, for instance, in the description of cell-body and axonal dynamics. The conductivity may also depend on the frequency and magnitude of fields, but ordinarily these effects are secondary and ignorable in first-order approximations. I will assume henceforth that the scalp conductivity tensor, $\boldsymbol{\sigma}^{\mathrm{scalp}}$, is a diagonal tensor with a normal component $\sigma_r^{\mathrm{scalp}}$ and two equal tangential components $\sigma_t^{\mathrm{scalp}}$:

$$\boldsymbol{\sigma}^{\mathrm{scalp}} = \sigma_r^{\mathrm{scalp}}\hat{\mathbf{r}}\hat{\mathbf{r}} + \sigma_t^{\mathrm{scalp}}\hat{\boldsymbol{\theta}}\hat{\boldsymbol{\theta}} + \sigma_t^{\mathrm{scalp}}\hat{\boldsymbol{\varphi}}\hat{\boldsymbol{\varphi}}. \qquad (4)$$

Both $\sigma_r^{\mathrm{scalp}}$ and $\sigma_t^{\mathrm{scalp}}$ are real-valued quantities.

2.3 Boundary Conditions

The sum of $\mathbf{J}^i$ and $\mathbf{J}^v$ gives the *total current density* $\mathbf{J}$ which in the quasistatic regime is divergenceless (or *solenoidal*). This means that the flow lines of $\mathbf{J}$ are continuous and form closed loops in space at each instant. Since the impressed current vanishes outside the active region, the continuity of $\mathbf{J}$ implies the continuity of $\mathbf{J}^v$ in the extracellular space. Consequently, due to (3) any abrupt change of $\boldsymbol{\sigma}$ at the interface between adjacent macroscopic layers implies a reciprocal change of the normal component of $\mathbf{E}$. At the scalp-air interface this condition can be written in the form

$$\sigma_r^{\mathrm{scalp}}E_r^{\mathrm{scalp}} = \sigma^{\mathrm{air}}E_r^{\mathrm{air}}, \qquad (5)$$

where the subscript r indicates radial (normal) component. Typical values for σ^{air} and $\sigma_r^{\mathrm{scalp}}$ are: $\sigma^{\mathrm{air}} = 3\times10^{-15}\,\mathrm{S/m}$, see Pawar et al. (2009) and $\sigma_r^{\mathrm{scalp}} = 0.44\,\mathrm{S/m}$, see Horesh (2006). This means σ^{air} is negligibly small as compared to $\sigma_r^{\mathrm{scalp}}$ ($\sigma^{\mathrm{air}}/\sigma_r^{\mathrm{scalp}} \sim 10^{-14}$), so that E_r^{scalp} is null along the scalp-air interface. But the negligibility of σ_{air} and the vanishing of E_r^{scalp} do not imply that E_r^{air} vanishes as well. Instead, E_r^{air} is expected to increase sharply at the scalp-air interface and then fall off rapidly with increasing distance from the head.

The vanishing of E_r^{scalp} greatly simplifies computations. Because $\mathbf{E}$ becomes tangential it can be fully estimated from the electric potential using a numerical technique.

In practice, however, things are much more complicated. When recording EEG the current actually comes out of the head driven by a non-zero field component E_r^{scalp}. A layer of conduction gel provides an interface to measurement electrodes through which the current exits. The lack of information about the potential distribution in the normal direction precludes a direct estimation of E_r^{scalp}.

2.4 An Approximation for E_r^{scalp}

The three components of the scalp electric field are not totally independent, but they must satisfy the conditions that $\mathbf{E}$ is curl-free and $\sigma^{\text{scalp}}\,\mathbf{E}$ is solenoidal. That is,

$$\text{curl}\,\mathbf{E} = 0, \tag{6a}$$

$$\text{div}\left(\sigma^{\text{scalp}}\,\mathbf{E}\right) = 0, \tag{6b}$$

at each time instant. One may raise the question of whether we can use one of these constraints to express E_r^{scalp} in terms of E_θ and E_φ to obtain the vector scalp electric field. The first constraint is definitely not suitable for this purpose, since it relates cross-derivatives. The second constraint E_r^{scalp} may be helpful. By combining (2a) and (6b) and assuming a homogeneous medium, we obtain

$$\frac{\sigma_r}{r^2}\frac{\partial}{\partial r}\left(r^2 E_r\right) + \frac{\sigma_t}{r\sin\theta}\frac{\partial}{\partial\theta}\left(\sin\theta\,E_\theta\right) + \frac{\sigma_t}{r\sin\theta}\frac{\partial E_\varphi}{\partial\varphi} = 0, \tag{7}$$

or equivalently,

$$\alpha\,\text{div}_\mathsf{T}\mathbf{E}_\mathsf{T}(\mathbf{r},t) = -2\,E_r(\mathbf{r},t) - r\frac{\partial E_r(\mathbf{r},t)}{\partial r}, \tag{8}$$

where $\alpha = \sigma_t/\sigma_r$ and $\text{div}_\mathsf{T}\mathbf{E}_\mathsf{T} = \text{div}\,\mathbf{E} - (\partial E_r/\partial r)\,\hat{\mathbf{r}}$ is the *tangential divergence* of the tangent field $\mathbf{E}_\mathsf{T} = \mathbf{E} - E_r\hat{\mathbf{r}}$. An approximate expression for E_r^{scalp} follows from a careful analysis of (8) at electrode sites. Let us assume for a moment that $\mathbf{r}^{\text{scalp}}$ is an arbitrary location at the outer scalp surface. Applying (8) yields

$$\alpha\,\text{div}_\mathsf{T}\mathbf{E}_\mathsf{T}^{\text{scalp}} = -2\,E_r^{\text{scalp}} - r^{\text{scalp}}\left.\frac{\partial E_r}{\partial r}\right|_{\text{scalp}}, \tag{9}$$

where $\partial E_r/\partial r$ is understood as a left derivative. An intuitive interpretation of this equation is that it shows how changes in E_r^{scalp} affect the lateral spread of current driven by $\mathbf{E}_\mathsf{T}^{\text{scalp}}$. This spread depends on how the normal field relates to its normal derivative, and eventually it vanishes somewhere at a certain time. If $\mathbf{r}_{\text{scalp}}$ lies on the scalp-air interface, where E_r^{scalp} is null all the time, then the lateral divergence is entirely determined by the rate of vanishing of E_r given by $\partial E_r/\partial r$. The higher this rate, the more significant the spread.

At the places where a measurement electrode is resistively connected to the scalp to record the signal the current can flow out of the head driven by the electric field. Presumably, this decreases the rate $\partial E_r/\partial r$ locally, since E_r is not attenuated to zero at electrode sites. With this assumption we obtain the following approximation for E_r^{scalp}:

$$E_r^{\text{scalp}} \approx -\frac{\alpha}{2}\text{div}_\mathsf{T}\mathbf{E}_\mathsf{T}^{\text{scalp}}. \tag{10}$$

Since the substitution $E_\theta = (1/r)\partial\Phi/\partial\theta$ and $E_\varphi = (1/r\sin\theta)\partial\Phi/\partial\varphi$ into $\text{div}_\mathsf{T}\mathbf{E}_\mathsf{T}^{\text{scalp}}$ yields the surface Laplacian of Φ, this approximation for E_r^{scalp} is essentially an estimation of the surface Laplacian of Φ.

3 Numerical Procedure

A detailed discussion of the problem of using splines on non-Euclidean space such as the spherical scalp model is presented in Carvalhaes and Suppes (2011). This applies equally well to the determination of the scalp electric field. One reason for using splines is because it is a reliable method with very low computational cost. Following Carvalhaes and Suppes (2011), we define three N-vectors, $\mathbf{e}_\lambda^r$, $\mathbf{e}_\lambda^\theta$, $\mathbf{e}_\lambda^\varphi$, giving the respective components of the scalp electric field at electrode locations $\mathbf{r}_1, \cdots, \mathbf{r}_N$ at a time t. These vectors are obtained via a linear transformation of the associated scalp potential distribution $\{V_1, \cdots, V_N\}$:

$$\mathbf{e}_\lambda^r = \mathbf{L}_\lambda \mathbf{v}, \quad \mathbf{e}_\lambda^\theta = \mathbf{E}_\lambda^\theta \mathbf{v}, \quad \mathbf{e}_\lambda^\varphi = \mathbf{E}_\lambda^\varphi \mathbf{v}, \tag{11}$$

where $\mathbf{v} = (V_1, \cdots, V_N)'$ and the transforming matrices $\mathbf{L}_\lambda$, $\mathbf{E}_\lambda^\theta$, and $\mathbf{E}_\lambda^\varphi$ are N-by-N matrices. The parameter λ is a regularization parameter to control the trade off between smoothing and interpolating the data. The matrix $\mathbf{L}_\lambda$ is the differentiation matrix that represents the surface Laplacian operator. Because $\mathbf{e}_\lambda^\theta$, $\mathbf{e}_\lambda^\varphi$, and $\mathbf{e}_\lambda^r$ are reference-free, the transforming matrices are all singular. Namely, the columns of $\mathbf{L}_\lambda$, $\mathbf{E}_\lambda^\theta$, or $\mathbf{E}_\lambda^\varphi$ sum to zero in order for (11) to be invariant under the transformation $\mathbf{v} \to \mathbf{v} + \alpha \mathbf{1}$, where $\alpha \in \mathbb{R}$ and $\mathbf{1} \in \mathbb{R}^N$ is a vector of N ones.

Obtaining the differentiation matrices requires the solution of

$$\begin{pmatrix} \mathbf{K} + N\lambda \mathbf{I} & \mathbf{T} \\ \mathbf{T}' & 0 \end{pmatrix} \begin{pmatrix} \mathbf{c} \\ \mathbf{d} \end{pmatrix} = \begin{pmatrix} \mathbf{v} \\ \mathbf{0} \end{pmatrix}, \tag{12}$$

where $(\mathbf{K})_{ij} = \|\mathbf{r}_i - \mathbf{r}_j\|^{2m-3}$, m is an integer larger than 2, and $(\mathbf{T})_{ij} = \phi_j(\mathbf{r}_i)$, where $\phi_1, \cdots, \phi_M$ is a set of linearly-independent polynomials in $\mathbb{R}^3$ of degree less than m. The constant M is given by $M = \binom{m+2}{3}$ and is subject to the condition $M < N$. The solution vectors $\mathbf{c} = (c_1, \cdots, c_N)'$ and $\mathbf{d} = (d_1, \cdots, d_M)'$ give the coefficients of the spline interpolant

$$f_\lambda(\mathbf{r}) = \sum_{j=1}^{N} c_j \|\mathbf{r} - \mathbf{r}_j\|^{2m-3} + \sum_{\ell=1}^{M} d_\ell \, \phi_\ell(\mathbf{r}). \tag{13}$$

Note that because $\mathbf{T}$ has a group of linearly dependent columns constrained by the geometric equation of the spherical head, the linear system (12) is singular. Carvalhaes and Suppes (2011) proposed to circumvent this problem by applying the pseudo-inverse matrix. Carvalhaes (2012) presented an approach accounting for the realistic case of a nearly spherical or nearly ellipsoidal scalp for which (12) is ill-conditioned.

4 Experimental Data and Results

By way of illustration this section shows the performance of the scalp electric field on the classification of two experiments carried out by my colleagues from the Suppes Brain Lab. The participants of each experiment performed specific tasks during which the EEG was continuously recorded from multiple scalp locations. The signals were offline preprocessed and numerically differentiated to obtain the electric field. The classification results shown below were maximized by seeking the optimal number of Principal Components (PCs) estimated from the data. The spatial components of the scalp electric field were further improved by spatially smoothing the data via λ regularization. Classifications were carried out by using 10-fold cross-validation on linear discriminant analysis. A detail description of all these procedures, along with further details of each experiment are in a separate publication, Carvalhaes et al. (2014).

In the first experiment the task for the 7 participants was to identify 9 two-dimensional images which were presented multiple times to each participant on a computer screen. Each image was a pairwise combination of 3 geometric shapes (*circle, square, and triangle*) and 3 colors (*red, green,* and *blue*). The signals were offline averaged to reduce temporal noise, and then classified according to the 9 features.

The second experiment was a two-class problem in which the stimuli consisted of a "stop" sign or the sound of the English word "go". The task assigned to participants was to create a mental image as vivid as possible of the stimulus that had just been presented. The goal was to recognize brain activity related to stimulus imagination only.

Figure (1) summarizes the classification results, showing mean recognition rates across all participants of each experiment. The rates were remarkably good, bearing in mind the chance level indicated with a dashed line. Improvements were specially significant in the recognition of the two-dimensional images. This better performance of the scalp electric field and its components in comparison to the scalar potential was confirmed in terms of positive effect sizes, which were generally greater than 0.5 and up to 2.1 standard deviations.

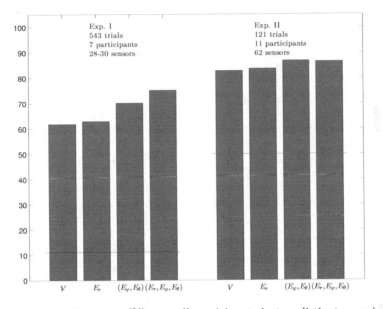

FIGURE 1 Mean recognition rates (%) over all participants in two distinct experiments. From left to right, the bars show the rates for the potential, normal component of the scalp electric field, tangential electric field, and total scalp electric field. The horizontal dashed line across bars indicates the chance level

5 Conclusion

This paper discussed the method of using the scalp electric field to improve the classification of EEG signals. The electric field trivially solves the reference-electrode problem, providing a local measure of the current density elicited by the brain activation. The practical effect of this measure on the challenging problem of brainwave recognition was exemplified by experiments on visual perception and mental imagery. The results were very satisfactory, offering an encouraging prospect for other applications and further development.

References

Abascal, J.-F. P. J., S. R. Arridge, D. Atkinson, R. Horesh, L. Fabrizi, M. D. Lucia, L. Horesh, R. H. Bayford, and D. S. Holder. 2008. Use of anisotropic modelling in electrical impedance tomography: description of method and preliminary assessment of utility in imaging brain function in the adult human head. *Neuroimage* 43(2):258–268. doi:10.1016/j.neuroimage. 2008.07.023.

Babiloni, F., C. Babiloni, L. Fattorini, F. Carducci, P. Onorati, and A. Urbano. 1995. Performances of surface Laplacian estimators: A study of simulated and real scalp potential distributions. *Brain Topography* 8(1):35–45. doi:10.1007/BF01187668.

Carvalhaes, C. G. 2012. Spline interpolation on nonunisolvent sets. *IMA J Numer Anal* 33(1):370–375.

Carvalhaes, C. G., J. A. de Barros, M. Perreau-Guimaraes, and P. Suppes. 2014. The joint use of tangential electric field and surface Laplacian in EEG classifcation. *Brain Topography* 27(1):84–94.

Carvalhaes, C. G. and P. Suppes. 2011. A spline framework for estimating the EEG surface Laplacian using the Euclidean metric. *Neural Computation* 23(11):2974–3000.

Griffiths, D. J. 1999. *Introduction to Electrodynamics*. New Jersey: Prentice Hall, 3rd edn.

Hämäläinen, M., R. Hari, R. J. Ilmoniemi, J. Knuutila, and O. V. Lounasmaa. 1993. Magnetoencephalography: theory, instrumentation, and applications to noninvasive studies of the working human brain. *Reviews of Modern Physics* 65(2):413–497. doi:10.1103/RevModPhys. 65.413.

Haus, H. A. and J. R. Melcher. 1989. *Electromagnetic Fields and Energy*. Englewood Cliffs, New Jersey: Prentice Hall.

Hjorth, B. 1975. An on-line transformation of EEG scalp potentials into orthogonal source derivations. *Electroencephalography and Clinical Neurophysiology* 39(5):526–530.

Horesh, L. 2006. *Some Novel Approaches in Modelling and Image Reconstruction for Multi-Frequency Electrical Impedance Tomography of the Human Brain*. Ph.D. thesis, Department of Medical Physics, University College London.

Huiskamp, G., M. Vroeijenstijn, R. van Dijk, G. Wieneke, and A. C. van Huffelen. 1999. The need for correct realistic geometry in the inverse EEG problem. *IEEE Transactions Biomedical Engineering* 46(11):1281–1287. doi:10.1109/10.797987.

Lotte, F., M. Congedo, A. Lecuyer, F. Lamarche, and B. Arnaldi. 2007. A review of classification algorithms for EEG-based brain-computer interfaces. *Journal of Neural Engineering* 4(2):R1–R13. doi:10.1088/1741-2560/4/2/R01.

Nunez, P. l. L. and R. Srinivasan. 2006. *Electric Fields of the Brain: The Neurophysics of EEG*. New York: Oxford University Press, 2nd edn.

Pawar, S. D., P. Murugavel, and D. M. Lal. 2009. Effect of relative humidity and sea level pressure on electrical conductivity of air over Indian Ocean. *Journal of Geophysical Research* 114(D2):D02205. doi:10.1029/2007JD009716.

Perrin, F., J. Pernier, O. Bertrand, and J. F. Echallier. 1989. Spherical splines for scalp potential and current density mapping. *Electroencephalography and Clinical Neurophysiology* 72(2):184–187.

Petrov, Y. 2012. Anisotropic spherical head model and its application to imaging electric activity of the brain. *Physical Review E: Statistical physics, plasmas, fluids, and related interdisciplinary topics* 86(1):011917. doi:10.1103/PhysRevE.86.011917.

Plonsey, R. 1969. *Bioelectric Phenomena*. New York: McGraw-Hill.

Tuch, D. S., V. J. Wedeen, A. M. Dale, J. S. George, and J. W. Belliveau. 2001. Conductivity tensor mapping of the human brain using diffusion tensor MRI. *Proceedings of the National Academy of Sciences of the United States of America* 98(20):11697–11701. doi:10.1073/pnas. 171473898.

18

Similarity Trees Derived from Pairwise Classifications

Marcos Perreau-Guimaraes

1   Introduction

In the last couple of decades, a growing field of research has been trying, with increasing success, to recognize brain activity from various kinds of recordings techniques, such as electroencephalogram (EEG), magnetoencephalogram (MEG) or functional magnetic resonance imaging (FMRI). These studies analyzed all kinds of activity, from motor Pfurtscheller et al. (1994); Waldert et al. (2008) to perception Cox and Savoy (2003); Ullman et al. (2002) or cognitive tasks Klimesch (1999); Schaefer et al. (2008); Perreau-Guimaraes et al. (2007). Some experiments were limited to the recognition of binary states, left/right finger twitching or absence/presence of some sensory input. These studies were usually set to establish the neurological pathways of a specific function, such as motor control of fingers. Other experiments pursued the study of a broader set of tasks or sensory inputs. In a previous experiment we recorded the EEG from participants reading or listening to a set of 100 geographical sentences, each made of half a dozen words. In another experiment we displayed to the participants circles, triangles and squares that could be either red, blue or green. The recognition of a richer set has the benefit of allowing insights beyond the recognition rate. In Suppes et al. (2009) we proposed a theoretical framework, based on partial order of similarity and similarity trees, for the study of the structures underlying the tasks. Using this framework we were able to show invariance between stimulus and brain waves of syntactic and semantic structures. In order to scale up these results to the thousands words of the English language, or to the many shapes and colors of geometry, or to the complexity of music, we have to surmount a number of difficulties. In our previous work we used an one-against-all strategy for polychotomous classification, predicting the class of an observation from the outcome of K-1 binary classifiers, where K is the number of classes. This works well for a small number of classes, but as the number of classes increases the design becomes increasingly unbalanced. Indeed, starting from a balanced design with N observation in each class, each binary one against all classification involves N versus (K-1)N classes. Other strategies, such as majority vote and probability coupling, predict the class of an observation from the K(K-1)/2 pairwise classifiers, keeping all classifications to N against

Foundations and Methods from Mathematics to Neuroscience.
Colleen E. Crangle, Adolfo García de la Sienra and Helen Longino.
Copyright © 2014, CSLI Publications.

N problems. Our goal is to propose and evaluate 3 methods to infer similarity trees, such as the ones we defined in Suppes et al. (2009), from a set of pairwise classifiers. The first method defines a naive metric from the pairwise classification rates. The second and third methods use the majority vote and coupling strategies. We start in the first section by extending the asymmetric similarity trees defined in Suppes et al. (2009) to a symmetric variant adapted to the naive metric. The second section describes the classifiers and in the third section we evaluate the three methods using simulations.

2 Symmetric Similarity Trees

We start by giving definitions and notations used in Suppes et al. (2009). Let $P(o_j^+ | o_i^-)$ be the conditional probability of a trial of class i to be classified as belonging to class j. Let also $\mathcal{O}^+$ define the set of classes prototypes and and $\mathcal{O}^-$ the set of test classes. Let $\mathcal{O} = \mathcal{O}^+ \cup \mathcal{O}^-$ be the merged set of all the prototypes and all the tests.

Let O_I and O_J be subsets of $\mathcal{O}$, the merged product is given by

$$O_I O_J = \{o_j^+ | o_i^- : o_j^+ \in O_I \ \& \ o_i^- \in O_J \ or \ o_i^- \in O_I \ \& \ o_j^+ \in O_J\}.$$

A set $O_I O_J$ is a valid merged product if and only if it contains at least one ordered pair $o_j^+ | o_i^-$. Any valid $O_I O_J$ can be ordered according to the original order $\prec$ defined by the conditional probabilities. From this ordering we define

- $\min O_I O_J$ = the least pair under the ordering $\prec$,
- $\max O_I O_J$ = the largest pair under the ordering $\prec$.

We then extend the order to the set of valid merged products using

$$O_I O_J \underset{min}{\prec} O_K O_L \ if \ \min O_I O_J \prec \min O_K O_L$$

$$O_I O_J \underset{max}{\prec} O_K O_L \ if \ \max O_I O_J \prec \max O_K O_L.$$

The leaves of the similarity tree are the partition made of the $2N$ singleton of $\mathcal{O}$,

$$\mathcal{P}_0 = \{\{+o_1\}, \ \cdots, \{+o_N\}, \{-o_1\}, \ \cdots, \{-o_N\}\}.$$

Let Q_0^* be the set pf all the valid elements of Q_0 ordered by the relation $\underset{min}{\prec}$.
The partition $\mathcal{P}_1$ is then obtained by replacing in $\mathcal{P}_0$ the singleton sets $O_{i'}$ and $O_{j'}$ such as $O_{i'} O_{j'}$ is maximum in Q_0^*, by their union $O_{i'} \cup O_{j'}$.

The inductive step from k to $k+1$ is given by merging in the same fashion the maximum element in

$$\mathcal{P}_k = \{O_1^k, O_2^k, \ \cdots, O_{2N-k}^k\}.$$

The root of the similarity tree is

$$\mathcal{P}_{2N} = \{\{-o_1, \ \cdots, \ -o_N, \ +o_1, \ \cdots, \ +o_N\}\}.$$

In the case of a metric the distances are commutative and we do not need to distinguish between prototype and test. We derive symmetric similarity tress trees from pairwise distances. Now we have only a set $\mathcal{O}$ of classes and

$$O_I O_J = \{o_j | o_i : o_j \in O_I \ \& \ o_i \in O_J\}$$

We still have $O_I O_J = O_J O_I$ and all sets $O_I O_J$ are valid.
We order $O_I O_J$ with the distance order $\prec$ and we define

- $\min O_I O_J$ = the least pair under the ordering $\prec$,
- $\max O_I O_J$ = the largest pair under the ordering $\prec$.

We then extend the order to the set merged products using

$$O_I O_J \underset{\min}{\prec} O_K O_L \; if \min O_I O_J \prec \min O_K O_L$$

$$O_I O_J \underset{\max}{\prec} O_K O_L \; if \max O_I O_J \prec \max O_K O_L.$$

We start constructing the similarity trees from the partition made of the N singleton sets of $\mathcal{O}$,

$$\mathcal{P}_0 = \{\{o_1\}, \cdots, \{o_N\}\}.$$

Note that $\mathcal{P}_0$ is obviously a partition of $\mathcal{O}$, since, (i) each element of $\mathcal{P}_0$ is a nonempty set, (ii) the intersection of any two elements of the sets Q_0 of all the merged pairs in $\mathcal{P}_0$ are ordered by the relation $\underset{\min}{\prec}$.

The partition $\mathcal{P}_1$ is then obtained by replacing in $\mathcal{P}_0$ the singleton sets $O_{i'}$ and $O_{j'}$ such as $O_{i'} O_{j'}$ is maximum in Q_0^*, by their union $O_{i'} \cup O_{j'}$.

The inductive step from k to $k+1$ is given by merging in the same fashion the maximum element in

$$\mathcal{P}_k = \{O_1^k, O_2^k, \cdots, O_{N-k}^k\}.$$

The root of the similarity tree is

$$\mathcal{P}_N = \{\{o_1, \cdots, o_N\}\}.$$

3 Deriving Polychotomous Conditional Probabilities from Pairwise Classification

The general problem is to predict the class of an observation x among K possible classes. When $K = 2$ the prediction is performed by projecting $x \to h(x)$ into a real line and using a decision function $D(h(x)) = class$

$$\mathbf{x} \longrightarrow D(h(\mathbf{x})) \longrightarrow j. \tag{1}$$

In a linear model the projection function corresponds to a vector of coefficients $\mathbf{h}$ with the same length as trial $\mathbf{x}$. The function is then the inner product $h_i(\mathbf{x}) = \mathbf{h_i} \cdot \mathbf{x}$. The decision function is usually (not always) a thresholding function such as

$$D(h(x)) = \begin{cases} 0 & if \; h(x) < thr \\ 1 & if \; h(x) \geq thr \end{cases}.$$

The rate of correct classification for the pair of classes (i,j) is given by

$$r_{ij} = \frac{m_{00}^{ij} + m_{11}^{ij}}{m_{10}^{ij} + m_{10}^{ij}} \tag{2}$$

where m_{00}^{ij} and m_{11}^{ij} are the number of trials correctly classified and m_{01}^{ij} and m_{10}^{ij} the number incorrectly classified.

When $K > 2$ several methods exist. Linear Discriminant Analysis (LDA) defines $K - 1$ projections from the eigenvectors corresponding to the $K - 1$ non zeros eigenvalues of the cost function $J(S_w^{-1} S_b)$ where S_w and S_b are the within and in-between scatter matrices estimated from the sample. The function J is usually the trace function, $J(S_w^{-1} S_b) = tr(S_w^{-1} S_b)$. The prediction of the class of a test observation x is given by either a maximizing decision function or a multivariate model built with the $K - 1$ projections of the data. Instead of one projection function, $K - 1$ projection functions $h_1, h_2, \ldots, h_{K-1}$ are applied to each trial $\mathbf{x}$, and the decision function

$D(h_1(\mathbf{x}), h_2(\mathbf{x}), \ldots, h_N(\mathbf{x}))$, depending on the $K - 1$ scalars, gives the predicted class.

$$\mathbf{x} \longrightarrow D\left(\begin{bmatrix} h_1(\mathbf{x}) \\ h_2(\mathbf{x}) \\ \ldots \\ h_N(\mathbf{x}) \end{bmatrix}\right) \longrightarrow j.$$

From the number m_{ij} of test trials of class i classified as class j we build the confusion matrix $\mathbf{M} = (m_{ij})$. The correct classification rate r is the sum of the diagonal divided by the number of test trials

$$R = \frac{\sum_i m_{ii}}{\sum_{ij} m_{ij}}.$$

This method works well when the number of classes is not to high. With a large number of classes LDA is outperformed by classification methods deriving the prediction of a test observation from the $K(K - 1)/2$ pairwise decisions $D_{ij}(h_{ij}(x))$, where ij denotes the classification of class i against class j.

In Friedman (1996) Friedman proposed a "max wins" or majority vote method where the predicted class is the class that "wins" the largest number of pairwise classifications. Lets take a 3-class problem as an example. If for a test observation x the results of the $3(2)/2 = 3$ classification is

$$\begin{aligned} D_{12}(h_{12}(x)) &= 0 \quad \rightarrow \text{class } 1 \\ D_{13}(h_{13}(x)) &= 0 \quad \rightarrow \text{class } 1 \\ D_{23}(h_{23}(x)) &= 1 \quad \rightarrow \text{class } 3 \end{aligned}$$

then class 1 wins and the observation is predicted as being from class 1. This uses only the predicted classes from the $K(K - 1)/2$ pairwise classifiers. We can also infer from the 2x2 M^{ij} pairwise confusion matrices the estimated class probabilities for each pair of classes. For $i \neq j$

$$\begin{aligned} \hat{p}^{ij}(0|0) &= \frac{m_{00}^{ij}}{m_{00}^{ij} + m_{01}^{ij}}, \; \hat{p}^{ij}(1|0) = \frac{m_{01}^{ij}}{m_{00}^{ij} + m_{01}^{ij}} \\ \hat{p}^{ij}(0|1) &= \frac{m_{01}^{ij}}{m_{10}^{ij} + m_{11}^{ij}}, \; \hat{p}^{ij}(1|1) = \frac{m_{11}^{ij}}{m_{10}^{ij} + m_{11}^{ij}} \end{aligned} \tag{3}$$

It is not generally possible to find a set of probabilities $p(i) = \{p(1|i), p(2|i), \ldots, p(K|i)\}$ compatible with the $\hat{p}^{ij}$ probabilities. Using an approach similar to the Bradley-Terry method for paired comparisons, Tibshirani proposed a coupling method to find the set of probabilities $\hat{p}(i)$ that best fits the $\hat{p}^{ij}$ probabilities. He defines the matrix $r_{ij} = \hat{p}^{ij}(1|0)$ if $i < j$ and $r_{ij} = \hat{p}^{ji}(0|1)$ if $i > j$. The diagonal values r_{ii} are not defined. The method fits iteratively the r_{ij} to the model given by

$$\mu_{ij} = E(r_{ij}) = \frac{p(i)}{p(i) + p(j)} \tag{4}$$

Starting from some guess of the $p(i)$ values, it uses the iteration

$$p(i) \leftarrow p(i) \frac{\sum_{j \neq i} n_{ij} r_{ij}}{\sum_{j \neq i} n_{ij} \mu_{ij}} \quad \text{if } \sum_{j \neq i} n_{ij} \mu_{ij} > 0 \tag{5}$$

where n_{ij} is the total number of observations in the pair of classes i and j. At each step the $p(i)$ are re-normalized such as $\sum_i p(i) = 1$.

Lets get back to the similarity trees. Here we are interested in deriving similarity matrices from the pairwise classifiers. The simplest method is to use the pairwise classification rates as a distance between classes. The greater the classification rates, the farther apart are the classes. We define the naive metric matrix by $s_{ij} = 1 - R^{ij}$ if $i > j$, $s_{ij} = 1 - R^{ji}$ if $i < j$ and $s_{ii} = 1$, where R^{ij} is the proportion of correct classification for the pairwise classifier i versus j. We also derive a similarity matrix from the majority vote method by estimating the conditional probabilities $\hat{p}(i)$ from the majority vote confusions. Finally we define the coupling similarity matrix by averaging the $p(i)$ obtained for each test observation in the coupling classification method.

4 Simulation and Results

We defined 9 classes from 2 features and 3 possible values each. For each class, the observations x_{ij}, $i, j \in \{1, 2, 3\}$, were defined by the model

$$\begin{aligned}
x_{ij} &= a_1 f_{ij} + b_1 g_{ij} + \epsilon \\
y_{ij} &= a_2 f_{ij} + b_2 g_{ij} + \epsilon
\end{aligned} \tag{6}$$

where (f_{ij}, g_{ij}) follows a bi-variate normal distribution with mean m_{ij} and covariance matrix Σ_{ij} and the noise ϵ follows a normal distribution with mean 0 and standard deviation σ. We sampled 100 observations in each class for a total of 900 observations. We set

$$m_1 = \begin{bmatrix} 1 & 2 & 3 \\ 1.1 & 2.1 & 3.1 \\ 1.2 & 2.2 & 3.2 \end{bmatrix}$$

$$m_2 = \begin{bmatrix} 1 & 1.1 & 1.2 \\ 2 & 2.1 & 2.2 \\ 3 & 3.1 & 3.2 \end{bmatrix} \tag{7}$$

in order to simulate two 2-level feature sources with the two sources having the features in orthogonal directions. The observations (x, y) simulate measurements at some point on the scalp. By varying a_1, b_1, a_2, b_2 we can get closer to either source. We fixed all the standard deviations σ_{ij} to a constant. The noise standard deviation is $\sigma = 0.1$. Any classification method could be used to perform the pairwise classification, we chose a simple Bayesian classifier. For each one of the 100 simulation we split the 900 observations into two groups, half of the observations for fitting a multivariate normal model in each class and half of the observations were used as test in order to estimate the classification rates and conditional probabilities we need to build our similarity matrices. The training set is used to fit the model

$$X_{ij} \sim N(\hat{m}_{ij}, \hat{\Sigma}_{ij}), \ i, j \in 1, .., 3. \tag{8}$$

Once $\hat{m}_{ij}$ and $\hat{\Sigma}_{ij}$ are estimated, we obtain the probabilities for each test observation using Bayes rule

$$P(class = ij|X) = \frac{P(X|class = ij)P(class = ij)}{P(X)} \tag{9}$$

where $P(X|class = ij)$ is estimated from the training sample using the multivariate normal model $N(\hat{m}_{ij}, \hat{\Sigma}_{ij})$. Here $P(class = ij)$ is the constant $1/9$ as we have a balanced design, and we do not need to know $P(X)$ as it does not depend on the class. We can get the result by normalizing $P(class = ij|X) = P(X|class = ij)P(class = ij)/\alpha$, setting α such that $\sum P(class = ij|X) = 1$.

We will define the truth similarity matrix from the euclidean distances between the mean:

$$P_{truth}(o_{ij}, o_{kl}) = ((a_1(m_1(i,j) - m_1(k,l)) + b_1(m_2(i,j) - m_2(k,l)))^2 +$$
$$(a_2(m_1(i,j) - m_1(k,l)) + b_2(m_2(i,j) - m_2(k,l)))^2)^{1/2} \quad (10)$$

In order to compare the simulated similarity trees with the truth we will use the Kendall τ distance Kendall (1990, 1938) between ordered sets. This distance is defined as the number of disagreements between pairs of elements, normalized to one for a total disagreement. If $O_1 = \{o_1(1), \ldots, o_1(n)\}$ and $O_2 = \{o_2(1), \ldots, o_2(n)\}$ are two sets with an order $<$,

$$K(O_1, O_2) = (|\{(i,j) \in \{1, \ldots, n\}^2, i < j, ((o_1(i) < o_1(j)) \wedge (o_1(i) > o_1(j))) \vee$$
$$((o_1(i) > o_1(j)) \wedge (o_1(i) < o_1(j)))\}| / (1/2n(n-1)) \quad (11)$$

In our simulation, for the wins-all and the coupling similarity trees, the number of pairs of classes in $\{1, \ldots, 9\}^2$ is $n = 9 \times 9 = 81$, with 3240 such as $i < j$. In the case of the naive similarity tree the similarity matrix is symmetric, which leads to $n = 36$ with 630 such as $i < j$.

The similarity trees are uniquely determined by the ordering of the pairs of classes, so the distance between the orderings defines by isomorphism a distance between trees. The 100 simulations lead to an accurate estimation of the mean values of the distances, which are normally distributed. We also report 95% confidence intervals.

We first investigate the case where there is a clear choice in favor of the source 1. We set $a_1 = a_2 = 1$ and $b_1 = b_2 = 0.1$. Table 1 shows the mean distances between the truth and the 3 estimated similarity trees for values of $\sigma_{ij} = cte$ varying from .1 to 10. With the large number of simulations the confidence intervals are very small and do not overlap, suggesting the mean differences between the distances are statistically significant. We do not make the leap here to assume they would necessary be scientifically significant, as this would depend on the actual entity being measured, which is not the point investigated in this paper. Figure 1 shows the means in function of σ_{ij}. For small values of the dispersion the coupling method is outperformed by the naive and wins-all methods. The denominators in Equations 4 and 5 become zero for $i \neq j$ in the case of perfect classification which leads to poor estimates when the dispersion is small. The naive method compares well against the others for large values of the dispersions. The method performing the best in the most useful rage of dispersions is the coupling method. For dispersions greater than 10 the classifiers are not able to discriminate the classes and the distances reach a plateau for all three methods.

Now let us consider the case where sources 1 and 2 are equally important, with $a_1 = a_2 = b_1 = b_2 = 1$. Table 2 and Figure 1b show the mean differences from the truth in this case. We note the trends are very comparable the the first case, with the overall distances being slightly smaller.

Our simulation results suggest that the choice of the method to define the similarity matrix depends on the overall classification performance. When the classification rates are high, corresponding in our simulations to a small dispersion value, the naive metric outperforms the two more sophisticated methods. If the classification rates are signifi-

σ	Naive		Wins All		Coupling	
0.1	0.001	[−0.001, 0.002]	0.000	[−0.000, 0.001]	0.195	[0.191, 0.199]
0.5	0.249	[0.245, 0.253]	0.127	[0.125, 0.129]	0.192	[0.190, 0.194]
1	0.259	[0.254, 0.265]	0.206	[0.203, 0.209]	0.207	[0.204, 0.209]
2	0.286	[0.277, 0.295]	0.336	[0.330, 0.342]	0.248	[0.244, 0.253]
5	0.354	[0.342, 0.365]	0.422	[0.417, 0.427]	0.370	[0.362, 0.377]
10	0.414	[0.400, 0.429]	0.417	[0.410, 0.423]	0.431	[0.425, 0.437]

TABLE 1 Mean Kendall τ distances between the true and the 3 estimated similarity trees, with their 95% confidence intervals with $a = 10b$

σ	Naive		Wins All		Coupling	
0.1	0.001	[0.000, 0.001]	0.000	[0.000, 0.000]	0.148	[0.146, 0.149]
0.5	0.167	[0.163, 0.170]	0.081	[0.080, 0.083]	0.141	[0.140, 0.142]
1	0.175	[0.171, 0.180]	0.150	[0.147, 0.152]	0.150	[0.148, 0.152]
2	0.188	[0.181, 0.195]	0.273	[0.268, 0.279]	0.187	[0.183, 0.191]
5	0.262	[0.250, 0.273]	0.343	[0.336, 0.350]	0.293	[0.284, 0.302]
10	0.315	[0.301, 0.329]	0.335	[0.327, 0.344]	0.340	[0.332, 0.348]

TABLE 2 Mean Kendall τ distances between the true and the 3 estimated similarity trees, with their 95% confidence intervals with $a = b$

cant but well below perfect, then the coupling method will result in the most accurate similarity trees. The results seems to be invariant to the value of the linear coefficients defining the measure in function of the sources, which suggests that in an actual experiment the choice of the method is not influenced by the actual position of the measuring devices.

References

Cox, D. D. and R. L. Savoy. 2003. Functional magnetic resonance imaging (fMRI) "brain reading": Detecting and classifying distributed patterns of fMRI activity in human visual cortex. *NeuroImage* 19(2):261–270. doi:10.1016/S1053-8119(03)00049-1. URL http://linkinghub.elsevier.com/retrieve/pii/S1053811903000491.

Friedman, J. H. 1996. Another approach to polychotomous classification. *Technical report Stanford University* .

Kendall, M. G. 1938. A new measure of rank correlation. *Biometrika Trust* 30(1):81–93. doi:10.2307/2332226.

—. 1990. *Rank Correlation Methods*, vol. 3. London: Edward Arnold, a division of Hodder & Stoughton, 5th edn. ISBN 0195208374.

Klimesch, W. 1999. EEG alpha and theta oscillations reflect cognitive and memory performance: a review and analysis. *Brain research Brain research reviews* 29(2-3):169–195. ISSN 01650173. doi:10.1016/S0165-0173(98)00056-3. URL http://www.ncbi.nlm.nih.gov/pubmed/10209231.

Perreau-Guimaraes, M., D. K. Wong, E. T. Uy, L. Grosenick, and P. Suppes. 2007. Single-trial classification of MEG recordings. *IEEE Transactions on Biomedical Engineering* 54(3):436–443.

Pfurtscheller, G., D. Flotzinger, and C. Neuper. 1994. Differentiation between finger, toe and tongue movement in man based on 40 Hz EEG. *Electroencephalography and Clinical Neurophysiology* 90(6):456–460. doi:10.1016/0013-4694(94)90137-6. URL http://www.sciencedirect.com/science/article/pii/0013469494901376.

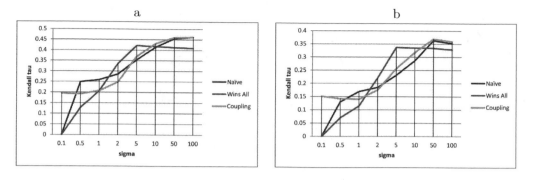

FIGURE 1 Mean Kendall τ distances between the true and the 3 estimated similarity trees in function of the dispersion of the class distributions for $a = 10b$ in a and $a = b$ in b

Schaefer, R. S., M. Perreau-Guimaraes, P. Desain, and P. Suppes. 2008. Detecting Imagined Music from EEG. In K. Miyazaki, Y. Hiraga, M. Adachi, Y. Nakajima, and M. Tsuzaki, eds., *Proceedings of the 10th International Conference of Music Perception and Cognition*. Causal Productions.

Suppes, P., M. Perreau-Guimaraes, and D. K. Wong. 2009. Partial orders of similarity differences invariant between EEG-recorded brain and perceptual representations of language. *Neural Computation* 21(11):3228–3269. doi:10.1162/neco.2009.04-08-764.

Ullman, S., M. Vidal-Naquet, and E. Sali. 2002. Visual features of intermediate complexity and their use in classification. *Nature Neuroscience* 5(7):682–687. doi:10.1038/nn870. URL http://www.ncbi.nlm.nih.gov/pubmed/12055634.

Waldert, S., H. Preissl, E. Demandt, C. Braun, N. Birbaumer, A. Aertsen, and C. Mehring. 2008. Hand movement direction decoded from MEG and EEG. *Journal of Neuroscience* 28(4):1000–1008. doi:10.1523/JNEUROSCI.5171-07.2008. URL http://www.ncbi.nlm.nih.gov/pubmed/18216207.

19

Representation, Isomorphism and Invariance in the Study of Language and the Brain

Colleen E. Crangle

Patrick Suppes is the quintessential philosopher scientist. I know of no other scientist who conducts research with such philosophical verve and no other philosopher who embraces data—lots of it—with such enthusiasm. As a computer scientist completing a PhD in Philosophy under Suppes in the 1980s, during the early and reckless days of Artificial Intelligence, I found the integrity Suppes brought to studies of cognition and my particular interest, language, irresistible. It has been an honor to work with him many times over the years since.

1 Introduction

One of the defining characteristics of Suppes' approach to both philosophy and science is the essential role he assigns to computation. Elsewhere in this volume, Michael Friedman describes Suppes' vision of the history of science as the ever richer accumulation of empirical data and the development of increasingly powerful methods of computation for making sense of those data. Brain research entails surely one of the most breathtakingly complex data collection and analysis activities yet encountered in science. As Suppes has trained his mind on this topic in recent years, he has brought to it his unique blend of the theoretical and the empirical. In this article, I discuss recent work I have done, based on earlier work with Suppes, that explores the question of how we can observe the activity of the brain experimentally, and the challenge of interpreting this activity. Specific computational methods had to be devised to decipher the data collected, a fact that keenly illustrates Suppes' view of computation's place in the history of science (Crangle et al. 2013; Crangle and Suppes 2013). These computations, moreover, are based on the three ideas of representation, isomorphism and invariance, notions that underpin and unify much if not all of Suppes' work. In this paper, I show how they provide the analytical means to better understand the processes and structures of the brain.

In the experiments I will describe, data on the electromagnetic activity of the brain, measured by electroencephalography (EEG), was collected from people while they were using language. But language in what context and for what purpose?

Foundations and Methods from Mathematics to Neuroscience.
Colleen E. Crangle, Adolfo García de la Sienra and Helen Longino.
Copyright © 2014, CSLI Publications.

A line of enquiry Suppes has pursued for many years is the much-analyzed problem of formulating a theoretical account of the truth or falsity of propositions. True to Suppes' disposition, he is not content with disembodied theories of propositional truth and falsity; rather he asks how the truth or falsity of ordinary empirical statements is computed in humans. Using the psychological concepts of associative networks and spreading activation, he has proposed in broad terms a mechanism by which people reach the conclusion of truth or falsity (Suppes and Béziau 2004; Suppes 2008, 2009, 2012). Neuroscience has given Suppes a new way to study this problem. Over a period of several years in a series of experiments, participants were asked to determine the truth or falsity of statements they saw or read while EEG recordings were made. These experiments furnish the data for the analysis described in this paper.

2 Representations of Words in the Brain

It has long been believed that since sentences appear to be understood incrementally, word-by-word, brain activity time-locked to the presentation of words in sentences can be analyzed to tell us something about the processing of that sentence. Recent work by Brennan and Pylkkänen (2012) has indeed confirmed this. They found specifically, for instance, that the activity 250 ms after word onset can be associated with the processing of words in sentences as opposed to words in lists.

In experiments described in Suppes et al. (1999a,b), 48 statements about the geography of Europe were presented to participants who were asked to determine their truth or falsity while EEG recordings were made. Presentation was both visual and auditory. The statements were in 10 randomized blocks, with all 48 sentences occurring once in each block, giving rise to 480 trials for each of nine participants. Half of the statements were true, half false, half positive, and half negative (e.g., *The capital of Italy is Paris and Paris is not east of Berlin*). These sentences, termed trigger sentences in the context of the experiment, had the following forms: *X is [not] W of Y, W of Y is [not] X, X is [not] Z of X, Y is [not] Z of Y*, where *X ∈ {Berlin, London, Moscow, Paris, Rome, Warsaw, Madrid, Vienna, Athens), Y ∈ {France, Germany, Italy, Poland, Russia, Austria, Greece, Spain}, W ∈ {the capital, the largest city}, Z ∈ {north, south, east, west}*, and *[not]* indicates the optional presence of *not*. Twenty one words were of interest in the analysis reported in this paper, namely the city and country names and the four words of direction or location. In the visual presentation of the sentences, words were displayed one by one on a computer screen. The same timing was used in the auditory presentation of the sentences, providing a precise start time for each word.

EEG measurements were collected from 15 sensors placed on the scalp.[1] On presentation of a sentence in a trial, the synchronized activity of millions of neurons in the brain at each sensor gives rise to a brain waveform that is recorded by EEG. From this brainwave data, consisting of 15 simultaneous recordings, we extract segments corresponding to the individual words from each recording.[2] We let each word's data sample begin approximately 160 ms after the auditory onset of the word, that is, 10 data points beyond the word's initial start point. This choice of start point was informed by work showing

---

[1] The sampling rate was 1,000 Hz and the data were down-sampled 16 times, giving an effective sample rate of 62.5Hz.

[2] Signal processing techniques such as Independent Component Analysis (ICA) are typically used to separate out electrocortical activity from data generated by artifacts such as eye blinks and to reduce the dimensionality of the data. It is from these component waveforms rather than the raw EEG data from the sensors that we extract data segments for each word.

the N400 component of the event related potential (ERP) to be associated with semantic processing and integration (Kutas and Hillyard 1984; Borovsky et al. 2012; Kutas and Federmeier 2011). To determine the end point, we extended each word's data sample by 50 data points (that is, approximately 800 ms) and calculated the average length of all the resulting data samples. This average length then determined the end point of each word's data sample. The data samples from all fifteen channels were concatenated, producing a single brain-data sample for each occurrence of a word.

TABLE 1   The trigger sentences containing the word *Paris*

> *The capital of Italy is Paris.*
> *Paris is not west of Berlin.*
> *Paris is east of Berlin.*
> *Moscow is north of Paris.*
> *Moscow is not east of Paris.*
> *Paris is south of Rome.*
> *Paris is not south of Athens.*
> *The capital of Spain is not Paris.*

As illustration, consider the word *Paris*, which appears in eight of the trigger sentences (shown in Table 1) and therefore produces 80 data samples for each participant. Suppose for the single participant S18 we extract from each data sample a brain-wave segment as described above. Averaging over each set of 10 samples, we get the eight brain waves shown in Figure 1 and averaging over all 80 samples the brain wave shown by the bold line. Figure 1 therefore gives approximate representations of the word *Paris* for S18.

Approximate invariance across the various occasions of the word's perception by S18 is evident. Such invariance between the brain representations of a word from one occasion of its use to the other by the same person is not surprising, whether that word is being read or heard. Also not surprising would be approximate invariance between the brain representation of a word in one person and its representation in another who speaks the same language. It is hard to imagine how communication would be possible without it. A number of specific representational questions were in fact posed in a series of experiments over more than a decade in Suppes' lab, (Suppes et al. 1997, 1998; Suppes and Han 2000; Wong et al. 2004, 2006; Perreau-Guimaraes et al. 2007; Suppes et al. 2009; Wong et al. 2008; Suppes et al. 1999b). The experiments entailed analyzing the brain-data samples obtained from individual trials using a statistical model to predict to which class a brain-data sample belongs.[3] The brain-data samples corresponded to words presented individually in the experiments, or words within sentences as with our 21 word types, or whole sentences, or even pictures of what the words referred to. Using such brain-data samples, the following series of representational questions were posed in the Suppes' lab, and answered affirmatively:

- Can we train a classifier of brain data so that we can predict which word the participant is reading or listening to?

---

[3] A linear discriminant model was used in a 5-fold cross-validation loop. The dimensionality of the data was reduced using a nested principal component analysis.

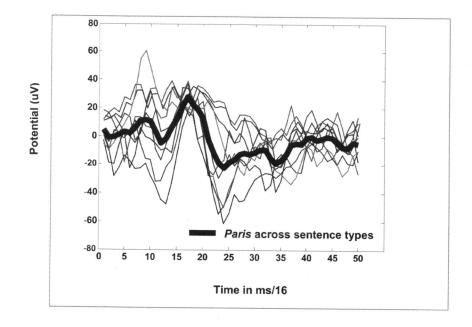

FIGURE 1 Approximate representations of *Paris* in and across eight sentences
for participant S18

- Can we train a classifier of brain data so that we can predict which sentence the participant is reading or listening to?
- Can we train a classifier of brain data so that we can predict which word within a sentence the participant is reading or listening to?
- Can we train the classifier using visually presented words (sentences, words within sentences) and use that classifier to make predictions for words (sentences, words within sentences) presented auditorily, and vice versa?
- Can we train a classifier on some participants and use it to make predictions for other participants?
- Can we train a classifier using words and use that classifier to make predictions for pictures depicting what the words refer to, and vice versa?

A fundamental result that follows from this work is that when brain-wave recordings are time-locked to the presentation of words singly or within sentences, under visual or auditory conditions, segments of those brain waves (after appropriate processing) can be identified as brain representations of those words. Furthermore, these representations have approximate invariance across test participants, across a variety of sentential contexts, across auditory and visual contexts, and across stimulus modality, verbal or pictorial.

3   Relations between Brain Representations of Words: Two Semantic Models

When a person determines the truth or falsity of a statement there must be some computation performed on the brain representations of the words in the sentence. Furthermore, this computation must take into account the fact that the relation between the brain representation of the word *Paris* and the brain representation of the name of the country Paris is the capital of, namely *France*, differs in some important way from the relation between the representations of *Paris* and, say, *Germany* or *Spain*. Even more so, the relations between the brain representations of any two country names, *Russia* and *Germany*, for instance, are surely different from the relations between the brain representations of any of the city names or the directional terms *north, south, east,* and *west.*

The question of interest I sought to answer was this: Can our brain data, collected while participants were computing the truth of statements about the geography of Europe, reveal anything about the relations between the brain representations of the words?

To answer this question I draw on the fundamental notion of isomorphism. Specifically, I ask if the relations between the brain representations of a set of words are isomorphic to the relations between those same words as represented in a semantic model.

But which semantic model? I describe two here that on the surface appear very different from each other. The first is WordNet, a lexical database of English compiled by experts and based originally on psycholinguistic principles. The second is a statistical model derived from a corpus, which is a body of language collected from written or spoken sources. The statistical model is based on the distribution of words within that corpus.

WordNet (Miller 1995; Fellbaum 1998; WordNet 2013) is based on three main ideas. First, words have one or more **senses**. For example, the word *country* has five senses.

#1. a politically organized body of people under a single government
#2. the territory occupied by a nation
#3. the people who live in a nation or country [in one of the other senses of the word]
#4. an area outside of cities and towns
#5. a particular geographical region of indefinite boundary (usually serving some special purpose or distinguished by its people or culture or geography).

Second, word senses are organized into **sets of related word senses**. A term of art has been defined for these sets: **synsets**. All the word senses in a given synset are synonymous with each other in some contexts; that is, the words can be substituted for each other in those contexts. This substitutability principle establishes that the words in a synset frequently occur in similar contexts or could occur in similar contexts without semantic discordance. The first synset that *country* participates in is **{state#4, nation#1, country#1, land#9, commonwealth#2, res publica#1, body politic#1}** and the second is **{country#2, state#7, land#5}**. (The specific sense of a word is indicated by the hash symbol followed by its number.) Each synset has a definition, although not in the formal sense of that word; WordNet definitions are called glosses, from the glossaries that were compiled for classical texts before there were modern dictionaries and definitions. All word senses in a synset have the same gloss, which is shared with the synset itself. Table 2 gives the five synsets that *country* is part of, along with examples of sentences using the word in each given sense.

The third principal of WordNet is that, in addition to the relation of synonymy that underpins its structure, **synsets and word senses are related** to each other. The prime relations are hypernymy and hyponymy. Hypernymy is the relation between two words or word senses when their extensions, or denotations or what they refer to, stand in the relation of class to subclass. Hyponymy is the inverse: word or word

TABLE 2 The five WordNet senses and synsets for *country*

{state#4, nation#1, **country#1**, land#9, commonwealth#2, res publica#1, body politic#1} (a politically organized body of people under a single government) "the state has elected a new president"; "African nations"; "students who had come to the nation's capitol"; "the country's largest manufacturer"; "an industrialized land"
{**country#2**, state#7, land#5} (the territory occupied by a nation) "he returned to the land of his birth"; "he visited several European countries"
{nation#2, land#8, **country#3**} (the people who live in a nation or country) "a statement that sums up the nation's mood"; "the news was announced to the nation"; "the whole country worshipped him"
{**country#4**, rural area#1} (an area outside of cities and towns) "his poetry celebrated the slower pace of life in the country"
{area#1, **country#5**} (a particular geographical region of indefinite boundary (usually serving some special purpose or distinguished by its people or culture or geography)) "it was a mountainous area"; "Bible country"

sense w1 is a hyponym of w2 when the extension of w1 is a subclass of the extension of w2. Colloquially, hyponymy is known as the *is_a* relation. For example, the word *ambulance* (which has only one sense) is a hyponym of the word *vehicle* (in its first sense) because an ambulance is a kind of vehicle, and we designate this relation as *ambulance is_a vehicle*. Conversely, *vehicle* is a hypernym of *ambulance*. Related to hyponymy is the relation *instance_of. France* is an *instance_of European country*, which *is_a country* in the second sense of the word. (Other relations in WordNet are part-whole or holonym, and member-of or meronym. The relation of antonymy or opposite-of is also present although it is a relation between words themselves not between synsets. *Hot* is an antonym of *cold*, for example, but not of the synsets {*cold, stale*} or {*cold, frigid*} nor of the individual words *stale* and *frigid*.)

WordNet can be visualized as a network in which nodes correspond to particular word senses or synsets. The five senses of *country* are labeled in Figure 2. Other higher-level organizing relations such as *is_a* that *country* participates in are not shown.

At its heart, WordNet organizes lexical items into word senses and then organizes those word senses into sets such that the word senses in a set "mean the same thing," with the common meaning captured in the gloss. These sets of word senses are often referred to as concepts or meanings and at some commonsense level that is what they are. But that commonsense view misses the essential point about WordNet, namely that words are defined by the contexts in which substitutability holds, their shared contexts.

In an informal sense, words are defined by the company they keep, and when a word keeps distinctly different sets of company it has distinct senses. An analogy would be a person being thought to have four "senses" or personas if she has colleagues at work, a group of people she regularly runs marathons with, people she sees each Sunday at church, and parents at her child's school with whom she arranges playdates, and she

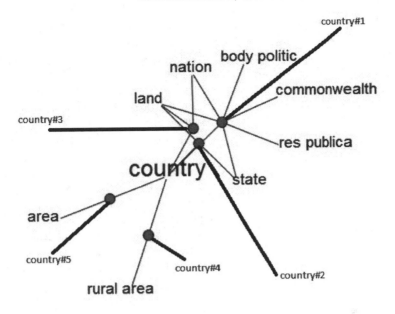

FIGURE 2 A visual snippet of WordNet for the word *country* and its five senses

behaves differently in important ways with each group. The word *country* under its first sense is in company with *state* under its fourth sense, *nation* under its first sense, *land* under its ninth sense, *commonwealth* under its second sense, *res publica* under its first sense, and *body politic* under its first sense. *Country* under its second, third, fourth and fifth senses keeps company as shown in Table 2.

Distributional semantics also has at its core the assumption that words are defined by the company they keep, that is, by their shared contexts. However, those shared contexts are not identified using human judgment but through statistical analysis performed on large corpora. Different results are obtained depending on the corpus chosen and what is deemed to define a word's context. If two words appear in the same document, they may be considered to share a context. Or they may be required to be within five words of each other in the same document, a more stringent notion of context. Additionally, a measure of the strength of that contextual connectedness is computed based on how often the two words share that context compared to how often they do not. If two words share a large number of contexts, we understand them to be more strongly related semantically than two words who share few contexts. A range of statistical techniques have been brought to bear on the problem of quantifying the strength of that relatedness, from simple co-occurrence frequencies to more complex techniques such as Latent Dirichlet allocation (LDA) (Blei et al. 2003). Similarly, a number of different corpora have been used in distributional semantic models. The Google Inc. dataset, consisting of English word n-grams and their frequencies, is one corpus. (An n-gram is a sequence of n words— typically two or four such as *balanced budget* or *United States of America*—that occur together more often than would be expected by chance.) Another corpus consists of a 16-billion-word set of English-language web-page documents. And another large corpus consists specifically of articles from Wikipedia (Mitchell et al. 2008; Brant and Franz 2006; Murphy et al. 2012; Pereira et al. 2010).

The choices made in distributional semantics have pronounced effects on the kind of semantic model you end up with. Briefly and broadly, you can produce a model that emphasizes topical relatedness such that *judge* is close to *court* and *referee* is close to *stadium*. Or you can produce a model that emphasizes categorical or taxonomic relatedness such that *judge* is close to *referee* and *court* is close to *stadium*. Distributional models can also be defined to reveal more general associative relations between words, such as that between *lipstick* and *lace*, for example, that almost certainly change over time as the world changes and different things are brought into juxtaposition in language (Deerwester et al. 1990; Lund and Burgess 1996; Landauer and Dumais 1997; Landauer et al. 1998).

Distributional models are said to reveal semantic relations that are hidden or latent in the texts, giving rise to the acronym LSA (latent semantic analysis) applied to some models. A well known latent semantic model is made available at `http://lsa.colorado.edu/` with a corpus consisting of texts of general reading up to 1st year college level. We used this model to compute pairwise relatedness scores for the words in our geography sentences, combining all the pairwise scores into a matrix that functions as a measure of how strongly related the words in a set are to each other (accessed January 15, 2013). (For simplicity, I will speak of similarity in the discussion that follows in place of the more general notion of relatedness.)

For WordNet, there are several ways to compute similarity for any pair of words. The most direct measure simply traverses the paths between words using the *is_a* and *instance_of* relations, noting the nature of the links and the lengths of the paths. For example, *ambulance is_a wheeled vehicle*, *bus is_a wheeled vehicle*, and *applecart is_a wheeled vehicle*. However, in the WordNet network, *ambulance is_a car is_a wheeled vehicle* and *bus is_a car is_a wheeled vehicle* but *applecart is_a pushcart is_a wheeled vehicle*. *Bus* and *ambulance* are therefore computed to be more similar to each other than they each are to *applecart*. Other more complex methods of computing relatedness have been devised. A detailed discussion can be found in Budanitsky and Hirst (2001) and Patwardhan and Michelizzi (2004) and online access is given to a program that computes several such measures (Pedersen and Michelizzi 2013). In our work, we pooled five different measures to take advantage of the strengths of each.

Using these two semantic models and the similarity measures they provide, we end up with two similarity matrices, one giving the pairwise similarity scores for the 21 words computed from WordNet and the other the similarilty scores derived from the distributional model of semantics. Let M_{WN} be the similarity matrix obtained from WordNet and M_D the similarity matrix obtained from the LSA distributional model. We see in the next section how these matrices are used to determine the relatively better fit of one semantic model over the other with the brain data.

4   Invariance between Semantic and Brain Representations

In analyzing the brain-data samples obtained from individual trials we used a statistical model, as stated earlier, to predict to which class a brain-data sample belongs. More generally, T brain-data samples $s_1, s_2, \ldots, s_T$ are classified into the N classes from $W = \omega_1, \omega_2, \ldots, \omega_N$, where W is the set of words the experimental participant was attending to, either seeing them on a screen or hearing them spoken. If test sample s_i is classified as ω_i then s_i and ω_i can be said to have a minimal similarity difference compared to the other possible classifications.

Let $M = (m_{ij})$ be the confusion matrix for a given classification task, where m_{ij} is the number of test samples from class ω_i classified as belonging to class ω_j. If our classification were perfect, that is, we correctly predicted for every brain sample which word the participant was attending to, the matrix would have on its diagonal the number of samples for each word from the experiment and there would be zeros everywhere else in the matrix. Such perfect prediction is unlikely. By computing the relative frequencies $m_{ij}/\sum_j m_{ij}$, however, we obtain a N-by-N estimate for the conditional probability densities that a randomly chosen test sample from class ω_i will be classified as belonging to class ω_j. Let these conditional probability estimates be given by the matrix $P = (p_{ij})$. Each element in this matrix gives in some fundamental way a measure of how strongly the two words, one from the row and one from the column, are related to each other in the brain of the participant. As with the semantic models, for simplicity we will use the notion of similarity in place of the more general relatedness.

It is tempting to hope that a straightforward comparison can be made of the brain representation similarities and those given by the semantic models. However, things are not that simple. Perfect element-by-element agreement between the brain-data matrix and the semantic-model matrix would constitute isomorphism between the two, an unlikely outcome under current experimental conditions. Failing isomorphism, it would be useful to know the *extent* of the agreement between the brain and semantic representations.

Here a generalization of isomorphism comes into play. For each class ω_i we define a quaternary relation R such that $\omega_i \omega_j R \omega_i \omega_k$ if and only if $p_{ij} < p_{ik}$, that is, if and only if the probability that a randomly chosen sample from class ω_j will be classified as belonging to class ω_i is smaller than the probability that a randomly chosen test sample from class ω_k will be classified as belonging to class ω_i. R is an ordinal relation of similarity differences, a partial order that is irreflexive, asymmetric, and transitive.[4]

Taking these N partial orders R (one for each ω_i) and the finite set W of classes ω_i, we have a relational structure (W, R) that provides a formal characterization of the brain data, specifically one that captures the relations between the elements of W.

We return to the semantic models, to M_{WN}, the similarity matrix obtained from WordNet, and M_D, the similarity matrix obtained from the distributional model of semantics. Just as with the brain data, we form from each of these matrices partial orders of similarity differences, R_{WN}) and R_D respectively (one for each ω_i). The partial orders simply capture the inequalities that two words ω_j and ω_i are more similar to each other than ω_j is to ω_k. Taking these partial orders and the finite set W of classes ω_i, we have relational structures (W, R_{WN}) and (W, R_D). By comparing (W, R) with both (W, R_{WN}) and (W, R_D) we can determine which model has a better fit with the brain data.

How is the comparison done? It is done in terms of the partial orders. For each semantic model, we ask how many partial orders are *invariant* with respect to the brain data and the semantic data. The *number* of such invariant partial orders is a measure of the *strength* of the structural similarity between the brain and semantic data for each semantic model. How do we determine the invariant partial orders? Details can be found in Crangle et al. (2013) but in outline the procedure is simple. We compute for each word a measure of the correlation between the two partial orders, one from the

---

[4]Specifically, the function defined by these inequalities is the function f defined on W such that $xyRuv$ iff $f(x) - f(y) < f(u) - f(v)$, with $f(x) - f(y)$ represented by p_{xy}. R can of course be simplified to a ternary relation.

brain data and the other from the semantic data. If there is a *statistically significant correlation*, we take the intersection of the two partial orders and thereby construct the partial order that is invariant with respect to the brain and the semantic data. (The Spearman rank correlation coefficient was the measure used.)

To obtain a robust measure of the structural similarity represented by these invariant partial orders, we performed the following computations. For each participant, we computed 60 single-trial classifications of the brain data for 10 words using random resampling with replacement. For half of these classifications we found the partial orders that were significantly highly correlated ($rho = .6485, p < 0.05$) and invariant with respect to the WordNet-based semantic data. For the other half we did the same for the LSA distributional model data. Four 10-word sets were used, ensuring that each of the 21 words appeared at least once in our computations. They were: {*London, Moscow, Paris, north, south, east, west, Germany, Poland, Russia*}; {*Paris, Vienna, Athens, north, south, east, west, Italy, Spain, Austria*}; {*Berlin, Rome, Warsaw, north, south, east, west, France, Greece, Poland*}; and {*Madrid, Rome, Vienna, north, south, east, west, Spain, Italy, Austria*}.

Figure 3 gives the results for the nine participants in the study. The Figure summarizes the results for the four 10-word sets and gives the total number of partial orders of similarity differences that were judged invariant between the brain data and the semantic data. As shown, a higher number of invariant partial orders was found for the WordNet data than the distributional semantics data. This pattern held for all the participants and for all but two of the 72 results represented in the figure (one for each of the two semantic models for the nine participants and the four 10-word sets).

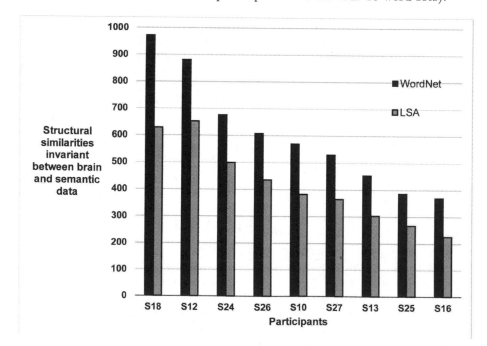

FIGURE 3 Significant invariant partial orders for the brain data and the semantic data from two models (Figure adapted from Crangle et al. 2013)

5 Concluding Remarks

What the results in Figure 3 suggest is that the organization of word representations in the brain is in some important way more like WordNet than that represented by the distributional model of semantics we used. As stated earlier, there are many different distributional models and below I describe a different one I used in follow-up work. However, the results in Figure 3 are robust and intriguing and open the path for more detailed investigations of semantic models and the brain.

Associative networks have long been discussed in the context of language processing. Simply put, an associative network is a set of nodes with links between them, where the nodes correspond to words or, more correctly for our purposes, brain representations of words, and the links correspond to associations between these brain representations. Early foundation work can be found in Collins and Loftus (1975) and Woods (1975) and more recent work on associative networks for natural-language processing and information retrieval can be found in Hirst (1988) and Crestani (1997).

What is still missing from discussions of associative networks in language processing is a coherent, general account of how associative networks are used in complex tasks such as determining the truth or falsity of an assertion. As stated earlier, Suppes sought to give a broad account of what that might look like, using activation and spreading activation, the principal operations proposed for associative networks. But he did not pay attention to the organization of the associative network. Many specific questions must be answered. What determines if there is a link between two words? How do the links give a measure of the strength of the relation between two words? The significant thing about WordNet is that it embodies a set of nuanced answers to these questions.

Distributional models of semantics have the potential to provide some such answers, especially since in recent years the Internet has presented itself as a rich source of semantic information. In Crangle and Suppes (2013) We examined the extent to which an Internet-derived model of distributional semantics can predict the brain activity for our set of geography words and the relations between them. I used a network of relations constructed from ukWaC, a large (>2 billion token) corpus of English constructed by crawling the .uk Internet domain, along with co-occurrence frequencies and point-wise mutual information to measure the strength of the relation between two words (Baroni et al. 2009; Turney 2001). That distributional model outperformed the distributional model reported on in this paper. This result is not surprising given the larger size of the ukWaC corpus. To date no systematic study has been done of the various corpus-based semantic models and WordNet to see if WordNet retains its apparent better match to brain data. Much could be learned from such a study, both for the brain and for assessing the adequacy of WordNet in the light of neurological data.

The investigation described in this paper was conducted in the context of asking what kind of semantic structures might underlie the computation of truth or falsity. It is a small step in our exploration of the brain but an important one in that it shows what is possible using the Suppes framework of representation, isomorphism and invariance.

6 Acknowledgements

I gratefully acknowledge the support provided by the Patrick Suppes Gift Fund and the Fulbright-Lancaster Science and Technology Scholar Award in the preparation of portions of this paper.

References

Baroni, M., S. Bernardini, A. Ferraresi, and E. Zanchetta. 2009. The wacky wide web: A collection of very large linguistically processed web-crawled corpora. *Language Resources and Evaluation* 43(3):209–226.

Blei, D. M., A. Y. Ng, and M. I. Jordan. 2003. Latent Dirichlet allocation. *The Journal of Machine Learning Research* 3:993–1022.

Borovsky, A., J. L. Elman, and M. Kutas. 2012. Once is enough: N400 indexes semantic integration of novel word meanings from a single exposure in context. *Language Learning and Development* 8(3):278–302. doi:10.1080/15475441.2011.614893.

Brant, T. and A. Franz. 2006. Web 1t 5-gram version 1.

Brennan, J. and L. Pylkkänen. 2012. The time-course and spatial distribution of brain activity associated with sentence processing. *NeuroImage* 60(2):1139–1148. doi:10.1016/j.neuroimage.2012.01.030.

Budanitsky, A. and G. Hirst. 2001. Semantic distance in WordNet: An experimental application-oriented evaluation of five measures. In *Proceedings of the NAACL 2001 Workshop on WordNet and Other Lexical Resources*. Pittsburgh: NAACL.

Collins, A. M. and E. F. Loftus. 1975. A spreading-activation theory of semantic processing. *Psychological Review* 82(6):407–428.

Crangle, C. E., M. Perreau-Guimaraes, and P. Suppes. 2013. Structural similarities between brain and linguistic data provide evidence of semantic relations in the brain. *PLoS ONE* 8(6):e65366. doi:10.1371/journal.pone.0065366.

Crangle, C. E. and P. Suppes. 2013. A web-based model of semantic relatedness and the analysis of electroencephalographic (eeg) data. In *Seventh International Corpus Linguistics Conference (CL2013)*. Lancaster University, Lancaster, UK: 8th Web as Corpus Workshop (WAC-8). URL https://sigwac.org.uk/wiki/WAC8.

Crestani, F. 1997. Application of spreading activation techniques in information retrieval. *Artificial Intelligence Review* 11(6):453–482. doi:10.1023/A:1006569829653.

Deerwester, S., S. T. Dumais, G. W. Furnas, T. K. Landauer, and R. Harshman. 1990. Indexing by latent semantic analysis. *Journal of the American Society for Information Science* 41(6):391–407.

Fellbaum, C., ed. 1998. *WordNet: An electronic lexical database*. Cambridge, MA: MIT Press.

Hirst, G. 1988. Resolving lexical ambiguity computationally with spreading activation and polaroid words. In S. L. Small, G. W. Cottrell, and M. K. Tanenhaus, eds., *Lexical Ambiguity Resolution: Perspectives From Psycholinguistics, Neuropsychology, and Artificial Intelligence*, pages 73–107. San Mateo, CA: Morgan Kaufmann Publishers.

Kutas, M. and K. D. Federmeier. 2011. Thirty years and counting: Finding meaning in the N400 component of the event related brain potential (ERP). *Annual Review of Psychology* 62:621–647. doi:10.1146/annurev.psych.093008.131123.

Kutas, M. and S. A. Hillyard. 1984. Brain potentials during reading reflect word expectancy and semantic association. *Nature* 307:161–163. doi:10.1038/307161a0.

Landauer, T. K. and S. T. Dumais. 1997. A solution to Plato's problem: The Latent Semantic Analysis theory of the acquisition, induction, and representation of knowledge. *Psychological Review* 104(2):211–240. doi:10.1037/0033-295X.104.2.211.

Landauer, T. K., P. W. Foltz, and D. Laham. 1998. Introduction to latent semantic analysis. *Discourse Processes* 25(2-3):259–284. doi:10.1080/01638539809545028.

Lund, K. and C. Burgess. 1996. Producing high-dimensional semantic spaces from lexical co-occurrence. *Behavior Research Methods, Instrumentation, and Computers* 28(2):203–208.

Miller, G. A. 1995. Wordnet: A lexical database for English. *Communications of the ACM* 38(11):39–41. doi:10.1145/219717.219748.

Mitchell, T. M., S. V. Shinkareva, A. Carlson, K. M. Chang, V. L. Malave, R. A. Mason, and M. A. Just. 2008. Predicting human brain activity associated with the meanings of nouns. *Science* 320(5880):1191–1195.

Murphy, B., P. Talukda, and T. Mitchell. 2012. Selecting corpus-semantic models for neurolinguistic decoding. In *Proceedings of the First Joint Conference on Lexical and Computational Semantics, Volume 1: Proceedings of the main conference and the shared task and Volume 2: Proceedings of the Sixth International Workshop on Semantic Evaluation*, pages 114–123. Stroudsburg, PA: Association for Computational Linguistics.

Patwardhan, S. and J. Michelizzi, eds. 2004. *WordNet: Similarity - Measuring the Relatedness of Concepts*, Proceedings of the Nineteenth National Conference on Artificial Intelligence - Intelligent Systems Demonstration (AAAI-04).

Pedersen, T. and J. Michelizzi. 2013. Wordnet: Similarity. Accessed 3 May 2013, URL `http://marimba.d.umn.edu/cgi-bin/similarity/similarity.cgi`.

Pereira, F., M. Botvinick, and G. Detre. 2010. Learning semantic features for fMRI data from definitional text. In *Proceedings of the NAACL HLT 2010 First Workshop on Computational Neurolinguistics*, pages 1–9. Los Angeles, CA: Association for Computational Linguistics. URL `http://www.aclweb.org/anthology/W10-0601`.

Perreau-Guimaraes, M., D. K. Wong, E. T. Uy, L. Grosenick, and P. Suppes. 2007. Single-trial classification of MEG recordings. *IEEE Transactions on Biomedical Engineering* 54(3):436–443.

Suppes, P. 2008. A revised agenda for philosophy of mind (and brain). In *Representation, Evidence, and Justification: Themes from Suppes*, pages 19–51. Frankfurt, Germany: Ontos Verlag.

—. 2009. Neuropsychological foundations of philosophy. In A. Heike and H. Leitgeb, eds., *Reduction. Between the mind and the brain*, pages 137–176. Frankfurt: Ontos Verlag.

—. 2012. Three kinds of meaning. In R. Schantz, ed., *Prospects of Meaning*, pages 567–579. Berlin and Boston: Walter de Gruyter GmbH & Co. KG.

Suppes, P. and J. Y. Béziau. 2004. Semantic computations of truth based on associations already learned. *Journal of Applied Logic* 2(4):457–467. doi:10.1016/j.jal.2004.07.006.

Suppes, P. and B. Han. 2000. Brain-wave representation of words by superposition of a few sine waves. *Proceedings of the National Academy of Sciences* 97(15):8738–8743.

Suppes, P., B. Han, J. Epelboim, and Z. L. Lu. 1999a. Invariance between subjects of brain wave representations of language. *Proceedings of the National Academy of Sciences of the United States of America* 96(22):12953–12958.

—. 1999b. Invariance of brain-wave representations of simple visual images and their names. *Proceedings of the National Academy of Sciences of the United States of America* 96(25):14658–14663.

Suppes, P., B. Han, and Z. L. Lu. 1998. Brain-wave recognition of sentences. *Proceedings of the National Academy of Sciences of the United States of America* 95(26):15861–15866.

Suppes, P., Z. L. Lu, and B. Han. 1997. Brain wave recognition of words. *Proceedings of the National Academy of Sciences of the United States of America* 94(26):14965–14969.

Suppes, P., M. Perreau-Guimaraes, and D. K. Wong. 2009. Partial orders of similarity differences invariant between EEG-recorded brain and perceptual representations of language. *Neural Computation* 21(11):3228–3269. doi:10.1162/neco.2009.04-08-764.

Turney, P. 2001. Mining the web for synonyms: PMI-IR versus LSA on TOEFL. In L. D. Raedt and P. Flach, eds., *Proceedings of the Twelfth European Conference on Machine Learning (ECML-2001)*, pages 491–502. Freiburg, Germany.

Wong, D. K., L. Grosenick, E. T. Uy, M. Perreau-Guimaraes, C. G. Carvalhaes, P. Desain, and P. Suppes. 2008. Quantifying inter-subject agreement in brain-imaging analyses. *NeuroImage* 39(3):1051–1063. doi:10.1016/j.neuroimage.2007.07.064.

Wong, D. K., M. Perreau-Guimaraes, E. T. Uy, and P. Suppes. 2004. Classification of individual trials based on the best independent component of EEG-recorded sentences. *Neurocomputing* 61:479–484. doi:10.1016/j.neucom.2004.06.004.

Wong, D. K., E. T. Uy, M. Perreau-Guimaraes, W. Yang, and P. Suppes. 2006. Interpretation of perceptron weights as constructed time series for EEG classification. *Neurocomputing* 70(1-3):373–383. doi:10.1016/j.neucom.2006.01.020.

Woods, W. 1975. What's in a link: Foundations for semantic networks. In D. Bobrow, A. Collins, and J. R. Carbonell, eds., *Representation and understanding*, pages 35–82. New York: Academic Press.

WordNet. 2013. About WordNet. Accessed 3 May 2013, URL http://wordnet.princeton.edu.

Part VI

Science in Practice

20

The Psychiatrist's Dilemmas

ANNE FAGOT-LARGEAULT

Abstract

Practising psychiatry means having to face conflicting options every day. Is psychiatric knowledge a branch of biology or a branch of psychology (or both)? Should the psychiatric patient be approached as someone who has a disease and wants to be relieved from a painful condition, or else as someone identified with the disease and who needs to be forced out of a cherished condition (or both)? In order to help the patient, should the psychiatrist prescribe drugs, or sessions of psychotherapy (or both)?

1 Introduction

Psychiatry and philosophy have a turbulent history of encroaching upon each other. I remember Patrick Suppes saying with pride: "I am simplistic". He was proud of it. That was 40 years ago. I want to be simplistic here. Psychiatry has to do with both psyche and body. On topics related to psychiatry, I interacted with Suppes at least twice. In the Fall of 2004, we had a "Mind and brain" seminar at Stanford (Philosophy 389: Dagfinn Follesdal and Jean-Pierre Changeux also cooperated). In the Fall of 2005 Professor Suppes delivered an invited talk at the College de France in Paris on the "Neuropsychological foundations of philosophy", and we had a small seminar with Prs Drs O. Dulac & C. Chiron, who specialize in the surgical treatment of severe epilepsy in children. I know that Suppes is interested in detecting the impact of psychotherapy on the brain. My modest contribution here will be to formulate schematically philosophical questions emanating from a practice of psychiatry at the hospital, in the emergency room, when I innocently wanted to reconcile the horns of a threefold dilemma.

2 The Theoretical Dilemma

When a medical student decides to specialize in psychiatry, he/she usually does it on the basis of one of his courses, or of some bright research he's heard of—he almost has not seen any real psychiatric patients yet. He has a theoretical view of the discipline that he is enthusiastic about. There is a common joke that says: being attracted to a theory is like being attracted to a woman. You fall in love, ignore the rest, become a believer, and think your belief is science. Let us take an example.

Foundations and Methods from Mathematics to Neuroscience.
Colleen E. Crangle, Adolfo García de la Sienra and Helen Longino.

Student A has just discovered a recent paper on autism, Torres et al. (2012). He believes in molecular psychiatry. He assumes that psychiatric disorders, such as infantile autism, have their cause in the brain cells. The field of his research will be on the genetics of autism and of other pervasive developmental disorders. In his publications you find statistical analyses eventually evidencing links between specific genetic polymorphisms and behavioral phenotypes (such as cognitive deficiencies).

Student B has devoured a book by Irvin Yalom, Yalom (1980). She believes in psychology-oriented psychiatry. She is interested in the historical antecedents of her patient's troubles. Language is her means of communication with her patients. She listens to them, she tries to empathize with their mental and emotional experience. She publishes case reports.

Neither of them will read the other's articles. Eclectism is rare. As a philosopher of science one would like them to communicate and work together. The patients they want to help have both a brain and an emotional experience. Assuming that their doctors must choose between the two sounds absurd. Yet the theoretical options split the community of psychiatrists. Should we think, as some do, that the unity of the discipline is a mere legacy of the past? The authors of a recent textbook announce that "in order to prevent future psychiatry from dissolving in a number of methodically defined subunits, and to further strengthen person-centered diagnostic approaches, we strongly need the historical perspective" in Hoff (2009, p. 12).

3 The Relational Dilemma

The psychiatry student eventually becomes a resident. How does he/she relate to the mentally or emotionally perturbed patient, for example in the emergency room of a general hospital?

Resident C addresses the patient as a person, who is suffering from an ailment that she is unable to deal with by herself, and the fact that she is here means that she is asking for help. The doctor tries to find out what kind of help is appropriate, he proposes possible solutions, and he lets the patient express preferences about what she thinks is best for her. Should the patient be in severe distress and/or unable to make sound decisions, resident C warns her that he himself will make the decision to institutionalize her now. That is at the risk of seeing the person run away before coercive measures can be taken.

Resident D is convinced that the important point is to make a sharp diagnosis, from which the therapeutic solution necessarily follows. She does not bother asking the patient what he wishes. The patient it out of his mind, he is intrinsically sick, discussing with him is a waste of time. She will have to choose for him. She does not talk to the patient. She talks about the patient to nurses, prescribes whatever drug or brace she deems necessary, and makes a decision: the patient will be locked in a psychiatric wards, or he will be put in a bed until he has sobered up, etc.

Saying that someone has a disease is very different from saying that someone is insane. Let the ailment be, for example, bipolar disorder. The relational dilemma (treating the patient as an equal partner or as someone to be protected and straightened up) refers to a duality of metaphysical presuppositions. Resident C assumes that the patient as a person does not identify with the disease: she has a bout of depression, she would enjoy getting rid of it, in no way can she be deemed responsible for being unwell. Resident D presumes that deep inside the patient is (ontologically) depressed, wants to be depressed, cherishes his depression, coincides with his depressed condition, and that it is nonsense to expect that he might empathize with the doctor's rational project to cure him. Whether the deep consent to his condition may be traceable to his genes, or to an existential choice, the patient here will be deemed to adhere to his depression, as though he were responsible for it; he must be sanctioned for his erroneous choice. This is reminiscent of an old debate in philosophy around the "initial choice of character", Plato and Turner (1889); Aristotle (1893); Kant and Abbott (1909); Adler (1912); Sartre (1957).[1] Henry Ey qualified mental illness as a "pathology of liberty", Ey (1960). Following Ludwig Binswanger and his notion of Daseinsanalyse, some psychiatrists bet on the possibility for psychoanalysis to have a patient return to the source of his very early traumas and life choice, realize that the melancholic posture he locked himself into was mistaken, opt out of his initial choice, and engage in a new existential attitude.

4   The Therapeutic Dilemma

Until 1949 there were practically no specific treatments of mental illness, except a few tentative shock treatments: malariatherapy (von Jauregg, 1917), insuline coma (Sakel, 1934), electroshock (Cerletti & Bini, 1938); or radical surgery like frontal lobotomy (Moniz, 1936). Then, within one decade (1949–1959), all potent psychotropic agents were discovered: antipsychotic (or neuroleptic) drugs, antidepressants, benzodiazepine compounds (anxiolytics). From then on, there is virtually no consultation in psychiatry that does not end with a prescription of some chemical. Therefore it might seem that there is no therapeutic dilemma. Prescribing drugs to sedate anxiety, stabilize the mood, or cast out delusions, is standard practice.

The Swiss psychiatrist Eugen Bleuer (contemporary with Sigmund Freud), who identified schizophrenia, and was the director of the Burghölzli clinic in Zürich, did not have any useful drugs to treat his patients. He is known for saying about them: "I couldn't restore them to health—then I listened to them". He meant that he was learning from them about the nature of their ailments. He did not claim that listening to them had any curative power. Freud however believed in the therapeutic efficiency of being able to express unconscious conflicts in language, although he did not disregard the possible effects of chemicals (he used cocaine himself, and studied its effects). Analysing one's dreams and memories in the course of a psychotherapy has become a routine way for psychiatric patients to be helped. As R.J. Kahana observed half a century ago: "In the past forty years, largely under the impact of psychoanalysis, dynamic psychotherapy has become the principal and essential curative skill of the American psychiatrist and, increasingly, a focus of his training", Kahana (1968, p. 458).

"The one hand does not know what the other is doing": as a matter of fact, psychiatrists commonly prescribe both psychotherapy and medication to the same patients, with no clear idea of how they interfere. Small empirical evidence has indeed been collected in favor of the belief that both together are better than either one alone. The mechanism is unknown. Note that psychotherapy in general (including cognitive remediation), being not a 'medical' treatment proper, is in many countries not covered by health insurance, while chemicals are. That makes it an elitist choice. The psychoanalytic school has argued that the high cost of psychoanalysis is a decisive element in

---

[1]For an overview, see entry 'character' in: Canto-Sperber Monique, et al., Dir., Dictionnaire d'éthique et de philosophie morale / *Diccionario de etica y de filosofia moral / Dictionary of ethics and moral philosophy*, 4th ed rev & augt, Paris, PUF, 1996 / Spanish transl.: Fondo de Cultura Economica USA 2001.

its effectiveness. How such an hypothesis may coexist with biological hypotheses on the causation of mental illness remains a mystery.

The psychiatrist Edouard Zarifian deplored the "religious war" that was going on in the nineteen eighties between exclusive believers in psychoanalysis ("brainless psychiatry of the past") and fanatic supporters of neurobiology ("mindless psychiatry of the future"), (Zarifian 1988, p. 204). The war has now calmed down, the dissociation remains, as if psychiatrists could only jump from one side to the other. Zarifian himself went from biological psychiatry, of which he was a specialist, to social and psychological psychiatry, when he was made responsible of the mental health of a large district. The psychiatrist Eric Kandel, Nobel prize in medicine 2000, went in the other direction. Here is the way how, in his autobiography, he summarizes his successive choices: "I entered Harvard to become a historian and left to become a psychoanalyst, only to abandon both of those careers to follow my intuition that the road to a real understanding of mind must pass through the cellular pathways of the brain", (Kandel 2006, p. 429).

References

Adler, A. 1912. *Über den nervösen Charakter*. Wiesbaden: Verlag Von J. F. Bergmann.

Aristotle. 1893. *The Nichomachean Ethics of Aristotle*. London: Kegan P., Trench, T. & Co. Translated by F.H. Peters, M.A. 5th edition.

Ey, H. 1960. *Manuel de psychiatrie*. Paris: Masson, 7th edn. Coll. Paul Bernard and Charles Brisset.

Hoff, P. 2009. Psychiatric diagnoisis: Challenges and prospects. In I. M. Salloum and J. E. Mezzich, eds., *Historical Roots of the Concept of Mental Illness*, pages 3–14. John Wiley & Sons, Ltd.

Kahana, R. J. 1968. Psychotherapy: models of the essential skill. In G. L. Bibring, ed., *The Teaching of Dynamic Psychiatry: A Reappraisal of the Goals and Techniques in the Teaching of Psychoanalytic Psychiatry*, pages 87–103. Madison, Conn: International Universities Press.

Kandel, E. R. 2006. *In Search of Memory. The Emergence of a New Science of Mind*. New York: W.W. Nortan & Company.

Kant, I. and T. K. Abbott. 1909. *Kant's Critique of practical reason and other works on the theory of ethics*. London: Longmans, Green & Co., 6th edn.

Plato and B. D. Turner. 1889. *The Republic of Plato*. London: Rivingtons.

Sartre, J. 1957. *L'être et le néant : essai d'ontologie phénomologique*. Paris: Gallimard.

Torres, A. R., J. B. Westover, and J. A. Rosenspire. 2012. Hla immune function genes in autism. *Autism Research and Treatment* 2012:13 pages. doi:10.1155/2012/959073.

Yalom, I. D. 1980. *Existential Psychotherapy*. New York: Basic Books.

Zarifian, E. 1988. *Les jardiniers de la folie*. Paris: Odile Jacob.

21

Illusions of Memory

ELIZABETH F. LOFTUS

1 Introduction

Some of the contributors to this volume of essays in honor of Patrick Suppes will have known him far longer than I have. But I would say knowing him for over 40 years is pretty long stretch. He was my PhD adviser back in the late 1960s, and I left my little shared office (that was not far from his considerably larger one Ventura Hall) in 1970 to take a teaching position on the East Coast. Our relationship was so different from the one I have with my own graduate students. They call me "Beth" from the get go. I only got the nerve to call him something other than Dr. Suppes after I successfully defended my PhD thesis. Suppes was intimidating to me as a young graduate student, not only because of the force of his personality, but because of what I knew about his academic ancestry... a descendant of William James and Wilhelm Wundt, among others, shown in Figure 1.

2 My Personal Memories

Perhaps some of my memories of working with Suppes expressed in my 1991 book "Witness for the Defense" Loftus and Ketcham (1991) were not fair to Suppes at worst, and were incomplete at best. I hope to rectify this now. I came to Stanford with a Bachelor's degree jointly Mathematics and Psychology. I had heard of the field "mathematical psychology" in which Stanford was tops, and entered as a graduate student in 1966. What could be more perfect than a field called "mathematical psychology" for a student with a joint degree? I ended up working on a Master's Thesis with Richard Atkinson, on computer assisted learning of verbal materials, but soon thereafter switched to working with "Dr. Suppes" whose projects seemed more relevant to a student with some proficiency in math. In "Witness" I would write about being halfway through my doctoral dissertation – eventually titled "An analysis of the structural variables that determine problem-solving difficulty on a computer-based teletype" - and being a bit bored with the whole thing.

By way of background, Suppes, by that time, had spent a number of years developing a computer-assisted instruction system for classroom use. Students would not only get instruction from the computer but would also be tested. Throughout the county, and in

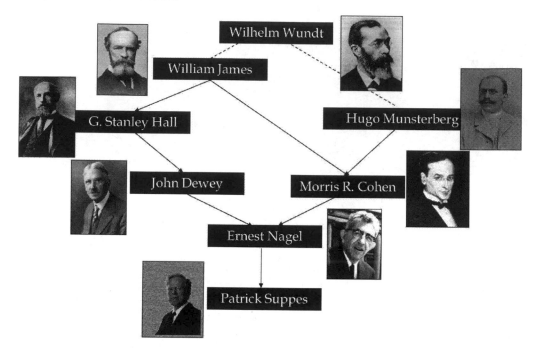

FIGURE 1 Academic ancestry for Patrick Suppes

other parts of the country, students were sitting down at little machines, trying to solve increasingly complicated word problems like this one:

"Tom collected 500 seashells and placed 43 of them in a showcase. How many shells were not placed in the showcase?"   Suppes et al. (1969, p. 1)

My dissertation project was based on data from sixth-grade students from a "depressed-area" who had completed the problem solving program. These students sat at teletype machines, connected by private telephone lines to a computer at Stanford, and tried to solve the word problems. My job, (over) simply put, was to analyze their data and make tentative conclusions about how adolescents solve problems, which problems are most difficult to solve, and why. We learned, for example, that various aspects of the math word problems contributed to their difficulty for students, such as the length and complexity of the sentences that described the problems and the number of different operations that had to be performed to solve the problem. In fact we were pretty successful at predicting the percentage of students who would go on to successfully solve the various word problems, achieving multiple correlations that exceeded .70.

Writing about these experiences two decades later, I reminisced:

"It was tedious work, no doubt about it. The theoretical model had been set up years earlier by (Suppes). and I was just one of several graduate students, each of us plugged into a specific slot, computer our statistical analyses, feeding our results into a common pot. It occurred to me that my particular job was a little like cutting up carrots to put in a soup. To the left and right of me were other students, equally frenzied and meticulous about cutting up their onions, celery, potatoes, chunks of beef, and then tossing them into the same huge pot. And I couldn't help thinking, All I've done is cut up carrots."

Loftus and Ketcham (1991, p. 5)

OK, lots of students seem to get a bit bored with their dissertation work as they are slogging through it. Missing from my brief description of research stew was any expression of just how much I had actually learned from working on that project with Suppes, and from a couple of later bits of work that we did together, now that he had become "Pat." I learned, for example, how to convert a doctoral thesis into a manageable scientific publication, Loftus and Suppes (1972a). I learned how to do multiple regression analysis, which was the primary analysis used in that research. More specifically, I learned how to do a stepwise multiple linear regression analysis to obtain regression coefficients, multiple correlations, and variance. I would end up using these methods in other research, Loftus and Suppes (1972b), and teaching them in various statistics courses, which I taught over the succeeding decades post PhD.

In the few years after I left Stanford, Suppes and I would meet periodically in New York where I was then teaching, to complete our other projects. Generally he was there on other business, and would squeeze me into his schedule. One unforgettable experience was the time he was so busy in New York, that our work together took place in the back seat of a taxi cab that was ferrying him from Manhattan to the New York airport. At the airport, we said goodbye, and he slipped some cash in my hand to be used to pay the taxi driver for the return trip back to Manhattan.

3 Life After Pat

After leaving Stanford, and wrapping up the projects with Patrick Suppes, my own research would take me down a path that never really intersected with the road on which he traveled. I recount this journey in an autobiographical chapter in Loftus (2007), but a more enjoyable version of my professional life was written by my former Stanford professor, Gordon Bower (2007). Here is a brief synopsis of what I tried to accomplish after leaving the Suppes nest.

My earliest studies demonstrated that innocuous procedures could alter memories of past events. Some of the more widely known findings were produced in the mid and late 1970s, starting with an experiment in which witnesses to an event were later interviewed with questions that insinuated novel (and incorrect) information into the memory record of the event. For example, after witnessing a slide show depicting an auto accident, the question "Did another car pass the red Datsun while it was at the stop sign?" might lead an interviewee who had seen a yield sign later to recall having seen a stop sign at the intersection. This was the modest empirical beginning of a research program that has helped to shift the way in which both scientists and educated laypersons understand the functioning of human memory.

In the current view (to which my research pointed), memory is no longer likened to a faithful, permanent recording device such as a camera or tape recorder. Rather, a more apt metaphor for the contemporary view is the rewritable memory of a computer. Further, with the aid of my research contributions, it is now well understood that the rewriting of memory can be precipitated by external agents, much as a worm or virus can precipitate changes in the computer's memory. I published hundreds of empirical articles and chapters on the malleability of memory, documenting the boundary conditions and individual differences variables that are associated with such distortions. The basic phenomenon of reduced memory after exposure to misleading information is now known as the "misinformation effect." (See Loftus (2005) for a decent summary of that "30-year investigation").

Not long after conceiving and starting to pursue this research program on the malleability of memory, I realized that this work had implications that called for consideration well beyond the covers of scholarly journals. The alterations of memory that I demonstrated in the laboratory had potential counterparts in criminal investigations, in which carelessly worded questioning by police and prosecutors might modify witnesses' memories of events about which they would later have to testify. I occasionally provided consultation or expert testimony on behalf of litigants in court cases that hinged on eyewitness testimony, in hopes of minimizing the likelihood of wrongful convictions or unfair verdicts. But, in hopes of reaching a larger audience than 12 individuals at a time, I made efforts to write about the findings for larger audiences. My book, *Eyewitness Testimony* (Loftus (1979/1996)), continues to see extensive use in training litigators as well as frequent citation in legal publications.

When news reports of adults discovering presumably long-repressed memories of sexual brutalization began to appear in the early 1990s, initial public reaction was predominantly to marvel at the malevolence of the accused parents or caretakers. Based on my own scientific work, and that of colleagues, I had concerns that some of these purported recoveries of memory had the potential to precipitate injustices predicated on mistaken assumptions about the veracity of memory, Loftus and Ketcham (1994). These concerns also led me to develop paradigms for exploring extreme instances of a malleable memory. One of my first studies, now referred to as the 'lost in the mall' study, demonstrated the ease with which one could create, in adults, 'recovered' unpleasant memories of having been, as a child, separated from a parent on a shopping trip, and after great distress being rescued and reunited with the family. My later studies showed how imagination, or exposure to other people's stories, or dream interpretation, or a multitude of suggestive techniques could lead people to have wholly false beliefs and memories about their past. Taken together, these studies show how very "rich false memories" can be planted in the minds of ordinary people. (See Loftus (2003) for more details).

My sustained research on the recovered memory problem has been important in shedding scientific light on deeply disturbing phenomena for which others in the profession were initially disinclined to consider taking a skeptical stance. I am referring to the "memory wars" – including the aforementioned epidemic of individuals who accused their parents and others of years of brutalization that has supposedly been repressed into the unconscious until therapists excavated the long-buried memories. Now, after numerous well publicized cases in which accusations of past abuse have been discredited, reports of 'repressed memory of abuse' are treated with greater balance by both the courts and by some members of the general public.

But studying memory distortion, and trying to use the science to change the way people think about memory, has not been without costs. I mentioned some of them when I proudly accepted the *Award for Scientific Freedom and Responsibility*, an award given to me at the 2011 meeting of the American Association for the Advancement of Science:

> "I feel grateful and privileged that the research I have done on memory in the past three decades has been honored for its contributions to science and human welfare. But of all of these awards, this one, in honor of scientific freedom and responsibility, has a special poignancy for me. I never set out to carry the banner for those glorious words *freedom* and *responsibility*; I was merely a scientist interested in the fallibility and malleability of memory—a subject that turned out to be central to the "repressed memory" moral panic that swept this nation in the 1980s and 1990s. If anyone had told me in advance that my

scientific commitment to knowledge would make me the target of organized, relentless vitriol and harassment (not to mention expensive litigation), I might have laughed at them—"Memory? Who gets angry over different memories?"

Every now and then I would find myself wondering: If I would known this in advance, would I have made the same decisions? Would I have decided to do the same kind of research, to spend countless hours in courtrooms testifying for the falsely accused, to write endless articles in rejoinder to dubious but persistent clinical ideas?

I do know that once faced with the choice between yielding to the wave of hostility and criticism that my research provoked or standing as strong as I could for science and justice, it was a no-brainer for me. But it was a decision that took an enormous personal toll, which is why this award is so meaningful and gratifying to me.

We live today in perilous times for science: conflicts of interest that taint research; pressures on scientists to cut corners to get fast results; a public culture that alternates between hostility to science and irrational expectations of what science can provide. If we as scientists want to preserve our freedom (and the welfare of others), now more than ever we have a responsibility.

And that responsibility is to bring our science to the public arena and to speak out as forcefully as we can against even the most cherished beliefs that reflect unsubstantiated myths.

During the most stressful of those trying times, I would sometimes say to myself "maybe I should have stuck with computer-assisted instruction. I could have helped people like he did, but without all the painful hassles. And maybe I'd be richer!" But his path was not path. Yet, I will always be grateful to 'Pat' for the role he played in launching a career that has been so satisfying. Most days I am thinking that I can't believe I get paid to do what I do every day.

References

Bower, G. H. 2007. Tracking the birth of a star. In M. Garry and H. Hayne, eds., *Do Justice and Let the Skies Fall: Elizabeth Loftus and her contributions to science, law and academic freedom*, pages 15 – 26. Mahwah, NJ: Lawrence Erlbaum Associates.

Loftus, E. F. 1979/1996. *Eyewitness Testimony*. Cambridge, MA: Harvard University Press.

—. 2003. Make-believe memories. *American Psychologist* 58(11):864–873. doi:10.1037/0003-066X.58.11.867.

—. 2005. Planting misinformation in the human mind: A 30-year investigation of the malleability of memory. *Learning and Memory* 12(4):361–366.

—. 2007. Elizabeth F. Loftus (Autobiography). In G. Lindzey and M. Runyan, eds., *A History of Psychology in Autobiography*, vol. IX, pages 199–227. Washington, DC: American Psychological Association Press.

Loftus, E. F. and K. Ketcham. 1991. *Witness for the Defense*. New York: St Martin's Press.

—. 1994. *The Myth of Repressed Memory*. New York: St. Martin's Press.

Loftus, E. F. and P. Suppes. 1972a. Structural variables that determine problem-solving difficulty in computer-assisted instruction. *Journal of Educational Psychology* 63(6):531–542. doi:10.1037/h0034074.

—. 1972b. Structural variables that determine the speed of retrieving words from long-term memory. *Journal of Verbal Learning and Verbal Behavior* 11(6):770–777. doi:10.1016/S0022-5371(72)80011-2.

Suppes, P., E. F. Loftus, and M. Jerman. 1969. Problem-solving on a computer-based teletype. *Educational Studies in Mathematics* 2(1):1–15.

22

EBP: Where Rigor Matters

Nancy Cartwright and Alexandre Marcellesi

Both authors would like to thank the Templeton Foundation's project 'God's Order, Man's Order and the Order of Nature' as well as the AHRC project 'Choices of evidence: tacit philosophical assumptions in debates on evidence-based practice in children's welfare services' for support for the research and writing of this paper. Nancy Cartwright would in addition like to thank the British Academy, which funded her project 'Evidence for Use', as well as the Institute for International, Comparative and Area Studies at UC San Diego for the 'Political civility and scientific objectivity' grant.

1 A Plea for Rigor in Evidence-based Policy

Patrick Suppes teaches two great lessons about rigor:

- Rigor matters.
- A little rigor can be a dangerous thing.

Our focus in this paper is on an area where rigor is badly needed and where it is highly touted but where Nancy is in trouble for insisting on it: the movements that go under the labels 'evidence-based medicine' and 'evidence-based social policy', which typically assert – in the interest of rigor – that randomized controlled trials (RCTs) are the gold standard in evidence. (We shall refer to both together as 'EBP' for short.) For instance, at a recent conference on evidence and causality, Sir Iain Chalmers, who was founding director of the UK Cochrane Collaboration (which oversees evidence-based medicine) and whose knighthood is for his contributions to healthcare, put some formulas of Nancy's, similar to ones we shall use here, up on the screen and made fun of them as both useless and unintelligible. But EBP is an area where rigor matters. The problem is that Suppes' second lesson is often ignored. There is too much rigor at one stage of the argument it takes to support a policy recommendation and very little thereafter – and that without mention. Yet it is well known that an argument is only as strong as its weakest premise. Downplaying the other premises that do not have solid support and stressing the rigor of the one premise gives a false sense of security. And this can be dangerous when policy decisions are at stake.

The EBP movement has generated a number of evidence hierarchies, grading systems for evidence, organizations and methods to review evidence pertaining to proposed treatments/interventions/policies, and warehouses where policies that pass review can be found. The evidence hierarchies rank not individual pieces of evidence but rather methods for the production of evidence, with well-conducted RCTs or systematic reviews or meta-analyses of well-conducted RCTs at the top. The hierarchies are then deployed by the various review organizations to evaluate how strongly supported treatments or policies are.

What you should have noticed right away is how vague our description is. Evidence is always evidence for some specific claim. Treatments are not the kinds of things that have evidence. So what are the claims about treatments or policies that evidence regarded as good by the EBP movement is supposed to be good for? There is the rub. It is really difficult to get a clear statement about this. We will give a short survey of some of the major sites and what they say in Section 2. In Section 3 we will look at the favored method in the EBP movement – the RCT. We will first propose one general form that an evidence claim that an RCT produces can take: "C causes E in (study population) A" or, to use a loose slogan: "It works somewhere". Second, using a version of Suppes' probabilistic theory of causality, we will sketch one rigorous argument that RCTs can certify this kind of claim in the ideal. Section 4 first proposes a form for the final hypothesis EBP is trying to provide evidence for: "C causes E in (target population) A" or, "It will work here". Second we will sketch a rigorous argument for establishing claims of this form taking the "It works somewhere" claims RCTs can establish as a starting premise. Thus we restore rigor by laying out an entire argument that starts with the assumption of a successful outcome in an ideal RCT and ends with the desired conclusion.

So... What's to make fun of about this?

2 What the 'Rigor' in EBP Frameworks Looks Like

2.1 Grading of Recommendations Assessment, Development and Evaluation (GRADE)

The GRADE working group is one of the most prominent advocates of the use of standardized grading schemes to assess the quality of evidence. It is particularly influential in evidence-based medicine: its grading schemes have been adopted by healthcare organizations such as the World Health Organization and the American College of Physicians, but also by vetting agencies such as Iain Chalmers' own Cochrane Collaboration.

GRADE offers two different grading schemes, one for the strength of recommendations to treat (strong, weak) and one for the quality of evidence (high, moderate, low, very low). A recommendation for a treatment is the output of a decision process that takes three elements as inputs: (i) an analysis of the (health-related) costs and benefits of this treatment, (ii) an assessment of the quality of the evidence supporting this cost-benefit analysis, and (iii) the values and preferences of patients. What the GRADE evidence scheme ranks is the quality of the evidence supporting the estimates of the benefits and harms of a treatment that are to feed into the cost-benefit analysis for this treatment.

These estimates of benefits and harms are estimates of treatment effects. Witness, for instance, what the GRADE authors says about the impact of deficiencies in RCTs on the quality of evidence:

"Our confidence in the evidence decreases if the available randomized controlled trials suffer from major deficiencies that are likely to result in a biased assessment of the *treatment effect.*"

(emphasis added, `http://www.gradeworkinggroup.org/FAQ/`, accessed December 2014)

GRADE's approach to grading evidence is detailed in two series of articles published in the *British Medical Journal (BMJ)* and the *Journal of Clinical Epidemiology (JCE)*. Throughout these articles, the authors of GRADE talk about 'the effects of treatments' in general, not in particular populations at particular times, implicitly assuming a 'narrow' conception according to which the effect of a treatment depends exclusively on what the treatment is and not on who receives it at what time and in what setting. The only place in which populations are discussed is the JCE article warning users of GRADE about the threats of 'indirect' evidence.

One way in which evidence for the effectiveness of an intervention can be indirect is by there being a difference between the target population and the study populations (Guyatt 2011b, §2.1). The GRADE authors make the following recommendation about how to react when a worry of indirectness arises:

> In general, one should not rate down [evidence] for population differences unless one has compelling reason to think that the biology in the population of interest is so different from that of the population tested that the magnitude of effect will differ substantially. Most often, this will not be the case." (op. cit., 1304–1305)

The GRADE authors do not provide a justification for the claim made in the last sentence of this quotation and, most importantly, they ignore potential behavioral and environmental differences between populations that may make an important difference to the effect of the treatment (just think of a case in which the condition targeted is high blood pressure, for instance). The problem created by the possible existence of relevant differences between study and target populations is thus dismissed without much argument. And users of GRADE are advised to make the default assumption that estimates of treatment effects obtained from a particular study population at a particular time can easily 'travel' to other populations and other times.

It is true that, without making this assumption, interpreting the results of meta-analyses becomes very difficult. Consider the example of a meta-analysis of the effect of antibiotics on acute otitis media in children given in (Guyatt (2011a, 387–388, Tables 1 & 2)). The GRADE authors report estimates for various effects of this treatment, e.g. pain at 24h, that are aggregates of the estimates produced by several studies, e.g. five RCTs for pain at 24h. Unless one assumes that these five estimates produced by RCTs conducted on five distinct study populations are all estimates of the same thing, something like the effect of antibiotics on pain at 24h for children suffering from acute otitis *in general*, i.e. in any population, then it makes little sense to aggregate them into a single estimate. The problem, of course, is that neither the problematic assumption that there is such a thing as the effect of antibiotics on pain at 24h *in general* nor the assumption that the five estimates that are aggregated are estimates of this particular effect are supported by either argument or evidence.

This does not prevent GRADE from rating the quality of the evidence supporting this aggregated estimate as 'High'. The reason for this rating is that the five studies from which this estimate is obtained all *individually* score highly on the GRADE criteria for quality of evidence, since they are all RCTs with no serious limitations, no serious inconsistencies, etc. According to the GRADE framework, to say that the quality of the evidence supporting this aggregate estimate is 'High' is to say that, "We are very confident that the true effect lies close to that of the estimate of the effect." (Balshem et al. (2011, 404, Table 2)).

But the "true effect" of what? Of the treatment in the particular population you are interested in treating? Of the treatment *in general*, i.e. in any population you might want to treat? Of the treatment in a superpopulation composed of the five study populations involved in the five RCTs from which the aggregate estimate was obtained? Of the treatment in a superpopulation composed of the populations which the five study populations were sampled from? Without a clear and principled answer to this question, it is difficult to interpret the aggregate estimate produced by the meta-analysis presented by the GRADE authors and, as a consequence, it is difficult to assess the quality of the evidence supporting this estimate. The evidence supporting the aggregate estimate *might* be good if what the "true effect" is happens to be the effect of the treatment in a superpopulation composed of the five study populations, but need not be if it is the effect of the treatment in the population you are interested in treating.

2.2 Oxford Center for Evidence-Based Medicine (CEBM)

The CEBM promotes EBP and produces evidence grading schemes. The most recent CEBM levels of evidence[1] offer different rankings of study designs, and of the evidence they produce, depending on the question one is interested in answering (e.g. Does this intervention help? What are the common harms? Etc.) The evidence that gets ranked by the CEBM levels of evidence thus is assumed to be evidence supporting particular answers to these questions.

Consider the question that most resembles ours: 'Does this intervention help?'. The CEBM rankings tell you that the best evidence for answers to this question is produced by systematic reviews of RCTs, systematic reviews that often take the form of meta-analyses. Just as in the case of GRADE, then, meta-analyses of RCTs are considered to produce the best evidence. But just as in the case of GRADE, the CEBM levels of evidence do not tell you exactly what the evidence produced by these meta-analyses of RCTs, or by RCTs individually, is supposed to be evidence for.

One will say: 'But they are evidence that the intervention helps (or doesn't help)!' What does it mean, however, to say that an intervention 'helps'? The same questions arise as before: Is it to say that it helped in some study population in which it was implemented? That it helps in every population in which it is implemented? That it will help in the population in which you intend to implement it? Again, as in the case of GRADE, it is not clear how one can rate the quality of evidence without a clear answer to this question. An individual RCT might provide very good evidence if the question asked is whether the intervention helped in the study population on which this very RCT was conducted, but not if the question asked is whether the intervention helps in every population in which one might implement it.

2.3 California Evidence-Based Clearinghouse for Child Welfare (CEBC)

The CEBC is a vetting agency that, unlike the Cochrane Collaboration, has its own evidence grading scheme. This grading scheme, which its authors call a 'Scientific rating scale', has five levels (from 'Well-Supported by Research Evidence' to 'Concerning Practice'), each level being defined by a list of criteria.[2] According to its authors, this

---

[1]http://www.cebm.net/index.aspx?o=5513

[2]http://www.cebc4cw.org/ratings/scientific-rating-scale/

'Scientific rating scale' is "a 1 to 5 rating of the strength of the research evidence supporting a practice or program." Of course, as for the EBP frameworks considered above, it is not made clear whether what is assessed is the evidence supporting the effectiveness of the program in the study population, in the target population, or in any population one might want to treat.

A quick look at the grading scheme, however, is enough to see that even their own vague description is not right. The CEBC grading scheme mixes together (i) whether a program's positive effects outweigh its negative effects and (ii) the strength of the evidence supporting this cost-benefit analysis. The fifth and lowest level, for instance, clearly is not a level of quality of evidence. This level, called 'Concerning Practice', is the level at which should be classified interventions such that "the overall weight of evidence suggests the intervention has a negative effect upon clients served". To present this 'Scientific rating scale' as ranking evidence thus is misleading and has the potential to confuse its users.

The tutorial video accompanying the CEBC evidence grading scheme does little to clarify the way this ranking scheme is supposed to work.[3] Consider, for instance, its use of the metaphor of the 'solidness of evidence': This video tells you to see evidence as a foundation, with the five levels of the evidence ranking scheme corresponding respectively to rock, gravel, sand, water, and gas foundations (from level 1 to level 5). It does not tell you, however, what this foundation is supposed to be a foundation for: What are you to build on top of your evidence? And the metaphor of the 'solidness of evidence' also fails to be faithful to the content of the CEBC grading scheme since, as we argued above, this ranking is not properly seen as a ranking of evidence. Consider again the fifth level of the ranking: If the overall weight of evidence suggests that a practice has negative effects, then the verdict that the practice has negative effects is presumably not based upon a gas foundation (otherwise why trust that any practice ranked at this level *really* has negative effects?).

The CEBC ranking scheme provides a striking example of pretend rigor: The use of expressions such as "Scientific Rating Process" or "Scientific Rating Scale" that is "Based on a Continuum" to describe the CEBC grading scheme stands in stark contrast with the lack of rigor in either the grading scheme itself or in the explanations and tutorial video accompanying it.

2.4 Substance Abuse and Mental Health Services Administration's National Registry of Evidence-based Programs and Practices (NREPP)

The NREPP is an agency that vets policies in the domain of mental health. Like the CEBC, it has its own system for grading the quality of evidence. This system grades six aspects (Reliability of measures, Validity of measures, Intervention fidelity, Absence of confounders, etc.) of studies on a 0.0–4.0 scale. The authors of NREPP's evidence grading scheme claim that, "NREPP's Quality of Research ratings are indicators of the strength of the evidence supporting the outcomes of the intervention. Higher scores indicate stronger, more compelling evidence."[4] As in the cases examined above, however, it is not made clear what is meant by "the intervention". Is it the intervention as it was implemented in the study population? The intervention as it might be implemented in

---

[3]http://player.vimeo.com/video/32101560
[4]http://nrepp.samhsa.gov/ReviewQOR.aspx

any population? The intervention as it might be implemented in the population you are interested in?

Some of the criteria that serve to grade studies, moreover, are stated in rather vague terms. To get a 4 on 'Appropriateness of [statistical] analysis', for instance, a study must satisfy the following conditions: "Analyses were appropriate for inferring relationships between intervention and outcome. Sample size and power were adequate."[5] What does it mean for sample size and power to be "adequate"? And adequate for what purpose?

In their presentation of it, the authors of NREPP's grading system do not state clearly whether the six different scores a study receives are to be aggregated (nor, if so, how) to give an overall 'Quality of Research' rating. Looking at the NREPP's database of evaluations, however, one notices that the six scores received by a study are in fact aggregated. How are they aggregated? Again, no explicit information is given regarding the method followed. In fact, the overall 'Quality of Research' rating attributed to a study simply is the average of the six scores it receives.

This is an odd choice, since not all the criteria determining the quality of the evidence produced by a study seem equally important. 'Intervention Fidelity', which requires that the intervention be implemented exactly as it was designed to be, does not seem nearly as important as (controlling for) 'Potential Confounding Variables' for instance. One would think a weighted average to be more appropriate. One might even think that a study that receives a score of 0 on 'Potential Confounding Variables' should receive an overall score of 0.

So, not only is it not clear what the evidence ranked by the NREPP's scheme is supposed to be evidence for, it is also not clear why anybody should believe that higher overall scores "indicate stronger, more compelling evidence."

2.5 Scottish Intercollegiate Guidelines Network (SIGN)

SIGN is a vetting agency that produces guidelines or recommendations regarding health-care policy for Scotland's National Health Services. It is a user, rather than a producer, of evidence grading schemes. SIGN, like the Cochrane Collaboration, has adopted (in 2009) GRADE's scheme for grading evidence. It is interesting to see how a vetting agency such as SIGN understands the evidence grading scheme it relies on. The authors of SIGN's handbook explicitly take the evidence grading scheme used by their framework to rank evidence for *effectiveness predictions*:

> It is important to emphasise that the grading does not relate to the *importance* of the recommendation, but to the strength of the supporting evidence and, in particular, to the predictive power of the study designs from which these data were obtained. Thus, the grading assigned to a recommendation indicates to users the likelihood that, if that recommendation is implemented, the predicted outcome will be achieved."
> SIGN (2011, 34, emphasis original)

This passage illustrates the assumption that seems to underlie EBP frameworks in general, including the ones examined above, namely that the 'best' study designs (i.e. systematic reviews of RCTs, or individual RCTs) automatically and straightforwardly produce evidence that is relevant, and sufficient, to warrant predictions regarding the effectiveness of policies that have yet to be implemented.

Unfortunately, this assumption is mistaken and one needs an argument of the kind to be presented in Section 4 in order to go from the result of an RCT to a well-supported

---

[5]http://nrepp.samhsa.gov/ReviewQOR.aspx

effectiveness prediction. There is thus little sense in talking about the "predictive power" of systematic reviews of RCTs, for instance, and in interpreting evidence grading schemes as ranking studies according to their predictive power.

2.6   So What?

The five EBP frameworks considered above all value methodological rigor highly. This is why they systematically rank RCTs (and systematic reviews of RCTs) at the top of the evidence grading schemes they use and expert opinion at the very bottom. What is more rigorous than a well-conducted RCT? And what is less rigorous than unchecked opinion, even if that of an expert? What an examination of a sample of these frameworks reveals, however, is that they lack rigor in key places. If you put forth a scheme for grading evidence, then, unlike the CEBC scheme, it should rank evidence, and evidence only. If you give an overall numerical score to studies, as the NREPP rating system does, then you should clearly explain how this score is computed and justify the choices involved in this computation. Most importantly, your evidence grading scheme should state clearly what the evidence it ranks is evidence for.

All the evidence grading schemes considered above equivocate on this last point. They never clearly answer the questions we keep underlining: What is the evidence ranked evidence for? Is it evidence that the intervention was effective in the study population? Is it evidence that the intervention will be effective in most population? Is it evidence that the intervention will be effective in every population? Is it evidence that the intervention will be effective in the population in which you want to implement it? We repeat this point once more not at the risk of boring the reader because it is of crucial importance. Ranking evidence without a clear answer to this question is vain. You cannot evaluate how good the evidence for a particular claim is unless you are clear on what this claim says.

What we do below is to explain what claims the evidence ranked by evidence grading schemes is generally in fact evidence for and explain why RCTs are thought to be very good at supporting claims of this kind. We also explain how to bridge the gap from the kinds of claims supported by RCTs, i.e. claims of the form "It works somewhere", to predictions of the effectiveness of interventions, i.e. claims of the form "It will work here".

3   The Starting Point: RCTs

3.1   The Probabilistic Theory of Causality

What is so good about RCTs? They are supposed to be a very good way – some insist, the only way – for controlling for unknown confounders. See for instance what the webpage of MIT's Abdul Latif Jameel Poverty Action Lab (J-PAL) has to say about RCTs: The reader is told that "randomized evaluations do the best job" at controlling for unknown confounders because they "generate a *statistically identical* comparison group, and therefore produce the most accurate (unbiased) results" while "other methods often produce misleading results."[6] And how did 'confounders', known or unknown, enter the discussion? Let us start back, way behind where the usual defence of RCTs begins, to get a more rigorous account. Confounders enter when we are trying to establish causal claims. So we shall begin with the probabilistic theory of causality. We shall not, though, use the theory in exactly the form Suppes first put it, but in a modified version Nancy

---

[6]http://www.povertyactionlab.org/methodology/why/why-randomize

has developed building from Suppes' account (Cartwright (1979)). For simplicity we will consider only yes-no variables.

Suppose then that the notion of causality at stake satisfies the following constraint:

Probabilistic causality: For any population A, C causes E in A iff for some $A(i) \subseteq A$ every member of which satisfies K_i, $P_{A(i)}(E/C \& K_i) > P_{A(i)}(E/\neg C \& K_i)$ where K_i is a state description[7] over a full set of causes of E, barring C itself.

The expression 'a full set of causes' takes some further paraphernalia to characterize. It can be done either relative to a formulation of the causal principles that govern A or to a set of causal pathways into E that obtain for A, where a full set of causal factors for E will contain one node from every pathway into E. We shall here leave it undefined. The factors that go into the Ks are just the 'confounding factors' that RCT advocates are concerned about. We shall use the term 'causal structure' from now on. A causal structure for outcome E in A is a set $\{\mathscr{C}_A, P_A\}$, where $\mathscr{C}_A$ is a full set of causal factors for E in A and P_A is a probability measure that holds in A over the space generated by $\mathscr{C}_A \bigcup \{E\}$.

Note first that this theory uses the notion of causality on the right-hand-side and hence cannot provide a reductive definition for causation. It does however provide an important constraint between probability and causality, which is a good thing for our enterprise since the immediate results of RCTs are statistics. Second, a direct application of the formula requires a huge amount of antecedent causal knowledge before information about probabilistic dependencies between C and E can be used to determine if there is a causal link between them. The RCT is designed specifically to finesse our lack of information about what other causes can affect E. Third, the theory allows that C may both cause E and prevent E (i.e., cause $\neg E$) in one and the same population, as one might wish to say about certain anti-depressants that can, it seems, both heighten and diminish depression in teenagers. This is especially important to note when it comes to RCTs since the effect size measured in an RCT averages over different arrangements of confounding factors so that the cause may increase the probability of the effect in some of these arrangements and decrease it in others and still produce an increase in the average.

3.2   Ideal RCTs

We shall describe the simplest basic structure, to make the argument outline clear. RCTs have two wings – a treatment group where every member receives the cause under test and a control group, in which any occurrences of the cause arise 'naturally' and which may receive a placebo. In the design of real RCTs three features loom large:

a. *Maskings* of all sorts. The subjects should not know if they are receiving the cause or not; the attendant monitors should not know; those identifying whether the effect occurs or not in an individual should not know; nor should anyone involved in recording or analyzing the data. This helps ensure that no differences slip in between treatment and control wings due to differences in attitudes, expectations, or hopes of anyone involved in the process.

b. *Random assignment* of subjects to the treatment or control wings. This is in aid of ensuring that other possible reasons for dependencies and independencies between

---

[7]A state description over factors $A_1, \ldots, A_n$ is a conjunction on n conjuncts, one for each A_i, with each conjunct either A_i or $\neg A_i$.

the cause and effect under test will be distributed identically in the treatment and control wings.

c. Careful choice of a *placebo* to be given to the control, where a placebo is an item indiscernible both for subjects of the experiment and for those administering the experiment from the cause except for being causally 'inert' with respect to the targeted effect. This is supposed to ensure that any 'psychological' effects produced by the recognition that a subject is receiving the treatment will be the same in both wings.

These are in aid of bringing the real RCT as close as possible to an ideal RCT. An RCT is ideal for testing "C causes E in A" iff the probability of all combinations of causal factors in A of E are the same in both wings except for C and except for factors that C produces in the course of producing E, whose distribution differs between the two groups only due to the action of C in the treatment wing. Suppose for simplicity $P_A(C)$ in the treatment wing $= 1$ and $P_A(C)$ in the control wing $= 0$. An outcome in an RCT is positive if $P_A(E)$ in the treatment wing $> P_A(E)$ in the control wing.

As before, designate state descriptions over a full set of causal factors other than C for E in A by K_i. In an ideal RCT each K_i will appear in both wings with the same probability, w_i. Then $P_A(E)$ in treatment wing $= \sum w_i P_A(E/C\&K_i)$ and $P_A(E)$ in control wing $= \sum w_i P_A(E/\neg C\&K_i)$. So a positive outcome occurs only if for some $i, P_A(E/C\&K_i) > P_A(E/\neg C\&K_i)$. Thus a positive outcome in an ideal RCT for C cause E in A occurs only if C causes E in some $A(i) \subseteq A$, and hence only if C causes E in A by the probabilistic theory of causality.

The RCT is neat, at least in the ideal, because it allows us to learn causal conclusions without knowing what the confounding factors are. By definition of an ideal RCT, these are distributed equally in both the treatment and control wing, so that when a difference in probability of the effect between treatment and control wings appears, we can infer that there is an arrangement of confounding factors in which C and E are probabilistically dependent and hence in that arrangement C causes E. It is of course not clear how closely any real RCT approximates the ideal.

Notice that a positive outcome does not preclude that C causes E in some sub-population of the experimental population and also prevents E in some other. Again, certain anti-depressants are a good example. They have positive RCT results and yet are believed to be helpful for some teenagers and harmful for others.[8]

4 The Destination: A Prediction of Effectiveness

Out of the morass of vague expressions reviewed in Section 2 of what the evidence in EBP is supposed to be evidence for, let us take SIGN's formulation to express the basic idea: "Thus the grading assigned to a recommendation indicates to users the likelihood that, if that recommendation is implemented, the predicted outcome will be achieved." We take it that this is meant to be some kind of causal claim; and also that the users constitute some new population A' different from any experimental population A. Let us suppose then that the target claim that we aim to produce evidence for is C causes E in A'. This is a fairly weak claim recall, since it is consistent with it being true that C also causes $\neg E$ in A'. Why should a positive RCT result for C causes E in A speak in any way for the truth of C causes E in A'?

[8]See for instance the U.S. Food and Drug Administration medication guide at `https://web.archive.org/web/20070702044521/http://www.fda.gov/cder/drug/antidepressants/`

It will do so if these conditions for RCT relevance are both satisfied:

R1. Populations A and A' have the same causal structure for E.

R2. One of the K_i that picks out a subset of A such that "C causes E in $A(i)$" holds also picks out a subset of A' that has members.

Of course it will also do so under weaker conditions. The weakest seem to be if both R3. and R4. are satisfied:

R3. C is in $\mathscr{C}_A \rightarrow C$ is in $\mathscr{C}_{A'}$.

R4. $P_{A'}(E/C \& K_i) > P_{A'}(E/\neg C \& K_i)$ for some i, where K's are state descriptions over $\mathscr{C}_{A'} - \{C\}$.

That is, C is a cause in A only if it is a cause in A' and there's at least one causally homogeneous subpopulation of A'—picked out by the causal structure *that holds* in A'—where C acts positively and that subpopulation has a non-zero probability.

The lesson to be learned is that although (ideal) RCTs are excellent at securing causal claims about the study population, there is a very great deal more that must be assumed – and defended – if those causal claims are to be exported from the experimental population to some target population. Advice on this front tends to be very poor indeed however. Recall GRADE's recommendation to take as a default the assumption that experimental and target populations are sufficiently similar unless there's good evidence to the contrary. Or consider the US Department of Education website, which teaches that two successful well-conducted RCTs in 'typical' schools or classrooms 'like yours' are 'strong' evidence that a programme will work in your school/classroom (USDE (2003, 10)).

This problem often goes under the label 'external validity'. A study has external validity when the claim established in the study population (here A) holds in a target population (A') as well. The Department of Education's advice about external validity is typical: A study will have external validity with respect to a given target if the two populations involved are sufficiently similar. The great advantage of a little rigor is that it can give content to this uselessly vague advice. From the point of view of the probabilistic theory of causality, 'like yours' can mean that R1. and R2. hold. At the very least, R3. and R4. must hold or the RCT results in A will be totally irrelevant to A'. Admittedly, these conditions are abstract so do not give much practical purchase on how to decide whether they obtain or not. But they are not, like the usual advice, without content. At least we know now just what kinds of similarity in what respects we need to look for.

5   Conclusion

We have briefly explained how to make up for the lack of rigor in EBP frameworks when it comes to justifying the relevance of RCT results to effectiveness predictions, that is, when it comes to bridging the gap between "It works somewhere" and "It will work here". The account sketched here using the probabilistic theory of causality that originates with Patrick Suppes has been developed in detail by Nancy together with Jeremy Hardie (2012). The argument taking one from RCT results to effectiveness predictions must be rigorous every step of the way in order for its conclusion to be properly supported by evidence. We urge practitioners and advocates of EBP not to focus solely on rigor in establishing the "It works somewhere" premise at the expense of rigor in establishing other premises that are equally necessary to yield the conclusion that "It will work here".

6   Acknowledgment

Both authors would like to thank the Templeton Foundation's project 'God's Order, Man's Order and the Order of Nature' as well as the AHRC project 'Choices of evidence: tacit philosophical assumptions in debates on evidence-based practice in children's welfare services' for support for the research and writing of this paper. Nancy Cartwright would in addition like to thank the British Academy, which funded her project 'Evidence for Use', as well as the Institute for International, Comparative and Area Studies at UC San Diego for the 'Political civility and scientific objectivity' grant.

References

Balshem, H., M. Helfand, H. J. Shünermann, A. D. Oxman, R. Kunz, J. Brozek, G. E. Vist, Y. Falck-Ytter, J. Meerpohl, S. Norris, and G. H. Guyatt. 2011. Grade guidelines: 3. rating the quality of evidence. *Journal of Clinical Epidemiology* 64(4):401–406. doi:10.1016/j.jclinepi.2010.07.015.

Cartwright, N. 1979. Causal laws and effective strategies. *Noûs* 13(4):419–437. doi:10.2307/2215337.

Cartwright, N. and J. Hardie. 2012. *Evidence-Based Policy: A Practical Guide to Doing It Better*. New York: Oxford University Press.

Guyatt, G. e. a. 2011a. Grade guidelines: 1. introduction—grade evidence profiles and summary of findings tables. *Journal of Clinical Epidemiology* 64(4):383–394. doi:10.1016/j.jclinepi.2010.04.026.

—. 2011b. Grade guidelines: 8. rating the quality of evidence—indirectness. *Journal of Clinical Epidemiology* 64(12):1303–1310. doi:10.1016/j.jclinepi.2011.04.014.

SIGN. 2011. *SIGN 50: A guideline developer's handbook*. NHS Scotland, Edinburgh.

USDE. 2003. *Identifying and Implementing Educational Practices Supported by Rigorous Evidence: A User Friendly Guide*. Coalition for Evidence-Based Policy, Washington, DC. URL http://www2.ed.gov/rschstat/research/pubs/rigorousevid/.

Part VII

Commentaries

23

Commentaries

PATRICK SUPPES

1 Michael Friedman

Michael's detailed characterization of my deep interest in both science and philosophy is the most accurate I have had the pleasure of reading. He has described very well my passion for insisting that philosophy, above all, should be concerned with the foundations of science as thought of and practiced by scientists. I much agree with his view that in many ways intellectually my work is closer to that of Carnap than to that of Quine. Superficially, it might be thought that my affinities would lie most closely with the pragmatism Quine endorsed on many occasions. Michael rightly recognizes that in spite of my acceptance of Quine's pragmatism, I am in fact, and in practice, closer to Carnap's concern to understand the detailed structure of science, especially that of physics. On the other hand, I have had little interest in Carnap's program for using logical methods of philosophy to eliminate any tendency of metaphysics to be developed as it was in the 19th century. My focus has been to understand the ways in which the philosophy of science can contribute to the understanding of science itself. I agree with Carnap on the dangers of opening the door for transcendental metaphysics, but that danger now seems remote. The pressing problem is to get a deeper grip on the complex nature of modern scientific work.

Michael rightly stresses my own emphasis on the need for greater appreciation of the role of scientific experimentation in almost every domain of science when fundamental problems are being analyzed. This is contrary to the tradition of emphasizing, especially by philosophers, the theory of a given science. Earlier in this century, physics was talked about as if we were going to discover the one grand theory of the origin and nature of the universe. This belief made the theory acquire a nearly religious state in its absolute claim of reaching such an understanding. Now we realize how hopelessly misguided this earlier conviction was.

One of the more ironic arguments for pointing this out is the current role of computers in physics. If I remember correctly, the great case for the importance and relevance of computers to physics was the fundamental role they would provide in making it possible to solve the important and fundamental partial differential equations that seem unmanageable. There is truth in this, but what was not foreseen is that computers would, above all, completely change the nature of data collection and analysis. It is

Foundations and Methods from Mathematics to Neuroscience.
Colleen E. Crangle, Adolfo García de la Sierra and Helen Longino.
Copyright © 2014, CSLI Publications.

generally recognized that in many parts of physics the experimental data (often purely observational in the case of astronomy and astrophysics) are so vast and complicated that any hope of giving a final theory of them now seems impossible. For me, this is a happy separation of a mistaken view of physics as establishing absolute truths about our world and the correct view of how schematic and fragmentary actual science is. Michael catches well this aspect of my current thinking, which is certainly not present in most of the early part of my career when I was more naïve about the nature of science and not serious enough about its limitations. I hasten to add that I do not have some nonscientific set of beliefs or attitudes to go beyond these limitations. What is important for me is to preach the doctrine of the necessarily fragmented character of our knowledge in every area of experience. I also want to say that even though my first serious mentor, Ernest Nagel, did not put the matter exactly this way, I am confident that if he were still around, he would mainly agree with me.

Michael also does a good job of recognizing the permanence of my deep interest in the history of science as an important part of good philosophy of science. Of course, like a variety of scholars, I find anecdotes about the past, and even more, careful recounts of conceptual controversies that were not easy to settle, irresistible. For a variety of reasons, we seem to have easier access to early controversies about the foundations of mathematics more than we do about physics or any other science. But what is important to emphasize, in my own view, is that the kind of arguments used to test simple hypotheses about the behavior of matter seem very similar to corresponding arguments in elementary mathematics about whether or not the square root of two is a rational number. The great variety of these sources provides wonderful conceptual examples for the philosophy of science. Ptolemy's *Almagest* reads like a modern textbook of physics and astronomy, but a little more rigorous in the proof of theorems. Moreover, it is a mixture of theory and observation that provided a model of how astronomy should be written for hundreds of years. This example is not an isolated one. It would be possible to give a very satisfactory course in the conceptual development of science and the accompanying philosophy by restricting the lectures and readings just to ancient astronomy.

I want to close in a more positive way in expressing my feelings about what is important in philosophy, especially in which relation to science. In my desire to emphasize the importance of science for almost every systematic question in philosophy, I sometimes give, I am sure, the impression that philosophy is no longer very useful. In so far as I have done this, it is a mistake. In some sense with the avalanche of scientific work now appearing in almost every area of experience, it is more important than in the past to have a philosophical perspective on scientific work. In many parts of science, casual claims of importance are made for findings that are not well supported by serious data. Recognizing almost automatically such failures conceptually is one of the things good philosophers of science do. This is just as true of mistaken claims made to the public about the significance of some findings. It is part of a good philosophical training to greet such claims with skepticism and to offer the way toward a more serious and sophisticated presentation of scientific results for the general public, which in these matters ranges from colleagues in other disciplines to university-level students. Finally, in these frantic times of too much of everything, young scientists, above all, do not have the time and energy to understand well the conceptual development in the past of their particular branch of science. It is a proper and important role for philosophy to provide in the general education of students this past background of such great significance in the development of modern science. Understanding, in a conceptual way, our past systematic and conceptual views of the world is a worthy enterprise for generations of philosophers to come.

2 Jens Erik Fenstad

Like many other participants in this volume, I have known Jens Erik Fenstad for many years and have enjoyed our discussions of a great variety of topics. I have benefited from these many discussions very much. In the same spirit, I have learned much from his paper entitled "What Numbers Are".

As is often the case, I find myself in general agreement with his viewpoint, but also find myself moving in a different direction to answer the many questions he raises. For example, I agree with his emphasis on the close relation between culture and knowledge, but I do not agree with what seems to be his desire to give the mental a separate kind of status, equal to that of physical substances.

This is perhaps the first important point to try to make my own viewpoint clear. I like, and continually adopt, Aristotle's distinction between form and matter, which he uses to discuss the science and knowledge of his time, from physics to psychology. Aristotle is also much opposed to Plato's efforts to make form alone have an independent eternal existence. With this I much agree. To put it simply, what exists in the real world are substances and processes made up of both form and matter. Matter without form does not exist, and similarly, form without matter does not exist. This Aristotelian view is also my own. It sets the standard of how we should talk about mental phenomena. I do not believe it is philosophically useful or correct to talk as if some mental states are on a par with and equal to physical states. I can accept that we should be careful in talking about physical states in too simple a way. It is perhaps better to talk about substances, some of which may have mental properties as well as physical properties. To take an example of great importance in current science, let us for a moment discuss the brain. Certainly almost everyone grants that brains are physical objects exhibiting complex and subtle physical processes of an electrical, magnetic or chemical nature. Yet, it is also true that these processes can be described and analyzed as mental processes. The fact that we can make such a mental analysis does not give the mental a separate independent status. This truth I insist upon as an Aristotelian principle of great general importance. Other kinds of examples bring this point out in a way that is easy to understand. We agree an apple is red, but does not mean we believe in the independent status of redness. Being red is a property of many physical objects and processes, but redness itself is not a physical object or physical process.

So, in this spirit, I do not accept the independent, or nearly independent, status that philosophers like Husserl would like to give processes exhibiting in some way intentionality. But my objection in this direction still leaves a place, and an important one, for culture. Different individuals and different societies know different things and think about the world in different ways. We can also recognize that there is a concept of correctness that permits an objective comparison about many of these different ways of thinking and knowing. Yet, there are other aspects of culture that can vary without one part or one view being the true or correct one, in a way that invalidates the correctness or acceptability of the view another culture. I have in mind in making this comparison, habits and actions that are in themselves neither true nor false. We may like the table manners of one culture more than another, but this does not mean in any sense that the table manners of one person are in some sense more true than those of another. It is the case that we can refer to a standard and then have a relative notion of correctness. So I can say of someone that she eats soup in a style that was taught in boarding schools for young women many decades, and is therefore correct with reference to that style, now no longer so widespread. Instead of talking about eating soup, I could just as well talk about different cultures having different styles of painting or representations in other forms of the human body. Each of these we could reference to a standard and talk about correctness, or relative correctness with reference to such a standard, but this does not imply in any sense that there is one true standard of painting or one true style for eating soup. Cultures have variety. It would be a great mistake to try to eliminate their variety by claiming there should be some universal standard. On the other hand, there is another line of talk that leads to a different standard of correctness, and that is being correct about a property of some object or process in the natural world. Here, cultural relativism is out of place. It is out of place to claim it is a cultural matter to determine the temperature in the morning at a given place and date. Or to take two other examples, it is not a matter of cultural variation to say that all kinds of physical properties are invariant under certain kinds of operations and not others. This is a matter of how the physical world is put together. To me, such questions about simple physical properties have just the same status as questions about number.

I do not agree with Jens Erik about assigning a special permanent abstract status to mathematical objects, but rather to treat them as forms in the sense of Aristotle. Such forms can be realized in matter as substances in the real world, but also not so realized and only be known as forms or structures that have not been realized, and perhaps can never be realized in the real world. To repeat what I said above, this applies especially to mental states, which only have real existence as properties of some special substances such as brains, or more generally, human beings or other animals. In this same spirit, I reject any sort of status of an independent kind for Popper's objects that belong to a Third World.

What I said does not mean I defend some Aristotelian view adequate to explain or describe, without modification and further development, many properties of the real world, especially properties that we think of as mental or psychological in nature. What I am discussing here is the basic framework within which science and culture should be placed and analyzed.

The comments I have made are general, and necessarily so Jens Erik has written a paper that is rich in references and details of many kinds that I cannot possibly consider with care and accuracy in this brief commentary. What I have tried to do is to indicate my sympathy with much of what he has to say, and yet to argue for a different formulation of the general framework in which more detailed discussions take place. I believe that we can have a rather strong difference on questions of general philosophy, but still agree on many of the particular things he has to say about mind and culture.

3 Jaakko Hintikka

It is a pleasure to comment on Jaakko Hintikka's article. He and I have a friendship that now goes back about 50 years. I still remember how pleased I was to be able to recruit him part-time to be a member of the Department of Philosophy at Stanford. I have had the pleasure as well of visiting Finland at his invitation several times, not to speak of numerous meetings elsewhere in Europe and the United States. Moreover, the topic of this paper is one that takes me back very quickly and naturally to the 1950s, when I was going to Alfred Tarski's seminar in Berkeley and writing thereafter my textbook on axiomatic set theory (1960). Like Jaakko, I realize there are many fascinating ways to think about the foundations of mathematics that do not involve just a simple statement in clear set-theoretical terms. I wrote something about it then, but he has written much more than I have, and recently certainly has thought about the matter more than I have.

So, I just want to make a few informal comments of my intuitive perception of some of the possibilities for axiomatizing set theory, or one of its near equivalent. The heart of Jaakko's proposal in his article is to use Skolem functions instead of existential quantifiers, especially in formulating some form of what is customarily called the axiom of choice. So Jaakko wants to alter the underlying standard logic, not set theory itself, in formulating the axioms of set theory. In recent years, he has written a number of articles in this direction. I have not read them all carefully, but I found his ideas suggestive and original in many ways, perhaps too original for people who have strong convictions about how foundations should be formulated.

This is not the place to enter into a detailed discussion of his bold suggestions for some radical changes in the underlying logic. I do want to mention some of the ideas that occurred to me in those earlier years and continue to think about off and on. These are also ideas that interact with what Jens Erik Fenstad proposes in his own approach to the foundations of mathematics, which he discusses in another chapter of this volume.

My own main intuition is that the concept of abstraction is in many ways more natural than the general concept of set. The reason is that ordinary language is full of such use of abstractions. In reporting cognitive ideas or emotional feelings, we must be able to give brief and abstract descriptions of them in order to make conversation work well. Such abstraction is necessarily found everywhere in ordinary talk, so the idea of abstraction is one that is intuitively easy to grasp. In contrast, the notion of set is much less familiar, and in its more exotic forms seems strange to the uninitiated.

I suppose my favorite example is to compare the complexity of the purely set-theoretical definition of a cardinal number, with the simple creative definition by identity that two sets that are equipollent have the same cardinal number. In this simple framework, it is undecidable whether or not a cardinal number is a set. A favorite conjecture of mine is that with some work and some modifications, all the standard axioms of ZF set theory can be replaced by creative definitions of identity. Each such definition creates a new class of entities, which in general is not the same as the class created by another such definition. This approach takes a different attitude toward introducing classes of entities as needed, rather than trying to put all objects, at least all mathematical objects, in the framework of one kind of object. Perhaps the main argument in favor of this kind of construction is that it mirrors in mathematics what takes place in ordinary language all the time. It also encourages less attention be paid to the type of entity, rather than to what properties a given entity possesses.

I doubt that Jaakko will be satisfied with this suggestion of how to move away from the standard framework, and I do not pretend that I have fully worked out the necessary details. As I have stressed, it does exemplify a kind of approach that seems natural, both from the standpoint of ordinary use of language and from the creation of new theoretical entities in all parts of science.

What I have ignored here is something important. I have not dealt with the relative strength of such axioms as those consisting of creative definitions of identity, but this can probably be done more or less along standard lines. What I do like about Jaakko's proposal, here and in other recent papers of his, is the attempt to investigate more deeply the fundamental foundations we want to start with.

I suppose I cannot resist one more subversive remark. It seems to me that great era of clarifying, in the spirit of Hilbert Gödel, Tarski, and others, the foundations of mathematics, is now coming to some sort of close. One of the related, but unfamiliar topics to mathematicians, is a new concentration on the nature of mathematical thinking. From conversations with various mathematicians, I realize that many think that this topic is now more important than the extension or modification of classical foundational axioms. So, I end my remarks with a question to Jaakko, "What do you think of this new direction?"

4   Harvey M. Friedman

I have known Harvey since he was a young man of 18. He was a prodigy who had just received a PhD in mathematical logic. I was pleased to appoint him at this very young age to be an assistant professor in logic in the Department of Philosophy at Stanford. He was at that time the youngest assistant professor in the university. Harvey liked such a record, and it has continued to be a mark of his career that he has set many similar ones.

What is really important about Harvey is that he is one of the few logicians today who have both superb technical command of the modern mathematical developments in logic and at the same time has retained a deep interest in the philosophy of logic and mathematics. Harvey has had a career of more than forty years of splendid research and is now, almost at retirement age, carrying on as actively as ever.

I will not try to go into details of his contribution to this volume, because it is too technical for this context. However, I think it is easy to explain his motivation and goals, as I understand them and as they are clearly expressed in the introduction to his contribution. One problem he is deeply into is whether or not there is a system of axioms that is really suitable as a basis for the entire body of what mathematicians think of as more or less standard mathematics. It is important to emphasize this phrase "standard mathematics", because there are many proofs in logic of such questions as the existence of very large cardinal numbers and their properties that cannot be proved in the standard framework of Zermelo-Fraenkel set theory with the axiom of choice (ZFC).

He is equally concerned with the general problem of the incompleteness of ZFC. To prove the continuum hypothesis was the first problem on Hilbert's famous list of open problems given at the International Congress of Mathematics in 1900, and Hilbert assumed it could be proved. Gödel proved in the 1930s that the continuum hypothesis was consistent with ZFC. The conceptually more important problem of its being independent of ZFC was proved by Paul J. Cohen in 1963. The incompleteness of ZFC was clearly established by Cohen's results.

The proofs and speculations that Harvey gives in his contribution to this volume display well the depth and range of his thinking about these problems. To show the inadequacy of the ZFC system of axioms as a basis for standard mathematics, he would very much like to find a proof of a simple statement in the arithmetic part of standard mathematics that (by demonstration) can be proved only by going well beyond the resources that ZFC can provide. He provides a simple statement that he argues lies in the arithmetic part of standard mathematics, and gives a proof using resources well beyond those of ZFC. He claims that ZFC is not sufficient, with a proof to appear elsewhere. If he makes good on that claim, and the mathematical community accepts his statement as lying in the arithmetic part of standard mathematics, this will be a stunning result.

5   Thomas Ryckman

Tom knows my views rather well, and most of what he has to say about my skeptical attitude to many philosophical isms is correct. Rather than making detailed replies to his many suggestive summaries of my views, I think it will be more useful to comment in a general way on how my views are continuing to develop in an evermore skeptical manner.

I think I understand what is driving this continuing development of my skeptical attitude toward the generalizations about science and its methods that are affirmed by many philosophers. Too many of my own generalizations of the past have been shattered by the intense effort I have made in the last two decades to get into research on the brain. Several decades earlier, I found it fairly easy to understand and use the experimental methods of psychology in the study of learning in a variety of settings and from a variety of theoretical viewpoints. The overwhelming difference between standard psychological experiments and those of neuroscience is above all the size of the data sets, often a difference of many orders of magnitude. Let me give a simple example that some psychologists will find too simple, but that I think will be useful for comparative purposes.

In many learning experiments, where only behavioral data are recorded, the response of the participant in the form of choosing one of a small number of responses is recorded, and often the latency of the response. A similar experiment on the brain, with 128 sensors records an electric or magnetic quantity measured to three or four decimals every millisecond. The quantity of brain data collected and the corresponding effort to understand the data are incomparably more difficult. In fact, the lesson for me has been that the fragmentary nature of science is increasing at a rapid rate.

Let me try to explain what I mean by this remark. Like other philosophers, I often tell students a much too simple story about science. It consists of some simple examples of Newtonian mechanics, or perhaps now, some simple biological experiments. The talk is very much in the spirit of everything being well understood and the theory telling us everything we want to know about the phenomena being studied. A single particle, acted upon by a total force that is null, moves in a straight line with a uniform velocity. This seems simple and one of the easiest things in the world to understand. It is remarkable that, in spite of the richness of ancient Greek astronomy, not to speak of other systems as well, ancient scientists did not really understand such uniform motion, did not recognize its dynamic equivalence to being at rest, and continued to consider the distinction between rest and motion of any kind to be the central distinction. When Newton developed clearly the dynamics of two isolated particles acted upon only by their mutual gravitational attraction, the elliptical orbits that then came out of the theory seemed obvious and natural as well, even though this result was not really well understood until the 17th century. This is where the simple story for students stops.

When we turn to the obvious case of three particles interacting only under the forces of mutual gravitation, we have almost nothing manageable to say to beginning students. Moreover, the mathematical difficulties of even the so-called restricted cases of three bodies turn out to be unmanageable in many cases, even for experts. Let me try to describe, in a fairly simple way, a class of cases of this kind that seem to make the elementary distinction between the determinism of Newtonian mechanics and random sequences mistaken.

We consider a three-particle system. Two of the particles of equal mass (m_1 and m_2) are moving in the standard elliptical orbits, because the only forces acting on them are their mutual gravitational attractions. Through the center of mass of the plane of these elliptical orbits, we introduce a third particle (m_3) with negligible weight moving only along the axis perpendicular to the plane. See Figure 1 for a diagram.

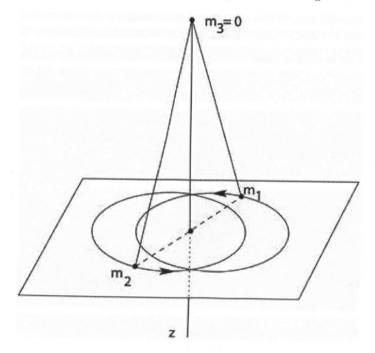

FIGURE 1 Restricted three-body problem

The symmetry of motion of this third particle is such that it never deviates from the perpendicular axis through the center of gravity of the two particle system. (For those who are concerned about introducing a weightless particle, the results introduced here can be approximated arbitrary closely by making the weight of the single particle arbitrarily small and the two other particles arbitrarily large.) Consider now a certain constant c that is greater than zero and can be estimated for the system of two particles moving in the plane. Introduce any random sequence of natural numbers all of which are greater than c. Then this random sequence can be reproduced by the Newtonian three-particle system introduced if the initial velocity of the nearly weightless particle is correctly chosen. What is called the symbolic dynamics of the mechanical system can exactly reproduce the random sequence. (The concept of symbolic dynamics was introduced by George Birkhoff in the 1930s to simplify mathematical attacks on the three-body problem.) Given any positive velocity greater than c, we take the largest integer equal to or less than that velocity to being the symbolic representation of it. At the level of the symbolic representation of the velocities, or even better, the time to go from the plane and to return (a half cycle), and then take the symbolic value of this time, we obtain through using the symbolic dynamics an exact reproduction of the given random sequence – the proof of this result is constructive, even though we conceptually think of the motion of the simple ideal and isolated three-particle Newtonian system as being fully deterministic. The lesson learned here is that a tidy simplicity will lead us to make mistakes in cutting up phenomena of motion into being deterministic or random, but not both. For detailed discussion and references to the literature of this example, see Suppes (2002, pp. 218–220).

Returning to the brain experiments, we have a similar problem of complexity. The behavior of the neurons in uttering the simplest sentence, or even in processing listening to that sentence as said by someone else, is too much for us. We will have important and useful fragmentary insights into how the brain works in dealing with such sentences, but an exact account comparable to our thinking about the two-body problem of gravitational motion seems out of reach for science in our time.

I take Tom's sermon to be a call for humility as well as skepticism in the face of the complexities of much current science. But I am persuaded that it has always been that way. For those in the know in ancient astronomy of every kind, the fragmentary nature of the results was surely understood, even if not emphasized.

One outcome of this kind of viewpoint about science is that it becomes more like the way we think about ordinary experience, fragmentary, momentary and often seemingly contradictory. The simplicities so loved by a long tradition in philosophy and in science itself must be recognized as what they are, useful on many occasions, but not fundamental in any deep sense.

6   Paul Humphreys

For several reasons, it is a pleasure to comment on the interesting paper of Paul Humphreys. First, many years ago I enjoyed very much having Paul as a graduate student in philosophy at Stanford. He came from England with a good education as an undergraduate in physics, and he has continued to use this physics background in a long series of papers in the philosophy of science. Second, Paul and I have always enjoyed sparring about important questions on the nature of science. We agree on most things, but have differed on the details. Third, in the present case I am delighted to see his returning to arguments about experimentation that go back to an old paper of mine in 1962 "Models of Data". At the time, the paper did not attract much attention, and was certainly considered by the model theorists I knew at the time, such as Tarski, as being too far afield to be of any real interest to the formal theory of models. (I am also pleased that Tom Ryckman referred to this old paper of mine in his contribution to this volume.) I look upon both of their papers as contributing to an effort to update my own thoughts of long ago on the importance of various kinds of formal models to the theory and practice of experimentation.

As the first main point of his paper, Paul gives an elaborate and interesting example from physics to show how there can be, even when a quantity to be estimated is not part of an explicit theory, many constraints that must be satisfied. I like very much what he has to say here. In any broad sense of the term, his physics example is more interesting than the rather simple model of reinforcement learning I gave in the original paper. Furthermore, in spite of my own already developed interest in the theory of measurement, I did not give it the place, indeed almost the central place, that Paul does in the analysis of experiments, especially in the physical sciences.

Paul may be less happy with one interpretation that I want to give to this emphasis of his. For me, it is another beautiful example of the fragmentation of science. In experimentation itself the use of theory, sometimes elaborate theory involving three or four basic concepts of physics or chemistry, is frequent and brought in without serious explanation, which is acceptable, but not always, when problems of approximation and accuracy are important. My critical point is that untangling this web of past theory and experimentation can often only be done properly by a few experts in the field. The rest of us have no alternative but to follow along and hope that what they are claiming to be true is indeed so. In other words what this fragmentation entails, necessary as it is, is the opaqueness in a deep intellectual sense of most advanced science. In ancient times, such opaqueness was often achieved by secrecy, but now the work has become so complex that secrecy is not needed. This kind of idea did not occur to me at all when I wrote the original article in 1962, but is inspired by Paul's current formulation of the reasons for having models of experiments or of data.

Another point of Paul's that I consider important is his insistence that data analysis in current times is often done without any guiding theory directly or hypothesis testing in the classical statistical manner. To a large extent this problem also can be attributed to the fragmentation of science. But it also encourages another remark about the consequences of this fragmentation.

Skepticism now exists everywhere in science about understanding in an absolutely fundamental way anything. This is a much greater change in scientific thinking that many philosophers are probably not yet ready to accept. But the evidence is everywhere, from the data on our expanding universe, to the mysteries of evolution and the beginning of life. These phenomena represent a timescale that is psychologically incomprehensible to most persons. Unfortunately, we understand the remote past no better that we understand the remote future. A possible fossil fragment here, a black hole there, may give us some limited insight, but little confidence of ever having an adequate theory.

Another thing I like about Paul's paper is the introduction explicitly of the concept of a model of the instrumentation, a subject mostly ignored in the philosophy of science, but of critical importance to interpreting data from the past and running experiments in the present. It is only in recent times that historians of astronomy have come to recognize that we do not have a very detailed understanding of the instrumentation used by various kinds of ancient astronomers. In many ways the story is even better about what we know of methods of moving and saving water or of building structures of surprising size and weight in any ancient civilization, not to speak of the instruments used in these large projects of the ancient past.

I want to emphasize that Paul has many excellent and more technical points about models of instrumentation that I will not elaborate on here. He has, I believe, introduced, however, a useful concept that needs treatment in most accounts of modern scientific experiments. I now turn to it.

Toward the end of his paper, Paul has returned to a classical topic still of great interest in physics and other parts of science. I refer here to what are often called in casual talk "inverse problems". I am especially reminded of Laplace's emphasis on reasoning in probability theory from effects to causes. Technically, the subject has come a long way since the end of the 18th century. Paul recognizes the importance of inverse methods in many parts of physics. It is a subject not adequately discussed almost anywhere in the philosophy of science. I am happy to see it in his paper here.

I began by saying that, in the past of many years ago, I had much enjoyed argument with Paul about problems in the philosophy of science. So far, I have almost entirely agreed with what he has had to say. Near the end of this paper, fortunately, I have found some grounds for argument. He does not really like my claim of many years ago, which I still maintain, that a sharp distinction between pure and applied mathematics is mistaken. I agree with him that some distinctions can be made here, but I think what is not usually considered in sufficient depth is how similar are the practices in these two parts of mathematics. Both kinds of mathematics depend upon intimate knowledge of what has been done in the past, and it is my belief that in their thinking there is a great congruence in their use of past mathematical findings, whether they be pure or applied in nature. But this is a dialogue he and I may be able to engage on a future occasion. For now, I want to express my pleasure in reading a paper that has done so much to make an old effort of mine seem once again relevant.

7 Adolfo García de la Sienra

I have known Adolfo for many years, first as a student, and then as a professor of the University of Veracruz in Mexico. In his contribution to this volume he has many things to say about the nature of measurement, and as might be expected, I agree with many of his opinions, but certainly not with all. I do like his focus on representational measurement in economics, because the most explicit theory of measurement in the social sciences has been developed by economists.

In the first part of his paper he has quite a bit to say about the difference between metrization and the representational theory of measurement. If there were time and space enough, I would have much to say about his proposed distinction, but in the present circumstances I will restrict myself to one remark that I do consider important. Adolfo's discussion of these matters in the early parts of his paper neglects entirely the second fundamental notion of the theory of representational measurement. This is the concept of invariance. What is meant by representation is never completely definite until the question is settled of how unique is a representation supposed to be. In some passages, Adolfo writes as if the only natural form of real representational measurement is the extensive form, which is usually considered unique up to the group of positive similarity transformations. Familiar examples are the transformations from pounds to kilograms, and from feet to meters. But this is too narrow a view, even for economists, who much prefer that utility not be an extensive quantity, but one subject to positive linear transformations that permit a change of the zero point.

At a more fundamental level, I much prefer a different way of talking about the relation between the natural world and mathematical structures. The issues here run across many types of problems in philosophy and in economics. Let me try to be down to earth and as simple as possible. Suppose we are running an experiment to determine the utility function of each participant. So we record their choices by familiar names or descriptions of objects chosen, from some given set of objects, such as kinds of food, a variety of records, CDs, or perhaps a choice of beers. We next try to create a pattern of restrictions on the choices given by each individual, so that these patterns of choices satisfy qualitative axioms of utility. When they do, we can then assign numbers as measures of utility, and these numbers as a pattern will be unique up to some given group of transformations, as discussed above. Only philosophers tend to want to make heavy weather out of this move from patterns of concrete choices to abstract patterns of numbers. The isomorphism between the two is natural and easy to understand, and this kind of measurement occurs in a somewhat simpler way in our daily visits to various kinds of stores to buy quantities of food or other products. We do not face any dilemma from the seeming abstractness of the numbers. They serve a useful and obvious instrumental purpose. I could, of course, take refuge in a standard formalism, and say that what is really important are the numerals, not the numbers. But I am happy with saying, as all of us do in the ordinary course of business, that many different numerical expressions can refer to the same number. We can grant philosophers, who desires, it that indeed this statement about references is correct. But it can be equally claimed that many of these different numerical expressions have a different meaning. So the expression "1+1" does not have the same meaning as the term "2". There is a big difference between our ordinary use of reference and meaning. Roughly put, references are fixed and stable, but meanings can wander over hill and dale. (For more on different kinds of meanings, see Suppes (2012).)

At the beginning of Section 1, Adolfo says the following, "the relationship between target systems, model systems and set theoretical structures is roughly the following: a model system is an idealized replica of a certain class of targets systems, a replica that can be described by means of axioms defining a set-theoretical predicate. Model systems are not set-theoretical structures but rather (imagined) physical, economic or geometric objects that represent their corresponding target systems."

I agree with much of what Adolfo has to say in this passage and in his amplification of it in the rest of the section, but I have come to feel that my former formulation, and Adolfo's as well, do not give the most natural account from a scientific standpoint. I now think that it would be much better to think of the model systems as being simplifying approximations of the target systems. For example, when we talk about the idealized two-particle system of Newtonian mechanics, it is natural to think of the relation of this idealized system to the one in the world, at least in Newton's day, as a nicely simplified approximation. This kind of approximation as a form of simplification is characteristic of almost all quantitative science. What is important is not a debate about the abstractness of the ideal system, but about the nature and quality of the approximation. I claim this has been true of astronomy from Ptolemy to Newton and Einstein. The magnificent achievements of astronomy in this long period have had nothing to do with issues of abstractness, but with the attainment of ever better methods of approximation. Of course, I do not want to suggest this is the whole story. Radical change in the notion of force and of its importance is equally significant. But this again is introduction of a new concept, but not of a new level of abstraction.

I am in closer agreement with what Adolfo has to say in Section 2 on empirical structures. But I do not agree with this statement in this section that a natural or social system (a target system in his usage) "instantiates" a mathematical structure if and only if it is exactly described by the axioms defining the structure. It is evident enough that this kind of characterization is not in the spirit of my ever greater emphasis in my own thinking on approximation. My choice would be to put the matter this way. Mathematical structures, except in a few trivial cases, can only approximate the real systems of the natural world. It is a problem, and a fundamental one, for the theory of representational measurement to take account of this fact in a deeper way than has been true of the past work in constructing such representations. It is important to note that it is much more difficult to axiomatize theories of measurement that introduce in a transparent way the kind of axioms that are easily modified to produce ever better approximations, especially probabilistic ones. From this perspective, the simple models of classical economics that assume perfect memories and perfect calculations by the agents being characterized are just that, simple models of limited use, waiting to be improved upon, not something to be proud of. I believe I stated my views on measurement in such a way that is easy to see where Adolfo and I agree, and where we do not.

Finally in Section 3, Adolfo discusses models of data in a way that I agree with on many points. But my effort has been to comment on where we disagree, and so this will be true of my final comment on this last section. In his comment on models of data, Adolfo emphasizes the distinction between finite data and the use of calculus and continuous quantities in the theory. I certainly agree that in any serious sense the data are finite, but the use of the calculus is really only accidental, as we see now so commonly in physics, where continuous theoretical quantities are mainly computed as numerically discrete, to meet the requirements of the computer programs doing the calculations. In fact, physics offers a wonderful example. The theory of approximation of many quantities has become so precise, and at the same time the mathematical difficulties of reasoning about these quantities in the standard form of classical mathematical analysis have become severe. These two developments have created a veritable revolution in how physical data are used in theory and in practice. The ever refined approximations of discrete data and computations now dominate physics, and so it will soon be in economics, as economists learn how to use all of the massive data that are available to them. Surely, economics is bound to become one of the "big data" sciences of the future.

8 Stephan Hartmann

I have known Stephan since he was a very young man in graduate school. We have had so many conversations about the foundations of quantum mechanics and related topics that I feel that this commentary is just a natural extension of that past. Stephan gives a clear and informative account of nonmonotonic upper probabilities as one appropriate framework for analyzing entanglement problems in quantum mechanics.

I want to add one supporting argument to the case he makes for the usefulness of such upper probabilities. It is not well enough recognized that the Hilbert-space formalism of von Neumann does not provide a framework for analyzing the individual sample paths of quantum particles. Rather, it is only mean probabilities that can be brought within the Hilbert-space formalism. Moreover, this formalism is not a natural one for studying the behavior of particles, even in the mean, over time. But, it is exactly temporal processing in decoherence that is central to a proper account. In contrast, the extension of the theory of nonmonotonic upper probabilities to stochastic processes is relatively straightforward, and can follow the lines of development of proper stochastic processes, so thoroughly studied in modern probability theory since the middle of the last century. I cannot help feeling that the collisions of particles that play such an important role in classical mechanics can be successfully entirely ignored in the temporal decoherence of quantum particles. This means I am predicting that the detailed theory of decoherence will need, in the end, to adopt the methods used in the analysis of stochastic processes, perhaps especially Brownian motion, with negative probabilities used in analysis of quantum entanglement. There is much more to be said on these matters, but this is not the right occasion.

I thoroughly agree with Stephan's choice of using the apparatus of upper probabilities in the framework of classical Boolean algebra. It is notable that the quantum logicians, by which I refer to those who work on quantum logic that is not classical, mention so seldom temporal processes, especially those of decoherence or decay. Somehow I feel it is much easier to accept a continuous sample path that at some point instantaneously loses its quantum character, for example, its quantum entanglement with another particle, rather than to think of a change of logic or algebra at this point. This is not the only argument, of course, to be considered in this debate, but it is one that occurs to me naturally and is consonant with the continued emphasis, even in quantum mechanics on temporal processing.

Finally, I like Stephan's open questions, and look forward to the solutions he develops in the future. Earlier in his article, Stephan mentions that physicists as yet have paid little attention to upper probabilities. No doubt within physics itself, their future is uncertain, but most surely they are an important and interesting alternative to be thoroughly explored.

9   R. Duncan Luce

I have known Duncan Luce, I believe, longer than anyone else contributing a paper to this volume. But this is only the beginning. We worked together on many projects, especially in the theory of measurement, until the time of his death in 2012. During this long period, he moved around a great deal and was on the faculty of several different universities at different times, but most of his time was spent either at Harvard or UC Irvine. I have talked to him so much and so often about the theory of measurement that I almost started to formulate a question for him about another paper in this volume. This is no longer possible, but I cherish our many years of work together and still have vivid memories of the times we spent and worked together, including traveling and dining in many places. I dedicate my commentary to him, but with a certain irony, for I know my way of looking at measurement was not a favorite of his. I do emphasize that this is only a particular viewpoint about a particular problem in the theory of measurement on which we were not entirely in agreement.

My aim is to mitigate any sense of disagreement, by showing how my own viewpoint supplements, but does not contradict anything Duncan has to say in his generalization of Hölder's Theorem. I take the general problem of measurement for both of us, and many others, has been concern to justify that a given scientific quantity, such as mass or velocity in physics, or utility in economics or subjective probability in psychology, satisfies (i) certain structural qualitative axioms, (ii) a quantitative representation theorem, (iii) a unique group of transformations defining the quantity's scale type, and (iv) invariance of the scientific quantity under this group of transformations. There is no doubt that the article Duncan contributed here and others he has written are a significant addition to the literature on Hölder's Theorem and the more general problem I defined.

My own approach is to relate the formal properties of a scientific quantity more directly to the theories that use it. In spite of my dedication to empiricism in the philosophy of science and recognizing the great importance of experimentation, structural questions are usually dominated by general theory, not a particular theory of measurement. My proposal is one that I made much earlier in a paper written with J.C.C. McKinsey (1953). The basic idea is straightforward, but can easily lead to the complicated proofs. Characterize the set of all models that can map models of the theory into models of the theory. The main focus of investigation is to determine the formal properties of these mappings or, in other language, transformations. For example, what transformations carry models of classical mechanics into models of classical mechanics. I use the word "transformations" loosely, because in this setting a transformation of the entire system is an n-tuple of transformations of scientific quantities that are not independent, but ordinarily strongly related, by the laws of the given theory. Each quantity, such as mass or angular momentum, has a simple intuitive interpretation. I give here the example of classical particle mechanics taken from McKinsey et al. (1953, p. 274):

> "The transformations mentioned in the theorem have a simple intuitive interpretation. The transformation $t' = \alpha + \beta t$ amounts to changing the unit of time by an amount $1/\beta$ (also reversing the direction of time, if β is negative), and shifting the origin of time by an amount $-\alpha$. The transformation $m'(p) = \gamma m(p)$ amounts to changing the unit of mass by an amount $1/\gamma$. The terms $\mathscr{C}$ and $t'\mathscr{B}$ in the transformation for the position vector amount respectively to a translation, and the imposition of a uniform velocity for the new coordinate system with respect to the old. The multiplication of $s(p, [t' - \alpha]/\beta)$ by the matrix $\beta^2 \mathscr{A}$ amounts to subjecting this vector to a general affine transformation. Since every affine transformation can be factored into a series of orthogonal transformations and stretches, we therefore see that the transformation on the position vector consists of the following: a series of rotations about the origin, reflections, and stretches is made, followed by a shift in the origin, and the imposition on the coordinate system of a uniform velocity. The force function, finally, is subjected to a series of rotations about the origin, reflections, and stretches; the rotations are the same as the rotations to which the position vector is subjected, and the stretches are proportional to those imposed on the position vector."

I emphasize that I have jumped to the end of the proof and am announcing the transformations determined by the requirement of the theorem that an n-tuple of eligible transformations always carry a system of classical particle mechanics into another such system. As the quotation given above elaborates in some detail, without any other assumptions including linearity, the expected scale results, with the proper relation between quantities, can be proved. A similar result for the particle mechanics of special relativity was proved a little later by Rubin and Suppes (1954). In this latter case, the results were a little more complicated, because of the invariant speed of light. The set of eligible transformations does not form a group, but a Brandt groupoid. The technical details are not important here, except to show that complicated structures may have more complicated invariant features than would be expected.

In closing, I do want to emphasize again that the viewpoint I have advocated with the general study of scientific quantities and the way they are related in developed theories in no way conflicts with the different approach that Duncan developed. His is more useful in those situations where theories of given phenomena do not yet have agreed-upon formal structures to provide the basis for the kind of theorem I have been discussing. His work is full of good examples, starting with his highly original book *Individual Choice Behavior* (New York: Wiley, 1959).

I miss more than I care to say the opportunity to have a lively argument with Duncan on this and other topics we returned to on various occasions over many years.

10 Kenneth J. Arrow

In his introduction, Ken mentions our long friendship, going back to 1950. It continues today, more than 60 years later, with us usually having lunch on Monday of each week. We have a long history together, and in our weekly conversations, we return to one decade and then another. It is not quite the same thing, but it seems close to a lunch conversation to write this commentary on what Ken has to say about the role of information in economics. As might be expected, I agree with much and disagree with little.

What I shall do is mainly concentrate on reviewing the kind of ideas he considers about information, by comparing their use in the general theory of probability and decision-making. There are many similarities to what he has to say about information in economics, but also some interesting differences. A good place to start is with Jimmie Savage's Foundations of Statistics, published in 1954, which I commented on early (Suppes 1956).

Savage is particularly good at showing that the general theory of decision-making in a given decision situation requires that the subjective elements need to be ordered in several different ways, in the determination of the utility function (or what is its negative, a loss function), on the set of possible consequences. This set of possible consequences is troublesome, in the way Ken's forecasts of information are troublesome. We really do not ever have adequate knowledge of the set of all possible consequences. As it is sometimes put, we need to integrate over all possible future histories of the universe, and in doing so, use our probabilistic subjective beliefs about these future histories, to determine the decision that optimizes expected utility. What I have just said sounds complicated, but is in principle, theoretically straightforward. Its obvious unsatisfactory character arises in exactly the same way as the problem of the knowledge of forecasting future markets or prices in Ken's model was. We cannot possibly implement in serious detail such models, either his or mine, closely related as they are. We are not able to elucidate in some meaningful way future possibilities. In this sense, the general theory of decision-making under realistic probability assumptions is just as difficult as that of forecasting prices in future markets.

It is also true, however, that there are real differences in the empirical details between markets and forecasts of a more general nature. For example, if farmers did not have an essentially stationary and periodic repetition of weather conditions from one year to the next, farming as we know it would be impossible. In comparison, there are certainly markets in which one can come to understand how to make money even though those markets are gyrating rather widely from one month to the next, and perhaps even more from one year to the next. A case in point would be the great difference of the slow change in climate over the past hundred years compared to the great change in markets, due primarily to massive technological changes. I am not trying to make a serious set of comparisons here, but only to suggest that the efficacy of decision models for making choices varies greatly from one kind of situation to another. I do think that we have not spent enough time in trying to make clear the distinctions that can be made and are of great importance in actual decision-making.

Let me just give one example about which I have written a good deal since those early days when I was entranced more thoroughly than I am today with Savage-style decision- making. What I have in mind is the important role that habits play in much ordinary decision-making, including the domain of markets. In the kind of decision-making discussed so far, either by Ken or by me, no explicit place is given to the role habits play in affecting deeply the actual decisions made. Only now, really, are corporations becoming sensitive to the great individual differences in the tastes and habits of their customers. It is, of course, possible from some very general mathematical standpoint to see how one could incorporate habit into the general models, but the important point is that they were in fact not made evident in either the classic free-enterprise theory of markets or the psychological decision-making of optimizing expected utility (Suppes 2003).

A second notion that is also not made explicitly part of the formal structure of markets, or psychological models of decision-making, is a quantitative measure of freedom. Too much of the talk in either of these areas of discourse is about a simple all-or-none approach to the state of freedom. But this is not true of real markets, where everyone recognizes under ceteris paribus conditions a market that has six competitive suppliers is certainly more free than one with two. Corresponding examples hold for wider decision-making, as, for example, in the range of choices available for choosing a spouse, or deciding on which university to attend.

So, perhaps my main remark, in the context of Ken's elegant sketch of economic theory, is that more concepts need to be used in either theory. My typical strategy for saying this more formally is that I would expect to be able to formulate the qualitative conceptual aspects of either his kind of economic theory, or the psychological theory of decision-making I am considering, in such a way that the independence or non- definability of the concepts of habit and freedom can be proved. Just to give a sense of what I am talking about here, I mention that in a recent paper I was mainly concerned to prove that, given only the qualitative properties of the three fundamental notions, comparative probability, independence and comparative uncertainty, no one can be defined in terms of the other two. In making this statement, without going into any technical details, it is important to emphasize that by introducing strong structural conditions on, for example, comparative probability or comparative uncertainty, the other two notions become definable (Suppes 2014). It should be clear that the general philosophical point I am making is that future theories of complex behavior will probably not be successfully formulated in terms of a few beautiful concepts, but rather in terms of a rather larger set of concepts expressing nuance rather than beauty, especially the beauty of simplicity.

One final remark. Nothing of what I said removes in any serious way the severe limitation on our ability, now and in the future, to forecast that future in any detail. Not only whether or not particular events will occur in the future, but even their probability distribution, will seldom be known. This is one of the great problems of any aspect of science that requires temporal forecasts, whether it be in astrophysics, meteorology, economics or psychology. Paradoxically in a way, this includes forecasting what progress will be made on these problems in future decades.

11   Jean-Claude Falmagne

I have known Jean-Claude for many decades. I originally met him in Belgium, and was pleased when he had the opportunity to come to the United States and decided to stay. We have talked about the foundations of measurement and related topics in mathematical psychology for many years. In responding to his paper, I look upon this commentary as simply being another conversation with him, from which I expect to learn a good bit.

I certainly would not have ever encountered on my own his extension of Hölder's Theorem to the permutability equation he investigates so thoroughly. What he does here is an extension that goes beyond my own conception of the theory of measurement to the more general analysis of physical equations, and in many cases, their derivation from qualitative principles. My own view, but not a terribly well-settled one, is that physics, and other physical sciences as well, are full of such potential derivations and manipulations of physical quantities. Moreover, it is easy to see that it is quite often difficult to draw a sharp distinction between a mathematical and physical result in this area of research. The general definitions he gives seem purely mathematical in formulation, which means not closely tied to any physical ideas, and two of his four examples are purely geometrical. All the same, the other two examples are clearly physical ones, and have the distinction of bringing out similarities in different but related physical concepts. I refer here to his first two examples, the Lorentz-Fitzgerald contraction and Beer's Law. Both of these examples deal with questions of velocity, but of a rather different kind. In fact, to show my feeling of uncertainty about how far into physics Jean-Claude's methods will pay off, I find my own formulation of the similarity in two examples to be rather badly formulated. Indeed it may well be that Jean-Claude is not mainly focused on matters of physics but of fundamental questions about the relationship between various kinds of equations, without any real commitment to there being direct physical applications.

I want to emphasize that I in no way think that applications to physics in any direct way are a necessary condition for the results of the kind he has found in his paper to be interesting. His two purely geometric examples of satisfying the permutability equation reinforce, as he intended, the case for the generality of this equation. Because of the range of his examples it would be extremely interesting to see a purely qualitative derivation of the general equation, and what special assumptions of a qualitative nature would be required to cover the four different examples he gives.

I have long argued, more in conversation than in print, that the program of moving from qualitative assumptions to quantitative representations should be widely generalized from the theory of measurement alone to many parts of physics. I still feel that way, and believe that Jean-Claude's work as presented here is a step in that direction.

12   Hannes Leitgeb

As I have found out on several previous occasions, one of the pleasures of reading a paper by Hannes is the rich sense in which one can agree with many of his claims, and at the same time disagree with others. I especially like, in my own case, the strong response I have on both sides. For me, this means that it is going to be especially interesting to argue with Hannes about our areas of disagreement.

As you will see, what I call "disagreement" could perhaps better be described by "different emphasis". Almost at the very beginning of the present paper he distinguishes between normative and empirical analysis of beliefs, and announces that his theory is normative in the sense of describing an inferentially perfect rational agent. After long and unsuccessful attempts of philosophers and others, starting at least with Plato, to define what such a perfect agent should be like, I am resistant to any dependence on this notion in my theory of belief of subjective probability. Curiously enough, my research program, however, sounds in many ways like Hannes'. I am concerned with the relation between quantitative and qualitative beliefs or probabilities. But once details are considered, we take a different path. I look for those features of human behavior that empirically and naturally violate the unrealistic canons of traditional rationality.

Let me mention three features that immediately come to mind. First, human memory is limited and fallible. This is a fact of ordinary experience, and of many excellent experimental studies by psychologists over the past hundred years or so. Second, human perception is strictly limited in its ability to make distinctions that can be of an obvious kind when the perceptual apparatus is augmented by a telescope or microscope. This sounds as if the problems of limitation in this case were only ones of technical observation, but of course that is not the case. We cannot distinguish two close shades of red, and we have just as much or more difficulty in distinguishing two kinds of anger or two kinds of anxiety in someone to whom we are talking. Such limitations are present in every domain of experience, and it is foolish in my view to think that we can ever approximate a perfect believer or decision-maker. Rather, let us come to understand the subtle machinery represented by our brain computations involving memory or fine-grained perceptions. Third, and closely related to these first two, is the human use of natural language. In fact, the problem of understanding language in all its aspects is so difficult and elusive that no one really talks about a normative talker or a normative listener. Such a concept seems too artificial, even for philosophers of language.

What I find surprising and agreeable is that Hannes next introduces the kind of intellectual setup familiar from the theory of measurement, with the central place given to a representation theorem connecting, in a given domain, the qualitative and quantitative. This means that, as in geometry, we should be able to state qualitative axioms strong enough for someone to be able to prove a quantitative representation having these qualitative properties. For example, this may be regarded as the most direct approach to defending a quantitative concept in physics. What I just said is a little too simple. We can certainly introduce the concept of mass or velocity into a purely theoretical and quantitative formulation of classical mechanics, without any need for qualitative axioms. But the need arises as soon as we start to consider and attempt to execute an experimental program for actually measuring physical quantities. The foundational or philosophical question of the relation in physics between the qualitative and quantitative occurs most naturally and most often in the explicit consideration of problems of measurement. So I urge Hannes to give more emphasis to the empirical side of having beliefs and making decisions. I am sure he will not be surprised at this kind of comment coming from me, but I make it all the same, because I am sure he could end up producing some new and surprising axioms about actual behavior, once he turned in this direction.

In the last part of his article Hannes has a useful discussion of several aspects of the more or less standard qualitative axioms of comparative probability. He makes the point that there are many different ways to turn in extending the standard theory. He mentions his own hope for a move toward doxastic and epistemic logic. My own choice is in a very much different direction. As I have recently written (Suppes 2014) what I find missing and much-needed is a separate theory of uncertainty or randomness as developed in modern probability theory. (In Bayesian theories the term "uncertainty" is used and in ergodic theories the term "randomness" is used.) The definition is the same, and the concept does not occur anywhere in Hannes' discussion. Whichever definition is used, the concept is about the distribution of the values of the random variable. The actual values do not matter, but they define a partition of the set on which the random variable is defined. The ergodic measure of this partition is the standard Shannon definition of the entropy of this partition. This entropy is the measure of the uncertainty or randomness of the random variable. It is proved in (Suppes 2014) that this concept of uncertainty or randomness cannot, at the purely qualitative level, be defined in terms of comparative probability and qualitative independence. So at the conceptual level, in the standard framework of qualitative probability, uncertainty is a new notion. However, it is rightly claimed that already in the early days of Greek philosophy Epicurus had the qualitative notion correctly formulated. I look forward to Hannes' remarks to me on these closing comments.

13 Brian Skyrms

For several years Skyrms has been writing about signaling games. The present article is a continuation of that effort. Skyrms examines in some detail far too many different cases for me to consider them in any detail. But as is customary in literature in this area, each discrete example brings out some special point, which is often of general importance. Unfortunately, I must forgo detailed discussion of his intriguing special cases.

At the beginning of this paper Skyrms distinguishes between low and high rationality. Low rationality corresponds to the kinds of questions about learning much studied over many decades by experimental psychologists, and it is on this topic that Skyrms builds his theory of signaling games, with an emphasis on reinforcement learning. I do want to remark that all of the examples Skyrms so carefully explores use a much more restrictive concept of reinforcement than is characteristic of much of the literature in psychology. In some sense, this restriction by Skyrms is surprising, given his long-term interest in biological matters. It is characteristic of reinforcement in the natural world that reinforcements are not stationary, but evolve and change continually, usually as an explicit function of time. Of course, from a formal standpoint, reinforcement patterns that are defined as a function of time present much more difficult mathematical problems in characterizing their solutions, including their asymptotic behavior. First-order Markov behavior, i.e., dependence on only what happened on the previous trial or action, is much too simple as a behavioral assumption for most real behavior. Of course, the same kind of objection can be made to the use of discrete trials, standard as they are in most psychological experiments. In the real world, continuous-time processes are much more natural and obviously fit better most human and animal behavior. Here, I am riding a favorite hobby-horse of my own, with full knowledge that it is easier to talk about studying continuous-time processes that actually doing so. All the same, it is a natural philosophical question to ask of the scientific tradition of learning theory. Why not recognize more explicitly the restrictive nature, so artificial as it is, to learning outside the laboratory. I admit at once that a glance at the theory of continuous games provides a good explanation for staying with discrete trials, as is common in most games as well as learning theory.

Again, along biological lines, I want to mention a subtle feature of Estes' style associative learning that is not brought out in the current discussion. What I have in mind is that different variants of trial-and- error learning lead to very different learning algorithms and asymptotic behavior. For example, in Estes' associative learning involving more than a single stimulus, and a probabilistic reinforcement schedule, the basic stochastic process of such learning is not first-order Markov in the responses. Instead it is a chain of infinite order, for which in the ergodic cases the past fades away fast enough to permit asymptotic convergence, but in other cases, remains dependent on the initial probability distribution of states. This fading away of the past seems to be a natural qualitative feature from an evolutionary standpoint. It is worth noting that in the case of Estes' associative learning, in many cases there is a natural first-order Markov chain, but not in the set of responses. The Markov chain is in the unobservable states of conditioning when in general there is more than one stimulus.

The signaling games Skyrms introduces are easy to understand and are intuitively appealing. On the other hand, the concept of signaling in games is as old as the rattle of dice in ancient times, and play an important part in such popular games as bridge and poker. The formal application of game theory to markets is an application full of significant signaling to which much thought has been devoted for more than a hundred years. Having Skyrms' thoughts on this history would be fascinating, especially in relation to the way his own ideas about signaling are continuing to develop.

14 Willem Levelt

I have known Pim for about 40 years. We have met and talked in many places in Europe and in the United States. Our views about language have many elements in common, although he is a much more accomplished linguist that I could ever hope to be. After his many years of research, and also his accomplishments as the founder and director of the Max Planck Institute for Psycholinguistics in Nijmegen, I found his large and definitive work on the history of psycholinguistics, published in 2013, to be one of his most outstanding pieces of work. I am sure it will stand as the definitive work for what it covers for the rest of the century.

What I was delighted to find in his article for this volume is a very good history of my own efforts in the 1970s to introduce and use probabilistic grammars and associated semantics to analyze the speech of young children. As he aptly remarks, when I wrote these papers, often with collaborators, it was not at all the style of the linguistics of that day, then dominated by Chomsky's ideas. To consider for a moment the possible use in a serious fashion of probabilistic notions in analyzing speech performance was absurd. Fortunately the work was published anyway and was noted at least by a few. As Pim remarks, now such probabilistic grammars are widely used for many different purposes. The model-theoretic semantics, on the other hand, that I introduced at the same time, is still seldom deployed in current analyses of children's language.

I remain confident that such semantics, so closely associated with context-free grammars, will prove to be useful as well in the modern massive efforts to analyze corpora of sizes we could not have imagined in the 1970s. As is well known for many other subjects, large and complicated data sets almost always call for some kind of probabilistic or statistical analysis. Yet it is still recognized by people of all kinds who work with these large data sets that the problem of meaning or semantics is not much closer to any sort of practical large-scale solution than was the case a couple of decades ago.

It seems likely that even in the case of speech recognition, where the emphasis is just on hearing the spoken words correctly, questions of meaning will need to be considered to make any further substantial improvements in the best systems of today, which are around 95% accurate, worse in practice than the rate of recognition of informed listeners. Yet it is still the case that understanding what is said in all kinds of important and useful applications of speech is absolutely necessary. Here are a few simple examples, a driverless car must be able not only to recognize the words of instruction but to understand clearly their meaning. Instructions to our other devices are subject to the same constraint. In fact we would like even more than this, as I have mentioned in some other commentaries in this volume. We would like to be able to diagnose as well the emotional state of, someone who is asked a question, in order to give not only a correct answer, but one that has the proper emotional tone as well.

15 Dagfinn Føllesdal

My associations with Dagfinn began more than half a century ago. Much of that time he has been a part-time professor at Stanford, as well as continuing his position in Norway.

During these many years we have given more seminars together and on more subjects than I can begin to recount here.

Perhaps my memory is the most vivid of the seminars on Aristotle and Aquinas we gave over many years. It was in these seminars that I came to a much deeper understanding of Aristotle's theory of perception. I value this experience above all, because it clarified once and for all that the simple slogan "same form, different matter" was a succinct but pregnant way of describing the fundamental distinction between form and matter in Aristotle's philosophy. This view of perception, of course, did not completely originate with Aristotle. Many aspects of it are formulated by Plato, and to a lesser extent, by other Greek philosophers. But what was and is important is that I realized how fundamental this idea is in trying to face up to the nature of the accuracy and speed of human perception, and also of many animals as well. I reach into my pocket and hold up a key, I ask you, "What is this?" In a matter of milliseconds really, your brain has computed the answer. How could such rapid and accurate answers be computed when the number of objects I could exhibit of this kind is so large?

From a formal standpoint of systematic theory, especially of the kind of mathematical psychology that has interested me all my academic career, the depth and clarity of this discussion of how perceptual recognition of objects and processes is computed, that is, by recognizing the form of the object, in the way that Aristotle so clearly explains, has been a model of qualitative psychology and philosophy. A beautiful concept of isomorphism is behind this fundamental explanation.

Here is my sketch of a more complete theory. I recognize the object by quickly obtaining from the immediate present perception of the object an isomorphic form I can find or create in my brain, i.e., my memory. Of course, computing this isomorphism is another matter. Neither Aristotle nor Aquinas nor anyone else for a very long time even recognized that there was a problem of systematic computation to understand. Moreover, the physics of the brain must be such as to make this computation quickly and accurately. This problem of computation is still not fully solved, and, I am afraid, not even recognized as a major problem, perhaps the central problem, of perception, by many psychologists and neuroscientists.

Indeed, as much as I admire Husserl, based on Dagfinn's congenial tutelage, Husserl, I must say, does not offer a genuine computational solution. This is the problem that I personally work on and think about more than any other when my mind is fresh and ready to tackle it. I will try to say what I think the solution will be like, but I am sure that many details will still be an active problem of research at the end of the present century. I will not be around to celebrate what is known about computational perception in the year 2100, but I hope that the original brilliant contribution of Aristotle will be recognized.

In the many seminars that Dagfinn and I have held over the years, which includes the list of philosophers he gives in his paper, for all of them but one I felt confident, at ease and holding my own in the discussion. As is probably apparent from what I have already said, the grand exception to this rule was Husserl. The best way to put it is when it comes to Husserl I am Dagfinn's student, and I am not sure he thinks that I have been a very good one.

If I have come to understand something about Husserl's philosophy, it is solely due to Dagfinn. Otherwise I am sure I would have made little progress. At least as Dagfinn expounds him, Husserl had many deep and interesting ideas. Moreover they extended, in a natural and scientifically interesting way, many ideas of William James, in my judgment the most important psychologist of the 19th century.

In spite of Dagfinn's sustained and often brilliant expositions of Husserl's thoughts, his ideas are still far from being understood in any real depth by very many philosophers. It may be one of the tasks for the future, but I have to say I am skeptical that this will happen. Not because Husserl is lacking in merit, but because the direction and development of modern neuroscience is moving at such a pace and volume of work that the young scientists of today who work on the brain struggle just to keep up with the literature in their own special area, which is like a small torrent flowing rapidly into the great stream of all publications in neuroscience.

Those poor philosophers who try to grasp some idea of what is going on will find it ever more difficult, because of the pace, intensity and volume of the research now being done. I recognize, of course, I may be too pessimistic about these matters, and there will be a shift that makes possible a general understanding of the brain and the mental computations it supports. If this more optimistic view is correct, then Dagfinn's sustained and intense efforts to keep the ideas of Husserl still present and sound as a philosophical foundation of neuroscience and psychology will have been well spent.

16   Russell Hardin

The paper of Russell Hardin raises a really interesting question, one that is certainly not been much discussed in the large literature on Hume. This is how central is the concept of mirroring to Hume's theory of associative learning. I believe it is correct to claim that it is not mentioned explicitly in the Treatise. But this is not the only place in the history of associative learning in the more than 200 years since Hume's Treatise first appeared. In the long and complicated history of learning theory from, say, 1800 to 2000, it is hard to think of a major treatment by anyone of mirroring. Yet, the importance of mimicry in learning endless kinds of skills has been widely recognized from time immemorial.

On the other hand, the extensive writings of Aristotle on mental images, sensation, perception and learning do not mention, as far as I can determine, either mirroring or mimicry. There is one brief mention of imitation in the Rhetoric (I, xi, 23). In some sense, this is surprising, given the very extensive discussion of mental images by Aristotle in formulating his theory of memory, imagination and learning.

Here is an excellent explanation of some important distinctions Aristotle makes in the brief treatise On Memory and Recollection:

> "...just as the picture painted on the panel is at once a picture and a portrait, and though one and the same, is both, yet the essence of the two is not the same, and it is possible to think of it both as a picture and as a portrait, so on the same way we must regard the mental picture within us both as an object of contemplation in itself and as a mental picture of something else. In so far as we consider it in itself, it is an object of contemplation or a mental picture, but in so far as we consider it in relation to something else, e.g., as a likeness, it is also an aid to memory. Hence when the stimulus of it is operative, if the soul perceives the impression as independent, it appears to occur as a thought, or a mental picture; but if it is considered in relation to something else, it is as though one contemplated a figure in a picture as a portrait..." (450 b 20–30)

The introduction of a brief summary of Aristotle's psychology is not meant, in any sense, to question the originality of Hume's own thinking. It is rather to show the continuity. A serious criticism of what I have said about Aristotle is that it does not get into the main area in which a direct comparison should be made for present purposes. I have in mind here the comparison of Aristotle's and Hume's theory of morality. I accept this criticism, but try to mitigate its influence on the present remarks, because it is far too much to undertake such a comparison in the context of this commentary.

I do want to make a point about both Aristotle and Hume that many will disagree with. For a variety of reasons in my recent experience in brain research, I have come to be skeptical about trying to sharpen the division between the theoretical and practical, or between pure knowledge and the knowledge that is so critical to action of almost any kind. When I think about my own children when very young, it seems to me that they really made no distinction of any importance between cognition and emotion. These were abstract categories that did not influence to any great extent their early learning to recognize their mother, especially her face, seems to me to be on a very similar basis as learning the expression of emotion in her voice. The cognitive facts of life, so to speak, and the emotional facts of life were scarcely distinguishable. Moreover, Hume's treatise can be read this way without too much difficulty. As another example, in the 20th century when the theory of association and conditioning dominated learning theory and other parts of psychology, sharp distinctions of any kind between emotion and cognition were not often made. As is well known, Hume wanted to begin his Treatise with Book II On the Passions, but was advised by others, and finally decided himself, that it would be too difficult for the reader to begin with this much more original treatment of the passions than the epistemology of Book I. He was right about this, and I think he was right about the distinction he makes in the Enquiry concerning the Principles of Morals. But he was right on insisting on the empirical nature of the support for the doctrines of great theoretical import that lie behind both of these highly original first two Books of the Treatise. I join Hume in his great skepticism of the power of pure reason to settle any question of substantive import. I am perhaps more radical in wanting to reduce even further the difference between the truths of nature and the truths of morality.

Russell mentions briefly some work on the expression of emotion in the brain. I do not disagree with anything he has to say on this, but do want to say a few more detailed things, given my own research concentration on such matters. First, there has been some tendency in brain research to make the claim that the behavior of the brain in making computations about emotion are in principle different from the computations about matters of fact. This is not the place to try to make a detailed argument, but I do want to emphasize that we are still at the point in brain research with a variety of computations that seem to support either expressions of emotions or matters of truth almost equally. I may turn out to be wrong in this matter, but certainly the current evidence is no sense decisive. Our understanding of the physical mechanisms of computation about anything in the brain is still too primitive to want such an important distinction as the claim that computations about emotion are in principle quite different from computations about matters of fact. It is well-known that emotion is often expressed in language, and it is equally well recognized that the perception of the listener about the expression of emotion is more accurate than that of the speaker. In a general way, from the standpoint of the listener, the expression of emotion in speech is just another feature to learn, and to learn quite early in childhood. In saying this, it is not my aim to annihilate important distinctions, but to criticize distinctions that seem to play no important role, or have as yet no significant empirical support. It is a question unanswered in current research literature as to whether the young child perceives more quickly expression of emotion or of some standard cognitive command. Just to be contentious, I would bet on the perception of emotion often being faster than the cognitive recognition of a common command such as "Don't forget your coat".

Finally, I come back to Russell's useful emphasis on Hume's apparently original introduction of mirroring, a natural concept, which does have a growing contemporary literature in psychology, but as far as I can see, not yet in neuroscience. Broadly speaking, I think of this as just another way of making associations, which are already regarded as properly ubiquitous and prodigal in nature. This is in no way to belittle what Russell has done in bringing mirroring to broader attention, but only to emphasize how natural the concept fits in, and therefore, how surprising it is that it has not been discussed in more detail at various times in the past.

17 Acacio José de Barros

As in the case of many other commentaries I have written for this celebration of my 90th birthday, writing about the paper by Acacio and Gary is much like writing about a paper of my own. In fact, it is even more so in the present instance, because much of what they write is an expansion and clarification of a long paper we wrote together and published in 2012. But in the framework I have used so often, the problem here is trying to find a representation of stimulus response behavioral models in terms of oscillators as models of brain activity at the system level. What Acacio and Gary do well is explain in a careful way the concepts underlying neural oscillator models. Even in the recent history of mathematical psychology, there has not been much analysis or use of oscillators as models of neural computation. More generally, they are scarcely mentioned in most of the literature of psychology. Of course, the situation is very different in physics. As the saying goes, oscillators provide the go-to model when nothing else comes to mind. This is an exaggeration, but not if one restricts the analysis to electromagnetic phenomena. This includes quantum mechanics, where one of the simplest and computationally most manageable examples is the one-dimensional quantum oscillator. (My own first paper in quantum mechanics, published in 1961, was about such oscillators.)

The view that system computations in the brain are mainly made by oscillators is now widespread, even though the problem of giving realistic and detailed oscillator models of any important and significant aspect of behavior is regarded as difficult, and in most cases, not yet really solved at any level of depth at all. This is not surprising, for the accuracy and speed of neural computations in many species that are not mammals has been more than adequately shown in a great variety of experiments, now well over the first hundred years of work.

Watching the simple behavior of birds or fish, it is easy to make a mistake and think that the behavior is so simple it will be equally simple to model the neural computations that induce and guide the behavior. But this is far from being the case. Indeed, even at the behavioral level, it does not take many experiments to demonstrate how complicated much animal behavior is. This is true even before one begins to think about how the computations are made to guide and execute such behavioral responses to what appear to be simple stimuli. To some extent, the tradition in experimental psychology in the 20th century promoted the deception that such behavior is simple. Why is this so? I believe it is so because psychologists were anxious to put distance between their work and earlier work, such as that of William James, which emphasized insight and introspection in their full human complexity.

But this deception of experimental psychologists is no worse than what we have come to realize is also the same situation in elementary physics. It used to be the practice, as I like to emphasize, that in elementary physics courses it was a point of pride to show how simple and clear were the solutions to many elementary problems. Only recently has it been properly recognized how carefully and artificially, in some sense, these problems have been selected for teaching beginning physics. As I like to emphasize about elementary classical particle mechanics, even for the restricted case of gravitational forces, when we try to solve the interaction of more than two particles, we find ourselves lost, and can only manage some very special cases with strong restrictive assumptions. Given that this is the case in something with as simple a set of initial axioms as those of classical particle mechanics, it is not surprising that deception about psychological experiments, even of the apparently simplest sort, is easy.

Because of my conviction that collections of oscillators are the most promising idea at this time for neural computing, I applaud Acacio's and Gary's effort to make the physical intuitions about oscillators more evident and easier to understand. On the other hand, their paper does make clear as well that even the simplest representations of behavioral models is a complex task, when the effort is to be committed to a thorough use of the physics of oscillators in the representation. I would like to think that there will be some easier and more accessible route into this analysis of neural computation in terms of oscillators or related physical concepts, but it seems unlikely that this will be true. We already know that in principle such familiar phenomena as weather storms are in principle unpredictable in detail for even 48 hours. It may be the same with brain computations. This is not meant to be a forecast of how hopeless it will be to make progress in understanding the weather or the brain, but rather to emphasize how difficult it will be to go as far as we would like.

18 Claudio Carvalhaes

I have nearly daily conversations with Claudio about the application of physical mechanisms to the analysis of brain data, especially data concerned with the production or comprehension of natural language. There are many aspects to the problems that he and I discuss, and that he works on intently. The present paper is a good example. Everyone knows the direct interpretation of EEG signals is that there is a measure of the electric potential on the scalp. Claudio's quite technical paper has a clear objective. It is to show that by using the scalp of electric field rather than the electric potential, we should be able to do a better analysis of brain waves, as recorded by EEG electrodes on the scalp.

In his paper, Claudio provides a thorough analysis of this problem and offers a number of specific technical arguments why we should expect to get better results from observing the electric field rather than the electric potential. It is understood, of course, that not two different methods of direct experimental observation are involved here, but rather whether or not we should analyze the electric potential, or rather, first compute the electric field from the electric potential and then proceed to analyze of brain waves. He makes a good case for choosing the latter, but he is also realistic in analyzing what the computational difficulties are in doing so. Historically there is no doubt that the direct course of using just the electric potential has dominated in the past brain research that uses EEG recordings. He makes a good case for changing this past practice.

It is still too early to assess how much the additional cost of computing the electric field will be justified by the better results obtained. Of course in my own context "better results" means better predictions about the brain mechanisms of perception in language production. The results of the two experiments he reports are encouraging, but by no means decisive.

This paper of Claudio's represents nicely the new kinds of efforts that will be needed to bring the theoretical physics behind the brain mechanisms he discusses to bear on better understanding of computational processes of the brain. The kind of details he pursues seem to be inevitably necessary if we are to use in a really effective way insights from physics and the theory of oscillators to understand in a much deeper way how the brain does compute. Claudio's work is still in its early stages and it will be some time before it will be clear whether or not this kind of detailed application of classical electromagnetic theory will pay off. If only because it is so difficult to think of reasonable alternatives, I am betting positively.

19   Marcos Perreau-Guimaraes

The paper by Marcos, with whom I have worked for almost a decade, illustrates nicely the increasing complexity of the statistical models being developed to analyze system brain data, especially that generated from EEG recordings. The interesting and elaborate extension he makes in his paper to earlier analysis we worked on together is a fine example. The detailed developments are two special and complex to comment on in a serious way in the present volume.

I emphasize that there is nothing unusual about this situation. In fact, it is a common situation increasingly in all parts of science which collect large bodies of data. The data are too complicated to yield to a thorough analysis by a model regarded as manageable from the conceptual standpoint. Moreover, even with the most valiant statistical effort, the models that yielded significant results are very far from accounting for much of the detail that can be easily analyzed in the data. In this respect increasingly large-scale data-intense experiments on brain processes have come to resemble in this respect models of astrophysics or, to mention something less esoteric, models of the weather or stock markets. This comparison is in some way surprising. At least it is so for me. I originally thought of brain data as being a kind of extension of old-fashioned biological experiments. But the real truth is the other way round. I mean by this that modern biological experiments are much like modern brain experiments, too rich in data to be reduced in any way to a neat and manageable theory of the kind much glorified in 19th century physics.

As in all such matters, there is something else to be said about the topic. It was realized by many people in the 19th century that the solutions in physics, for example, that seemed clear and rather simple covered only very simple cases of physical behavior. The great classical example is celestial mechanics. In idealized form it is great in solving questions about the behavior of two interacting bodies, where by interacting I of course mean interacting only by gravitational forces. Just introduce a third body and chaos ensues, not to speak of a modest number like 10 or 12, which in general is totally unmanageable. This is true not only of mechanics, of course, but holds for electromagnetic theory and other parts of physics. So in some sense, we are just waking up to the fact that our science is much less adequate than we thought to determine detailed answers to almost any natural questions. As some physicists have remarked, it was a great mistake of the past to teach introductory physics to students as if problems were easily solvable. They only seemed to be, because the ones used in the textbooks and lectures were of the very simplest sort.

So the kinds of complications introduced by Marcos to statistical analysis of brain data are to be expected, and in great quantity, in the years ahead. There is, I believe, no realistic hope of some simple theories taking over and proving satisfactory in a detailed way. This is, I think, and as I have emphasized in numerous comments in this volume, a permanent change in the nature of science, at least for this century.

I have also emphasized in comments to other articles in this volume, my hopes for a deeper physical theory of how the brain computes even the simplest kind of natural perception. But I am under no illusion that we are likely to find a simple theory to account for these seemingly simple examples of accurate and fast perception. I also admit that I could be wrong about this and there are some real surprises lying in wait to be discovered. It would be foolish not to put in such a disclaimer, because above all, predicting the future in any detail is a hopeless enterprise.

I do want to mention that in thinking about Marcos' careful exposition of using more complicated models for classifying correctly what the brain is doing, perhaps what stands most in the way are theories that are too simple. The overly simplified temporal processes characteristic of current computational models of the brain, including all those we have worked on in my own lab, will probably be found to be far short of the mark. Too much of the brain activity reported and discussed suggests that neurons or dendrites that seem to show computational ability also satisfy something like the assumptions of Brownian motion for very small particles of any kind. This does not mean that we cannot understand at all processes that are obeying microscopically the laws of Brownian motion, but it is hard work, and not engaged in too often. Moreover, there is a large gap between the use of Brownian motion in physics and the study of such processes in pure mathematics. If I am right in suggesting that significant parts of brain activity resemble Brownian motion, I can only be hoping that it is an important goal shared by mathematics and physics, in the study of such motion, to obtain results useful for the study of the brain.

20 Colleen Crangle

I have worked with Colleen a number of years on a variety of questions about language, but with a special focus on how language is processed in the brain. The article for this volume reflects this collaboration, and so my remarks here are as much about my own work as hers. I do want to emphasize that quite apart from my own contributions, she has had many original ideas of her own, and has written about them separately. The common part that I want to focus on here is the difficult problem of recognizing meaning in brain data processing of language. The difficulties are too numerous to go into here, but I have no doubt that the problem of fully and completely understanding how the brain processes language will occupy a large number of researchers for some time. To say this in commenting on Colleen's article, does not mean that I am pessimistic about making progress. Quite to the contrary, I think this is a golden opportunity now and for some time to come.

The problem itself is more difficult than the problem of speech recognition, but is of a similar character. Some progress can be helpful, and at the same time, be recognized as a very partial solution. For example, we can do quite a good job of speech recognition in ideal circumstances, but in the hurly-burly of ordinary conversation in a noisy environment no recognizer can compete with a person recognizing the speech who is fully familiar with the speaker and the circumstances. This comparison is not completely an appropriate one, because we do not have a person somewhere who can give such a sound and satisfactory interpretation of brain waves generated in producing or recognizing speech. In other words, we do not have the advantage of having anything like the guide provided intuitively by a person listening or producing speech. Perhaps this is just another way of saying that the problem of brain-wave recognition and processing of language is much more difficult than the problem of speech recognition or production.

Another model that is also deceptive is the simple and clear semantic theory we can give for first-order logic expressions and a well-defined model of what the expressions are about. In this latter case, logical expressions and the semantic analysis of their meaning can be relatively easy and straightforward, at least for all kinds of cases of some interest.

So Colleen's article raises the intriguing question of what kind of intellectual apparatus for analyzing speech will turn out to be really helpful in understanding how the brain processes natural language. In many ways, the results Colleen presents for either one of two methods for analyzing the meaning of language is impressive. Moreover it suggests and, in fact, has guided current research on language in the brain. Colleen described the results of a large number of experiments that have been run over the past decade or so to give insight into how the brain is processing language. But the general semantic problem of having a theory to interpret the meaning of brain data assumed to represent language suffers directly and substantially from the fact that we have no adequate semantic theory of a concrete nature to interpret ordinary English sentences spoken or written by a person. In other words, at the most direct level of language use we are still very much in the dark about what an adequate semantic theory will look like. I am not saying anything original here, but only summarizing what I believe is a noncontroversial view about the current state of affairs in the study of the semantics of natural language.

The pessimistic part of what I am saying is that the useful methods Colleen has introduced, and that others have used as well, are still a long way from providing the semantic theory we need. Our brains do it, but we don't know how our brains do it. If we did, the problem of using speech in interaction with our many devices of modern technology would be relatively simple. It is still science fiction to imagine talking simply or naturally to your refrigerator or stove or car.

To say that it is still science fiction is in no way to deny the merits of the positive examples Colleen gives. This applies both to the use of Word Net or the statistical analysis of similarity of words. On the other hand, it seems very unlikely that either one of these methods can by itself be extended to give a workable and usable analysis of meaning, as required for asking and answering questions, or acquiring new knowledge. I only want to make the point that we face real difficulties of speech-enabling our devices or digitizing our own mental capabilities. I certainly am not saying that the problem is unsolvable, but I do want to emphasize its difficulty and to make sure it is understood by those who do not know the literature too well that the real problems we face are still far from being solved. No doubt, Colleen's work and the work she surveys by others represent a good start. It is simply my fundamental point that we have a good start, but are still far from a good ending in understanding in various essential ways the processing and interpretation of natural language.

21 Anne Fagot-Largeault

I guess I first met Anne in the late 1960s. So it has now been almost half a century since we first talked. At the time we met, she was registered in Paris as a student of philosophy and, and also as a medical student. Somehow she was working at the same time on a PhD in philosophy and an MD in medicine. I am not quite sure how she managed this, but she did very well. The surprising thing is that in the middle of this intense activity, she decided to also take a PhD in philosophy at Stanford. She wrote a thesis that I liked very much. After spending a little more time in the United States, she went back to France, where she has had a highly successful career. I like to mention the fact that she was the first woman to be appointed a member of the College du France in philosophy.

She also visited Stanford in later years as a member of the French – Stanford committee, so I had an opportunity on various occasions to meet her. When I myself was visiting France I also stayed in touch. Over these many years she is somehow managed to establish herself in Paris both in medicine and in philosophy. I am still not quite sure how she does it, but her short paper on the psychiatrist's dilemma reflects this split career.

I have other input from a different direction about the topic of Anne's paper. My oldest daughter, Patricia, has both a PhD in neuroscience, and an MD in psychiatry. She now lives in Palo Alto and is a senior member of the faculty of the School of Medicine at Stanford. We have discussed on several occasions the topic of Anne's paper.

Patricia also recognizes, that there is a strong movement to combine drug therapy and talk therapy, as psychiatrists like to refer to psychotherapy. There is also a burgeoning movement in the United States to be more watchful, especially of drug prescriptions for minor mental illnesses, of children or those of adolescent age. At the same time it is fully recognized that talk therapy is not sufficient for severe cases of schizophrenia and a number of other psychiatric illnesses.

I agree with just about everything that Anne has to say. She does not have much to say, and certainly it is premature to make any strong clinical claims that a third method is about to enter the arena of treatment possibilities, namely, computational neuroscience. In other commentaries I have discussed the computational problems we confront in understanding how the brain works. I am not optimistic about how soon we will have much confidence in our knowledge, especially that is detailed enough to make a strong claim of being a nearly complete account of the computational aspects of brain processing. However, surely we will make a great deal of progress in this complicated problem in the course of the present century. It is optimistic, to say so, but it is optimistic in a way characteristic of many other scientific and technological developments we can foresee happening in the 21st century. It also seems likely that addressing problems of mental illness will end up being one of the most significant objectives of medical research in much of the century. New kinds of understanding should be available for thinking about and diagnosing mental illness. This is a bold claim, and I do not want to make any strong assertions about the correctness of this claim, but it matches well with other prognoses for biological and medical progress in the more than 80 years left in the century.

22   Elizabeth Loftus

I enjoyed very much reading Beth's account of her years as a student seeking a PhD at Stanford, and then later many details about her very successful career. I wish I could claim some responsibility for the good scientific work she has done on the fallibility of human memory, and the social and legal problems that can arise from this fallibility. But I cannot make such a claim. I have in my long career examples of several very successful persons who were originally were PhD students of mine, but have gone on to be utterly independent of any substantial influence of mine and also very successful in their careers.

I recognize that Beth has had to endure hostile and almost dangerous threats to her personally in defending what are surely mostly correct views about the distortions of memory people can come to believe for all kinds of purposes and reasons.

Ironically, in the final years of my own career, I am following Beth's interest in human memory, but on a very particular and different set of problems. Since the late 1990s I have been thinking and writing about the system behavior, not the cellular behavior, of human, and to some extent, animal brains. Work in this area almost inevitably involves work on memory. The kind of behavioral work that Beth has done provides a good model to bring researchers on human memory to be careful about the claims they make. On my view, the brain has to compute memories from what is stored in its own memory. Understanding these computations is a task that has just begun, but that will surely occupy many scientists the rest of the century.

I am especially interested in building physical models of the methods of computation used by the brain. In my own current research a favorite current model is that familiar computations about ordinary features of perception are done by neural oscillators. As part of this theory, I and my colleagues also conjecture that these oscillators are built up from strongly connected phase-synchronized neurons. In the spirit of Aristotle's theory of perception as recognition of same form but different matter, oscillators that recognize various features are then weakly connected, but phase synchronized to recognize so astonishingly fast features of ordinary perception. The kind of problems of memory fantasies, as I like to call them that Beth studies provide something still more subtle to understand how they are formed and used by the brain. It may be some time before we can catch up and justify at the brain level the good advice her studies have provided, as a scientific defense of memory abuses.

23 Nancy L. Cartwright

I have mentioned in several other commentaries that they are very much like conversations I have had with the authors, often over many years. This is especially true of Nancy's contribution. I have known her for a long time, even before she was a professor at Stanford, and I have continued to stay in touch with her and to follow her work in philosophy of science over many years. The topic of probabilistic causality has been one that we have often discussed, possibly more than any other. I very much enjoyed reading her contribution and her sharp criticisms of the vagueness of many practical recommendations about evidence-based policies or decisions. I have three general comments to make.

First, I find it striking that in the large current literature on evidence-based policies, how little attention is paid to the systematic theory of decisions developed so thoroughly in the mathematical statistics of the 1950s. Nancy's extensive survey of the kind of criteria used by a variety of medical and health organizations in assessing evidence-based policies was what persuaded me that little rigorous attention was being paid to the extension in the 1950s of statistical assessments, purely probabilistic in nature, to the inclusion of loss or gain as measured by expected utility. Of course, vague remarks are made in this direction, but almost no organization she mentioned makes any serious technical effort actually to measure gains or losses. I make this as just one more criticism Nancy can add to her critique of the lack of rigor in evidence-based policies.

Second, I want to add another point of criticism that she can add to her critique. This is a matter of estimating effect sizes of differences between experimental and control groups, whether in randomized assignments to the two groups or not. The typical examples are very much in the spirit of the large data sets that are coming to dominate much of modern science and also the attention of mathematical statisticians. The classical methods of testing for a significant difference between an experimental and a control group in a purely probabilistic way, with no loss function involved, has an important failing for all or most applications. If the difference, for example, between the means of the two groups is statistically significant in the classical sense, then, the language much used in the past evaluation of experiments would justify saying that the difference in the means is scientifically significant. The problem is that when the sample sizes are very large, very small differences in the mean can be highly significant statistically, and therefore, for those not further concerned, scientifically significant. The existence of such results is one of the reasons that physicists have not really adopted classical statistics. With their large sample sizes in many kinds of experiments, it is not the assessment of statistical significance classically as described above, but assessment of scientific significance that is important. The number of standard deviations between the means of the two groups is the measure they often used.

A good many years ago, there was a great outcry in operations research for "six sigmas". This means that a proper standard for a good engineering or scientific result is six standard deviations' difference in the means of the experimental and classical control groups. (The use of the word "sigmas" comes from the use of the Greek letter "sigma" to denote a standard deviation.) This is, of course, an extraordinary standard to strive for, seldom achieved in either education or medicine, two great areas that make extensive use of statistical evaluations of results. All the same, there is much to praise in physicists' criterion. The modern movement is to adopt a form of this criterion by insisting on assessing the significance of effect sizes, which are really a measure in terms of standard deviations. (An additional desirable analysis is that of the 95% confidence interval of an effect size.) A hopeful sign is that the additional standard of effect size is now being used in applied statistical assessments in education, and also, to some extent, in medicine. The conservative organizations that Nancy cites have mainly not yet caught up with this new requirement of assessing effect sizes. But the tendency to ever more systematization of evaluation by government agencies, and other entities of a similar sort, make it very likely that soon we shall all be imitating the physicists and talk about effect sizes, even if only a few studies in many applied areas can hope to reach the standard of six sigmas.

My third and final point is my strong general agreement with Nancy that the real problem for assessing evidence-based policies or practices is the maddening vagueness of many of the requirements written down as if they were "Holy Writ". My complaints do not end there. I also find very unsatisfactory, as Nancy does, the casual use of some fixed number of categories, such as three or five, without any attempt to support the underlying assumptions needed to make statistical testing of such classification meaningful. Their casual use in education or medicine remind me of the way evaluations were given at the end of the 19th century and beginning of the 20th. It is hard to believe that, after all these years, anyone will take seriously such casual use of a small fixed number of categories. For many rough-and-ready casual purposes such unjustified categorization can be useful. But as a doctrine or a model of scientific methods, it is scarcely acceptable.

References

Cohen, P. J. 1963. The independence of the continuum hypothesis. *Proceedings of the National Academy of Sciences of the United States of America* 50(6):1143–1148.

Luce, R. D. 1959. *Individual Choice Behavior: a Theoretical Analysis*. New York: Wiley.

McKinsey, J., A. Sugar, and P. Suppes. 1953. Axiomatic foundations of classical particle mechanics. *Journal of Rational Mechanics and Analysis* 2(1):253–272.

Rubin, H. and P. Suppes. 1954. Transformations of systems of relativistic particle mechanics. *Pacific Journal of Mathematics* 4(4):563–601. doi:10.2140/pjm.1954.4.563.

Savage, L. J. 1954. *The Foundations of Statistics*. New York: Wiley.

Suppes, P. 1956. The role of subjective probability and utility in the decision-making. In *Proceedings of the Third Berkeley Symposium on Mathematical Statistics and Probability, 1954-1955*, vol. 5, pages 51–73.

—. 1960. *Axiomatic Set Theory*. Princeton, N.J.: Van Nostrand.

—. 1961. Probability concepts in quantum mechanics. *Philosophy of Science* 28:378–389.

—. 1962. Models of data. In E. Nagel, P. Suppes, and A. Tarski, eds., *Logic, Methodology, and Philosophy of Science: Proceedings of the 1960 International Congress*, pages 252–261. Stanford: Stanford University Press.

—. 2002. *Representation and Invariance of Scientific Structures*. Stanford, CA: CSLI Publications.

—. 2003. Rationality, habits and freedom. In N. Dimitri, M. Basili, and I. Gilboa, eds., *Cognitive Processes and Economic Behavior. Proceedings of the Conference held at Certosa in Pontignano, Siena, Italy, July 3-8, 2001. Routledge Siena Studies in Political Economy*, pages 137–167. New York: Routledge.

—. 2012. Three kinds of meaning. In R. Schantz, ed., *Prospects of Meaning*, pages 567–579. Berlin and Boston: Walter de Gruyter GmbH & Co. KG.

—. 2014. Using Padoa's principle to prove the non-definability in terms of each other of the three fundamental qualitative concepts of comparative probability, independence and comparative uncertainty, and related matters. *Journal of Mathematical Psychology* 60:47–57.

Suppes, P., J. A. de Barros, and G. Oas. 2012. Phase-oscillator computations as neural models of stimulus–response conditioning and response selection. *Journal of Mathematical Psychology* 56(2):95–117. doi:10.1016/j.jmp.2012.01.001.